modern
masonry
brick, block, stone

by
CLOIS E. KICKLIGHTER
Professor of Industrial Education
Eastern Michigan University
Ypsilanti, Michigan

South Holland, Illinois
THE GOODHEART-WILLCOX COMPANY, INC.
Publishers

Library of Congress Cataloging in Publication Data

Kicklighter, Clois E.
  Modern masonry.

  Includes index.
  1. Masonry.   I. Title.
TH5313.K5   693   76—49927
ISBN 0—87006—224—7

INTRODUCTION

MODERN MASONRY: Brick, Block, Stone provides a thorough grounding in methods of laying brick, block and stone. It will insure a broad understanding of materials and their properties. Students in trade schools, vocational and technical schools or wherever in training for the masonry trades, will find it a source book of near-encyclopedic dimensions. It is equally suitable for apprenticeship training.

Simply and clearly written, it covers the following important aspects of the masonry trade:
1. The makeup, properties, uses and sizes of every type of masonry unit.
2. Uses, descriptions and illustrations of tools and equipment.
3. Descriptions, illustrations and uses of anchors, ties and reinforcement.
4. Types of courses and bonds.
5. Properties and technical details of entire masonry systems such as foundations, floors, roofs and walls.
6. Accepted techniques for laying all kinds of masonry units in all kinds of bonds.
7. Construction details for masonry walls, foundations, pavement, steps, garden walls and masonry arches.
8. Concrete materials, procedures for placing and finishing concrete.
9. Concrete reinforcement and concrete design data.
10. Types of forms for concrete construction as well as their uses and how to build them.
11. Mathematics for masonry in English and metric.
12. Blueprint reading.
13. Methods of cleaning completed masonry structures.

A reference section of more than 30 charts and drawings includes useful information on such things as sizes of masonry units, mortar types and classes, metric conversions, an outline of a bricklayers' apprenticeship course, abbreviations used on blueprints, spans for concrete garden walls, and many other useful items.

Many of the hundreds of illustrations are in full color showing patterns and bonds used in various types of masonry construction. There are numerous individual color samples of brick, stone, terra cotta and manufactured products in a variety of bonds. Line drawings and photographs illustrate methods of bricklaying and provide technical details of masonry structures.

Clois E. Kicklighter

Creative and imaginative masonry.

CONTENTS

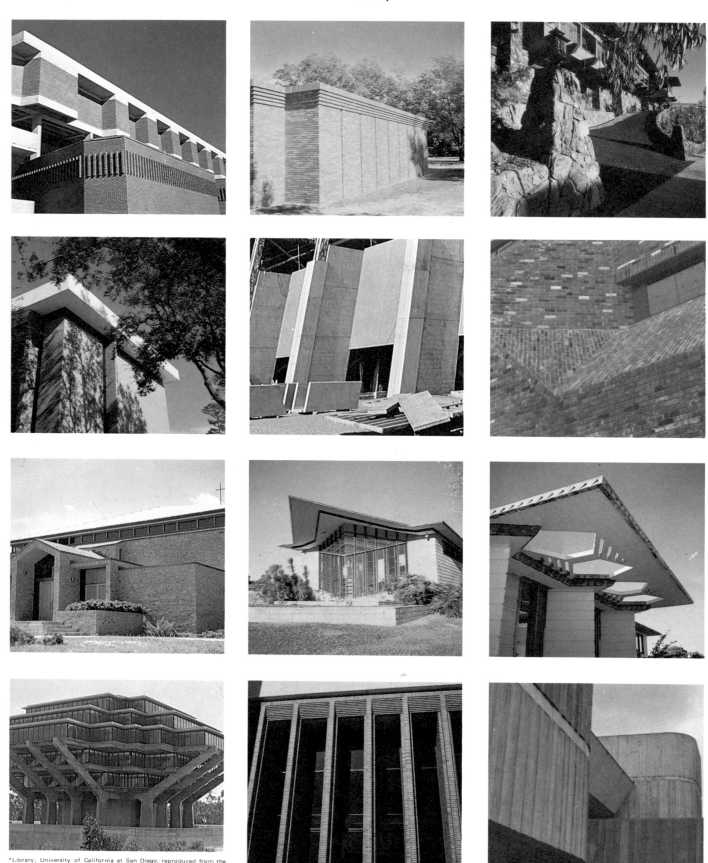

*Library, University of California at San Diego, reproduced from the slide set CONCRETE ARCHITECTURE IN NORTH AMERICA by KaiDib Fims, International, Glendale, California, USA. Photo: Dibble.

Fig. 1-1. These contemporary structures show only a few of the many textures, colors and shapes possible in modern concrete and masonry.

Chapter 1
CLAY MASONRY MATERIALS

The variety of masonry materials available for use in construction has greatly increased. There are many new shapes, textures, colors and applications. No doubt, this is due, in part, to the variety, versatility, durability and beauty of masonry materials. See Fig. 1-1. For instance, it is estimated that brick alone is made in at least 10,000 different combinations of colors, textures and shapes.

Masonry construction provides permanence. Few, if any, materials are as maintenance free and longlasting as brick, block and stone.

Masonry units may be grouped under several basic materials or products. These are:
1. Structural clay products (brick, clay tile and terra cotta).
2. Concrete masonry units.
3. Sand-lime brick.
4. Glass block.
5. Stone.
6. Mortar.
7. Masonry anchors, ties and joint reinforcement.

Each of these basic materials will be discussed in terms of classification, characteristics, size and applications.

STRUCTURAL CLAY PRODUCTS

Structural clay products are classified as:
1. Solid masonry units (brick).
2. Hollow masonry units (tile).
3. Architectural terra cotta.

These products, Fig. 1-2, are made from clay or shale. These materials are found throughout the country. Clays and shales suitable for the manufacture of brick fall into three groups:
1. Surface clays.
2. Shales.
3. Fire clays.

Clays are produced naturally as rocks weather. Pure clay is mainly silica and alumina. It is made up of very small crystals of certain rocks, chiefly feldspar. (Feldspar is a crystalline mineral made up of aluminum silicates with either potassium, sodium, calcium or barium.) Shales are a compressed clay. They are more dense than clay. Fire clays have higher percentages of aluminum silicates, flint and feldspar. The different characteristics of these materials makes possible the great variety of clay products that we have today.

CLASSIFICATIONS OF STRUCTURAL CLAY PRODUCTS

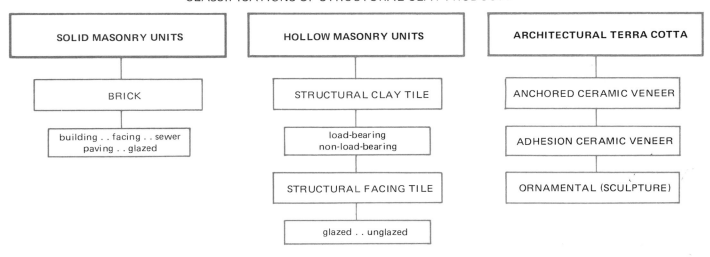

Fig. 1-2. Structural clay products are grouped under three basic types.

MANUFACTURE OF BRICK

Brick are commonly manufactured by three methods.
1. The soft mud process.
2. The stiff mud process.
3. The dry-press process.

SOFT MUD PROCESS

There are two variations of the soft mud process. They are the water-struck and sand-struck methods. Both methods prevent the clay from sticking to the mold. The clay is mixed wet and forced into a mold. The thickness of the brick is formed by scraping off the top of the mold.

In the water-struck process, the brick face is usually rather smooth, Fig. 1-3. A sand-struck brick has a sandpaper-like surface, Fig. 1-4. Particles of sand stick to the brick material during removal from the mold and create this roughness.

Fig. 1-3. The surface of this water-struck brick is smooth.

Fig. 1-4. This sand-struck brick has a sandpaper-like surface.

STIFF MUD PROCESS

In the stiff mud process, the clay is just wet enough to stick together. It is forced through a die and extruded (shaped) in the form of a ribbon or column. The thickness and width of the ribbon is the same as the dimensions of the brick being produced. Wires are used to cut the ribbon to lengths desired. Fig. 1-5 shows the brick produced by this method.

Fig. 1-5. Wire-cut brick has a rough surface.

DRY-PRESS PROCESS

In the dry-press process the clay is dry and loose. It is forced under high pressure into molds and forms the brick. These brick are very dense.

DRYING AND BURNING

"Green brick" are brick that have not been fired. They will contain from 5 to 30 percent moisture as they come from the molding or extruding machines. Most of this moisture must be evaporated before the brick are burned. Therefore, the brick are placed in a drying kiln for 24 to 48 hours. They may, on the other hand, be allowed to air dry.

When the moisture conditions are right, the brick are put into a kiln designed to maintain a certain temperature. Even distribution of heat is very important for uniform results.

The most popular types of kilns used are the PERIODIC and TUNNEL KILNS. The primary difference between the two is that the brick remain stationary in the periodic kiln and the temperature is changed for each stage of the burning. In the tunnel kiln, the brick are placed on moving cars which pass through different temperature zones.

Burning time may be as little as 50 hours in a tunnel kiln and as long as 150 hours in a periodic kiln. Cooling may require over 50 hours. This prevents checking (small cracks) and maintains the right color.

CLASSIFICATION OF BRICK

Brick are usually classified as:
1. Building brick (very often referred to as common brick).
2. Facing brick.

Fire brick is a special type of brick made from a refractory ceramic material which is highly heat resistant. It is used mainly in the construction of fireplace linings and hearths. It is not considered a main classification type.

BUILDING BRICK

Building or common brick are used chiefly as structural material where durability and strength are of more importance than appearance. Three grades of building brick are available.

1. Grade SW brick is highly resistant to frost action. It is used where brick are exposed to freezing weather. An example is brick used for foundation courses or retaining walls.
2. Grade MW brick is used where there may be exposure to temperatures below freezing, but where they are not likely to be permeated with water. An example is brick used in the face of a wall above ground.
3. Grade NW brick is designed to be used as a backup on interior masonry.

FACING BRICK

Facing brick are used on exposed surfaces where appearance is an important consideration. Such brick must also be strong and durable. Appearance is covered in three grades:

1. Type FBX. These brick are for general use in exposed exterior and interior walls and partitions where a high degree of perfection is required. They must be uniform in color and size.
2. Type FBS. This type is for general use in exposed exterior and interior masonry walls and partitions where wider color ranges and greater variation in sizes are permitted.
3. Type FBA. These brick are manufactured and selected to produce architectural effects resulting from nonuniformity in size, color and texture.

BRICK SIZES

For many years, brick were manufactured in only a few sizes, Standard, Roman and Norman. But today there are many more sizes available. We can separate them into two groups: modular and nonmodular.

A modular unit is one based on a measurement of 4 inches. It could include a measurement of 2 in. (one-half of the module) or 8 in. which is twice the module.

However, in brick, this size is only nominal. This means that it is not exactly 4 inches. Some allowance has to be made for the thickness of the mortar joints. Thus, a nominal 4 inches may be actually 3 7/8 in. or some other dimension.

Brick are available in sizes ranging in thickness from a nominal 3 in. to 12 in. Heights vary from a nominal 2 to 8 in. and lengths up to 16 in. are common.

Fig. 1-6 shows the most typical sizes of nonmodular brick. Sizes of modular brick are shown in Fig. 1-7.

Few manufacturers produce all of the sizes shown and other sizes are produced by some manufacturers. As design requirements change, new sizes may be added and less popular sizes dropped. For these reasons, it is recommended that the sizes available in a given region be checked before purchasing.

With the exception of the nonmodular "standard," "oversize" and 3 in. units, most brick are produced in modular sizes. *The nominal dimensions of modular brick are equal to the manufactured dimensions plus the thickness of the appropriate mortar joint.* See Fig. 1-8. Usually the mortar joint is 3/8 or 1/2 in. thick. The standard joint thickness (3/8 or 1/2 in.) is substracted from the length, height and thickness of the masonry unit to get the standard or specified dimensions of the unit. The actual dimensions, however, may vary from the standard usually by the following amounts:

1. Plus or minus 1/16 inch in thickness.
2. 1/8 inch in width.
3. 1/4 inch in length.

WHY MODULES ARE POPULAR

Modular brick sizes became popular because they provided a quick and simple way of estimating the number of brick needed. Since the nominal dimensions used in modular brick is the actual dimension of the brick plus the thickness of one mortar joint, the estimator can measure the area to be bricked and divide it by the nominal dimensions of the specific brick to be used.

For example, suppose that the "standard modular" brick is 4" x 2 2/3" x 8". If a mortar joint of 1/2 in. is used, the actual manufactured dimensions would be 3 1/2" x 2 1/4" x 7 1/2". These dimensions should be used when ordering the brick. Fig. 1-9 shows most common modular brick sizes.

COLOR AND FINISHES

Facing brick are produced in many colors, textures and finishes. In fact, no two brick are exactly alike. For this

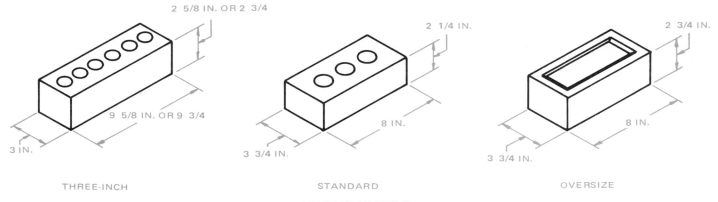

2 5/8 IN. OR 2 3/4   9 5/8 IN. OR 9 3/4   3 IN.

THREE-INCH

2 1/4 IN.   8 IN.   3 3/4 IN.

STANDARD

2 3/4 IN.   8 IN.   3 3/4 IN.

OVERSIZE

NONMODULAR BRICK

Fig. 1-6. These are nonmodular brick with actual dimensions shown.

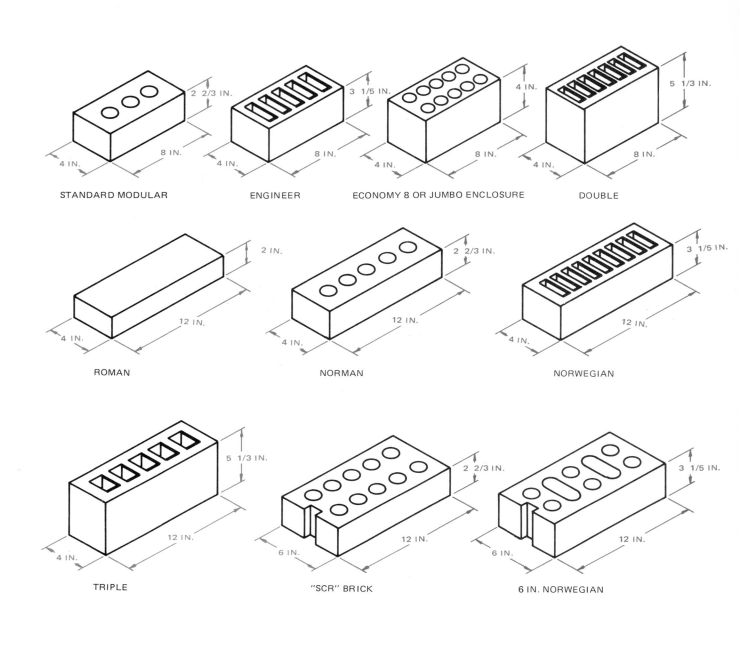

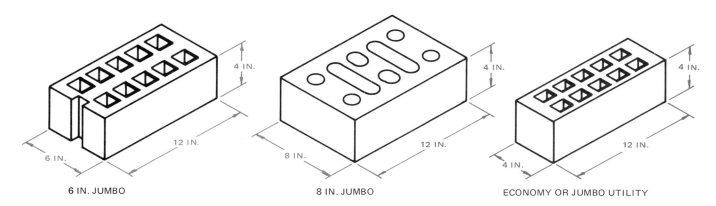

Fig. 1-7. Modular brick sizes are shown in nominal dimensions.

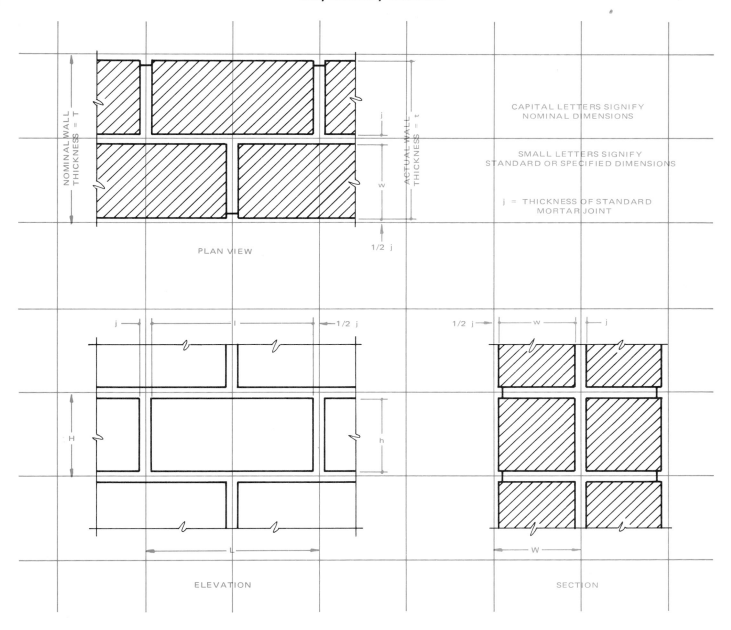

Fig. 1-8. The nominal dimensions of modular brick are equal to the actual dimensions plus the thickness of the mortar joint.

reason, ASTM (American Society for Testing and Materials) specifications for brick and structural clay facing tile state that texture and color shall conform to an approved sample showing the full range of color and texture. In general, from three to five samples will be required depending on the range of color and texture in the brick.

About 99 percent of all brick produced is in the range of the reds, buffs and creams. The actual color varies with the chemical makeup of the clay. Usually, this depends on the source. Color is also affected by the method of manufacture and the length and temperature of the burn.

Addition of chemicals and the method of molding and cutting the brick determine its texture. Much of the brick manufactured has smooth or sand finished texture. However, many other textures are produced on brick. Fig. 1-10 shows

several brick textures created during the production process.

A large sample of popular brick is included to illustrate the variety of color and texture available.

PROPERTIES OF BRICK

The term generally used to indicate burned clay units that are used primarily in building construction is STRUCTURAL CLAY PRODUCTS. They may help support the structure or they may serve only as a decorative finish. Technically, when assembled in a structure, they should support their own weight. They may be load-bearing or non-load-bearing.

Structural clay products must meet the following standards established by the Federal Trade Commission:

1. The composition is primarily of clay or shale or mixtures thereof; and

Sizes of Modular Brick

Unit Designation	Nominal Dimensions, in.			Joint Thickness in.	Manufactured Dimensions in.			Modular Coursing in.
	t	h	l		t	h	l	
Standard Modular	4	2⅔	8	⅜	3⅝	2¼	7⅝	3C = 8
				½	3½	2¼	7½	
Engineer	4	3⅕	8	⅜	3⅝	2¹³⁄₁₆	7⅝	5C = 16
				½	3½	2¹¹⁄₁₆	7½	
Economy 8 or Jumbo Closure	4	4	8	⅜	3⅝	3⅝	7⅝	1C = 4
				½	3½	3½	7½	
Double	4	5⅓	8	⅜	3⅝	4¹⁵⁄₁₆	7⅝	3C = 16
				½	3½	4¹³⁄₁₆	7½	
Roman	4	2	12	⅜	3⅝	1⅝	11⅝	2C = 4
				½	3½	1½	11½	
Norman	4	2⅔	12	⅜	3⅝	2¼	11⅝	3C = 8
				½	3½	2¼	11½	
Norwegian	4	3⅕	12	⅜	3⅝	2¹³⁄₁₆	11⅝	5C = 16
				½	3½	2¹¹⁄₁₆	11½	
Economy 12 or Jumbo Utility	4	4	12	⅜	3⅝	3⅝	11⅝	1C = 4
				½	3½	3½	11½	
Triple	4	5⅓	12	⅜	3⅝	4¹⁵⁄₁₆	11⅝	3C = 16
				½	3½	4¹³⁄₁₆	11½	
SCR brick	6	2⅔	12	⅜	5⅝	2¼	11⅝	3C = 8
				½	5½	2¼	11½	
6-in. Norwegian	6	3⅕	12	⅜	5⅝	2¹³⁄₁₆	11⅝	5C = 16
				½	5½	2¹¹⁄₁₆	11½	
6-in. Jumbo	6	4	12	⅜	5⅝	3⅝	11⅝	1C = 4
				½	5½	3½	11½	
8-in. Jumbo	8	4	12	⅜	7⅝	3⅝	11⅝	1C = 4
				½	7½	3½	11½	

Available as solid units conforming to ASTM C 216- or ASTM C 62-, or, in a number of cases, as hollow brick conforming to ASTM C 652-.

Fig. 1-9. The nominal and manufactured dimensions are shown for most common modular brick. Metric common brick will have a nominal size of 215 mm x 102.5 mm x 65 mm. Mortar joints will be 10 mm.

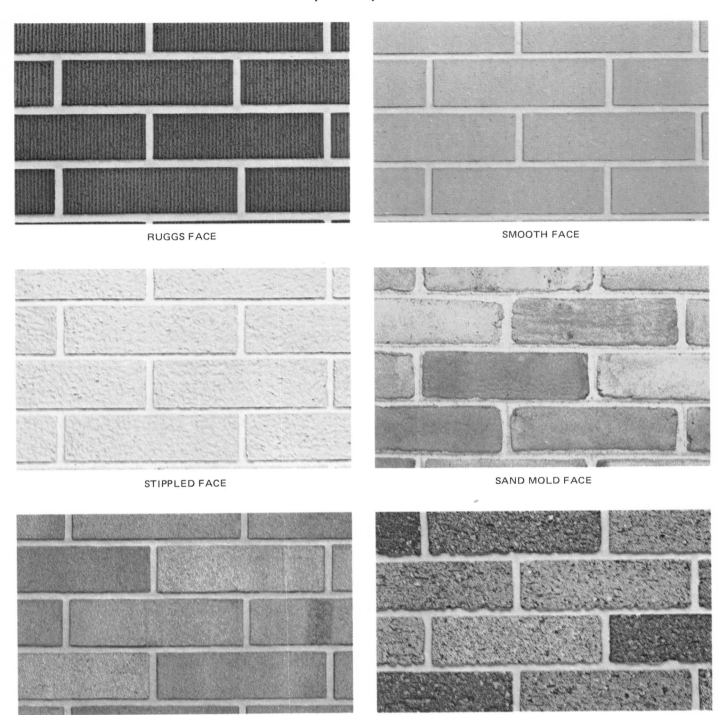

RUGGS FACE

SMOOTH FACE

STIPPLED FACE

SAND MOLD FACE

SAND STRUCK FACE

MATT FACE

Fig. 1-10. These brick textures are made during the production process.

2. The ingredients have been fused together as a result of the application of heat.

When products do not meet these requirements manufacturers must clearly show what the product is made of. Examples are: concrete brick, coral brick, plaster brick, sand-lime brick and concrete structural tile. These are not ceramic products.

Structural clay products may be separated into two groups based on how much solid material is in them:

1. Solid masonry units.
2. Hollow masonry units.

SOLID MASONRY UNITS: The American Society for Testing and Materials has set the standard. A solid masonry unit, they say, is "one whose cross-sectional area in every plane parallel to the bearing surface is 75 percent or more of its gross cross-sectional area measured in the same plane."

To know what this means we must understand some terms. The *bearing surfaces,* of course, are the tops and bottoms of

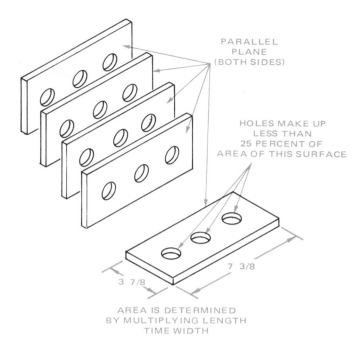

PARALLEL
PLANE
(BOTH SIDES)

HOLES MAKE UP
LESS THAN
25 PERCENT OF
AREA OF THIS SURFACE

7 3/8

3 7/8

AREA IS DETERMINED
BY MULTIPLYING LENGTH
TIME WIDTH

Fig. 1-11. Brick with holes in it are still considered "solid" brick if 75 percent of the area in any parallel plane is solid material. Imagine that the brick is sliced like the drawing above. Each cut surface is a parallel plane. To find area, multiply length by width. Area of the holes is found by multiplying 3.14 by the square of the radius.

the brick or block.

Now we need to know what a "parallel plane of the bearing surface" is. Just imagine that you have placed a brick on its side and have sliced it into a number of thin pieces. The sides of each slice would be parallel planes. If you measured the area of the slices, then subtracted the area of the holes in the brick you could tell what percent is solid material. See Fig. 1-11.

Many brick are solid. However, many facing brick have holes running through them. These cored brick are usually classified as solid masonry units because 75 percent of their area is solid material. The cored holes reduce weight, permit easier handling and make the mortar joint stronger.

HOLLOW MASONRY UNITS: A hollow masonry unit is less than 75 percent solid material in "its net cross-sectional area in any plane..." Structural clay tile are included in this classification.

WEIGHT OF BRICK: The weight of a brick unit will depend on materials used, method of manufacture, burning and the size of the brick. Since there are so many variables involved, you will have to check information secured from the manufacturer. Building brick weigh about 4 1/2 to 5 lb. each.

DURABILITY: Several conditions affect the weathering (durability) of a brick. These include:

1. Heat.
2. Cold.
3. Wetting.

NEGLIGIBLE WEATHERING

MODERATE WEATHERING

SEVERE WEATHERING

Fig. 1-12. Regions of the United States have been marked for severity of the weather.

4. Drying.

5. The action of soluble salts.

Fig. 1-12 shows the areas of the United States in which brick construction is subjected to severe, moderate and negligible weathering.

The weathering index for any locality may be found or estimated from material published by the Weather Bureau, U.S. Department of Commerce. This material is in the form of tables of "Local Climatological Data." A severe weathering region has an index greater than 500. A moderate weathering region has an index of 100 to 500. A negligible weathering region has an index of less than 100. The only weathering action that has any real effect on burned clay products is freezing and thawing when moisture is present.

EFFLORESCENCE: Efflorescence is a white powder or salt-like deposit on masonry walls. It is caused by water-soluble salts, Fig. 1-13. These salts collect on the wall's surface as water evaporates. The salts may have been mixed into the masonry units, the mortar or plaster.

Several things must happen for efflorescence to develop:

1. There must be soluble salts in the masonry wall.

2. Moisture must pass through the wall and carry the salts to the surface.

If either of these conditions are taken away, efflorescence will not occur.

COMPRESSIVE STRENGTH: *Compressive strength is the ability of a masonry unit to stand up under heavy weight without crumbling.* Four things will affect this strength:

1. Strength of the units (bricks) themselves.

2. Strength of the mortar.

3. Quality of workmanship.

4. The proportion of gross area that is given a bearing surface at bed joints.

High compressive strength is developed by using high strength brick and mortars and making well-filled joints.

In most construction today, compressive strengths do not exceed 100 psi (per square inch). Most brick have compressive strengths of over 4500 psi. Typical mortars have strengths

from 750 to 2500 psi. Therefore, we can say that most masonry walls have ample compressive strength for current construction practices.

BONDS AND PATTERNS

The strength, durability and appearance of a brick wall depends to a great extent upon the *bond* used to lay up the wall. The word "bond," when it refers to masonry, has three meanings:

1. Structural bond is the method by which individual units are interlocked or tied together to form a wall. See Fig. 1-14.

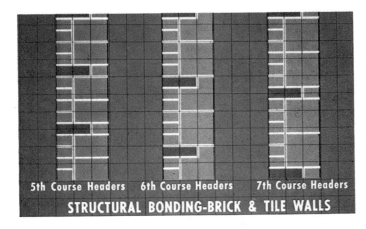

Fig. 1-14. Three typical brick and tile walls are shown. They illustrate various structural bonding methods using headers.
(Brick Institute of America)

2. Pattern bond is the pattern formed by the masonry units and the mortar joints on the face of a wall. The pattern may be the result of the structural bond used or may be purely decorative.

3. Mortar bond is the adhesion of mortar to the masonry units or reinforcing.

STRUCTURAL BONDS: Structural bonding in brick masonry may be done in three ways:

1. By arranging the brick in an overlapping fashion.

2. By the use of ties embedded (buried) in the mortar joints, Fig. 1-15.

Fig. 1-13. Efflorescence on masonry walls is caused by soluble salts in wet brick being carried to the surface as water evaporates.
(Portland Cement Assoc.)

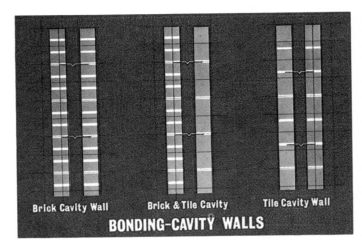

Fig. 1-15. Structural bonding of cavity walls using metal ties.

3. By the application of grout to adjacent *wythes* of masonry. (A wythe is a single vertical tier or stack of masonry.)

The overlapping method is best illustrated by two traditional bonds known as ENGLISH BOND and FLEMISH BOND, Fig. 1-16. The English bond consists of alternating courses of

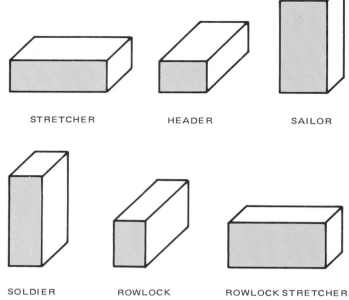

Fig. 1-17. Each brick position has a specific name.

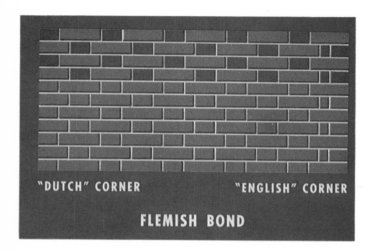

Fig. 1-16. These two traditional bonds are used in brick masonry.

headers and stretchers. (Alternating means to switch back and forth between two things. A *course* of brick is a continuous level row of brick.) The Flemish bond consists of alternating headers and stretchers in every course, so that the headers and stretchers in every other course appear in vertical lines.

Stretchers are the brick laid with the length of the wall. They develop longitudinal (lengthwise) bonding strength. Headers, laid across the width of the wall, bond the wall across its thickness. Fig. 1-17 identifies the name of six brick positions. Fig. 1-18 identifies the various brick positions in a wall.

Modern building codes specify that masonry-bonded brick walls must have no less than 4 percent headers in the wall surface. The distance between nearest headers should not exceed 24 inches vertically or horizontally.

Structural bonding of masonry walls with metal ties is alright for solid wall and cavity wall construction. See Fig. 1-19. No less than one 3/16 in. diameter metal tie should be used for each 4 1/2 sq. ft. of wall surface.

Ties in alternate courses should be staggered with the distance between adjacent (next to) ties not more than 24 inches vertically or 36 inches horizontally. Additional bonding ties should be placed around the perimeter of openings. These should be spaced not more than 3 ft. apart and within 12 in. of the opening. If smaller diameter ties are used, the spacing between them should be reduced.

Solid and reinforced brick masonry walls may be structurally bonded by pouring GROUT into the cavity between wythes of masonry. (Grout is a liquid cement mixture used to fill cracks.)

The method of bonding used will depend on the wall type, use requirements and other factors. The metal tie is usually recommended for exterior walls. It makes construction easier and provides greater resistance to rain penetration. The metal tie also allows for slight movements of facing and backing, which may prevent cracking.

PATTERN BONDS: Often, structural bonds, such as Flemish or English, are used to create patterns in the face of the wall. Patterns may be created by the way the mortar joint is handled. The arrangement of the brick remains unchanged.

It may also be desirable to produce a pattern bond by allowing certain brick to stick out beyond or sink below the plane (surface) of the wall. Pattern bonds can also be created by varying the texture of brick units.

FIVE BASIC BONDS: There are five basic structural bonds. They are:
1. The running bond.
2. The common or American bond.
3. The Flemish bond.
4. The English bond.
5. The block or stack bond.

Color and texture in both brick and joints create endless

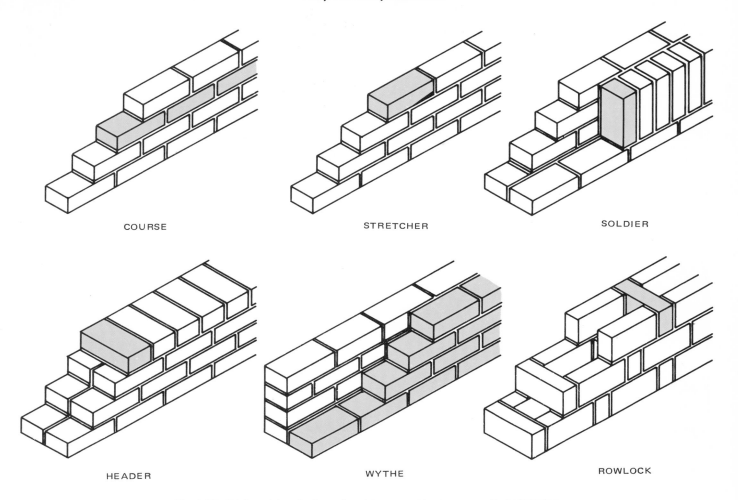

COURSE

STRETCHER

SOLDIER

HEADER

WYTHE

ROWLOCK

Fig. 1-18. Brick positions in the wall and terms used in masonry wall construction.

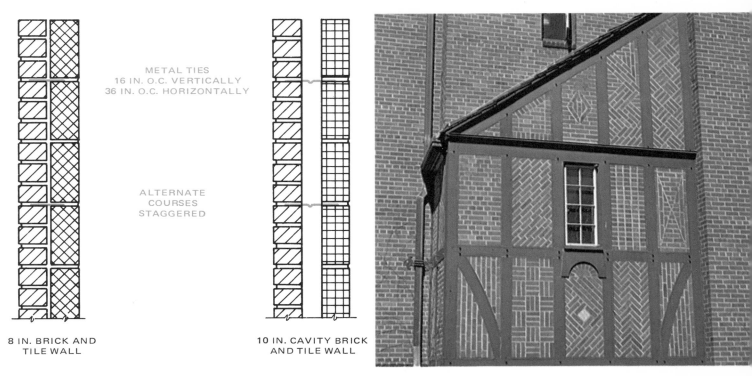

METAL TIES
16 IN. O.C. VERTICALLY
36 IN. O.C. HORIZONTALLY

ALTERNATE
COURSES
STAGGERED

8 IN. BRICK AND
TILE WALL

10 IN. CAVITY BRICK
AND TILE WALL

Fig. 1-19. A solid masonry wall and cavity wall bonded with metal ties.

Fig. 1-20. This wall is decorated with a collection of pattern bonds.

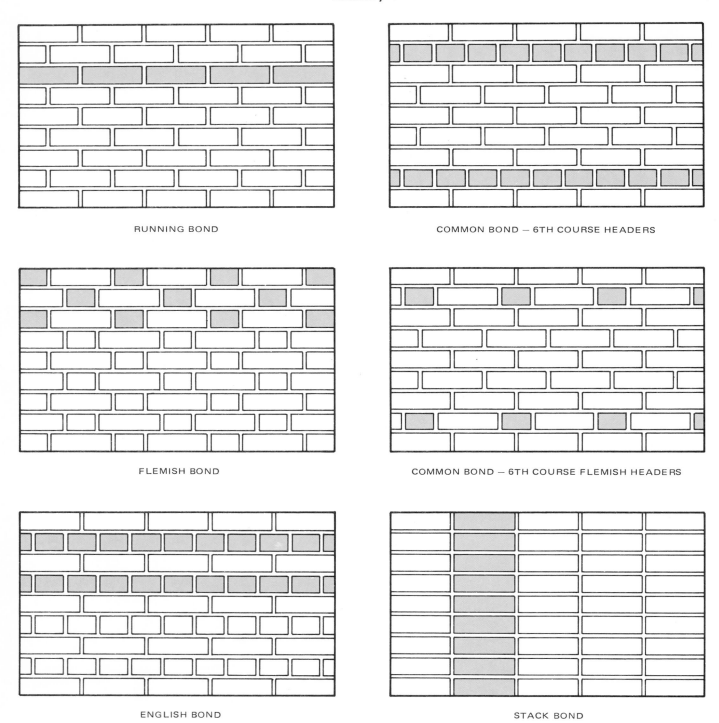

RUNNING BOND

COMMON BOND — 6TH COURSE HEADERS

FLEMISH BOND

COMMON BOND — 6TH COURSE FLEMISH HEADERS

ENGLISH BOND

STACK BOND

Fig. 1-21. The basic structure bonds.

variety. Fig. 1-20 shows how many different patterns can be arranged. Fig. 1-21 illustrates the five basic structural bonds.

RUNNING BOND is the simplest of the basic pattern bonds. It consists of all stretchers and is used largely in cavity wall and veneer wall construction. See Fig. 1-22.

COMMON or AMERICAN BOND is a variation of the running bond. It has a course of full-length headers at regular intervals to provide structural bonding as well as pattern. Header courses generally are used every fifth to seventh course, as shown in Fig. 1-23.

FLEMISH BOND alternate courses of stretchers and headers, Fig. 1-21. The headers in alternate courses are centered over the stretchers in the course between them.

If the headers are not needed for structural bonding, as in veneer wall construction, half brick may be used. These half brick are called *clipped* or *snap* headers.

The Flemish bond may be varied by increasing the number of stretchers between headers in each course. Three stretchers alternating with a header is known as a *garden wall* bond. Two stretchers between headers is called a *double stretcher garden*

Fig. 1-22. This reclaimed brick is layed in running bond.

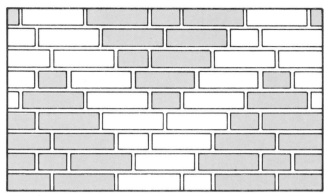

GARDEN WALL BOND WITH UNITS
IN DOVETAIL FASHION

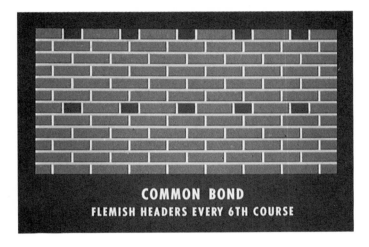

COMMON BOND
FLEMISH HEADERS EVERY 6TH COURSE

DOUBLE STRETCHER GARDEN WALL BOND
WITH UNITS IN DIAGONAL LINES

Fig. 1-24. Two variations of the garden wall bond.

Snap headers may be used when structural bonding is not necessary.

ENGLISH CROSS or DUTCH BOND is a variation of English bond. The only difference is that vertical joints between the stretchers in alternate courses do not align vertically. These joints center on the stretchers themselves in the courses above and below.

BLOCK or STACK BOND is purely a pattern bond. All vertical joints are aligned. There is no overlapping of units. See Fig. 1-25. Since the bond lacks the strength that other bonds

COMMON BOND
FULL HEADERS EVERY 6TH COURSE

Fig. 1-23. The common or American bond is a variation of the running bond. Two variations are shown here.

wall bond. Four or five stretchers between headers may be laid to form other garden wall bonds. See Fig. 1-24.

ENGLISH BOND is made up of alternate courses of headers and stretchers. The headers are centered on the stretchers and joints between stretchers in all courses are aligned vertically.

Fig. 1-25. In a stack bond, all vertical joints line up.

Fig. 1-26. Glazed brick of various colors were used to create this pattern.

provide, it is usually bonded to the backing with rigid steel ties. In load-bearing construction or large wall areas, it is advisable to reinforce the wall with steel. This is placed in the horizontal mortar joints.

Fig. 1-26 shows a pattern that was made through variations in color of the brick used. This is a common way to develop patterns.

WALL TEXTURE: Many new and interesting patterns and effects are created by projecting and recessing units, Fig. 1-27,

and omitting units to form perforated walls or screens, Fig. 1-28. These techniques greatly extend the traditional patterns and add a new dimension to brick masonry. Fig. 1-29 shows a modern brick sculptured wall which incorporates a variety of colors and textures.

MORTAR JOINTS: Mortar between brick units performs four functions:

1. Bonds units together and seals spaces between them.
2. Bonds to reinforcing steel and causes it to be part of wall.

Fig. 1-27. Pattern bonds may be produced by projecting or recessing certain brick from the plane (surface) of the wall as shown in these illustrations.

Fig. 1-28. Brick screens are highly decorative. They also help reduce noise and control light.

Fig. 1-29. A modern sculptured brick wall.
(Brick Institute of America)

3. It compensates for variations in the size of brick units.

4. It provides a decorative effect.

Mortar joint finishes are of two types: troweled and tooled joints. In the troweled joint, the excess mortar is struck (cut off) with a trowel and finished with the same tool. The tooled joint is made with a special tool to compress and shape the mortar in the joint. Fig. 1-30 shows typical mortar joints used in brick masonry.

CONCAVE and V-SHAPED joints are formed by the use of a steel jointing tool. These joints are usually rather small. They are effective in resisting rain penetration. Their use is recommended where heavy rains and high winds are likely to occur. This is the most popular mortar joint.

The WEATHERED joint requires care since it must be worked from below. It is a very functional joint because it is compacted and readily sheds water.

A STRUCK joint is common for ordinary brickwork. The joint is struck with the trowel and some compacting occurs,

Fig. 1-30. Typical mortar joints used in masonry work.

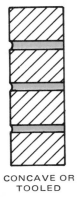
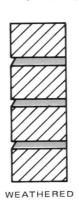
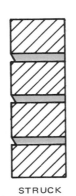
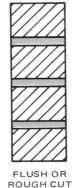
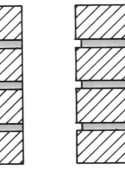

| CONCAVE OR TOOLED | V-SHAPED | WEATHERED | STRUCK | FLUSH OR ROUGH CUT | RAKED | EXTRUDED |

but it does not shed water as well as other joints.

The ROUGH CUT or FLUSH joint is the simplest joint to make. It is made by holding the trowel flat agains the brick and cutting in any direction. The joint is not compacted and leaves a small hairline crack where the mortar is pulled away from the brick by the trowel. This joint may not be watertight.

The RAKED joint is made by removing some of the mortar while it is still soft. This joint is difficult to make watertight even though slightly compacted. It is not recommended for areas with heavy rains, high winds and freezing temperatures. This joint is, however, very effective with irregular brick. It tends to enhance their character.

The EXTRUDED joint is made by squeezing out the mortar as the unit is laid. It is not trimmed off but is left to harden in its extruded form. This joint is best suited for dry climates, Fig. 1-31.

Fig. 1-31. The extruded mortar joint is best suited for dry climates because its ledges do not shed water very well.
(Portland Cement Assoc.)

Colored mortars may be used to accent brick masonry patterns, Fig. 1-32. It may be applied in two ways:
1. The entire mortar joint may be colored.
2. Where a tooled joint is used, tuck pointing is preferred.

When tuck pointing, the entire wall is finished with a 1 in. deep raked joint. The colored mortar is filled in later, with special care taken to keep the colored mortar off the brick.

HOLLOW MASONRY UNITS (TILE)

Hollow masonry units are a machine made product. They are extruded (squeezed) through a die and cut to the desired height or length. The raw materials may be surface clay, shale,

Fig. 1-32. Colored mortar may be used to accent the brick pattern.
(Brick Institute of America)

fire clay or combinations of these. They are pulverized (crushed), mixed with water and then extruded. After the moisture content is reduced to a satisfactory level, they are fired in a kiln at temperatures from 1750 to 2500 F.

CLASSIFICATION OF TILE

Unlike brick, tile is hollow. Hollow in this instance means that each masonry unit has a net cross-sectional area in any plane parallel to the bearing surface less than 75 percent of its gross cross-sectional area measured in the same plane. (Standards of the American Society for Testing and Materials)

Hollow clay tile may be separated into two groups:
1. Structural clay tile.
2. Structural facing tile.

STRUCTURAL CLAY TILE

Structural clay tile are produced as *load-bearing* and *non-load-bearing* types. Both load-bearing and non-load-bearing tile are produced with vertical or horizontal cells. (These are the open spaces through the tile.)

When the tile is to be laid with the cells in a horizontal plane, the unit is called *side construction* or *horizontal cell* unit. When the tile is to be laid with the cells in a vertical plane, the unit is called *end construction* or *vertical cell* tile. The size, number, shape and thickness of cells will vary from one manufacturer to another.

LOAD-BEARING TILE: Structural clay load-bearing tile are suitable for use in walls as backup or in partitions carrying superimposed loads (weight of other parts). Load-bearing wall tile are produced in two grades: LB and LBX.

Grade LBX tile are suitable for use in areas exposed to frost action. Grade LB tile are designed for areas not exposed to frost action.

NON-LOAD-BEARING TILE: Structural clay non-load-bearing tile are suitable for use as fireproofing, furring or the construction of partitions which do not support superimposed loads. Non-load-bearing tile is available in one grade, NB.

SIZES AND SHAPES: Structural clay tile is most often used in actual thickness of 3 in. and nominal thicknesses of 4, 6 and 8 in. However, nominal 2, 10 and 12 in. thick units are

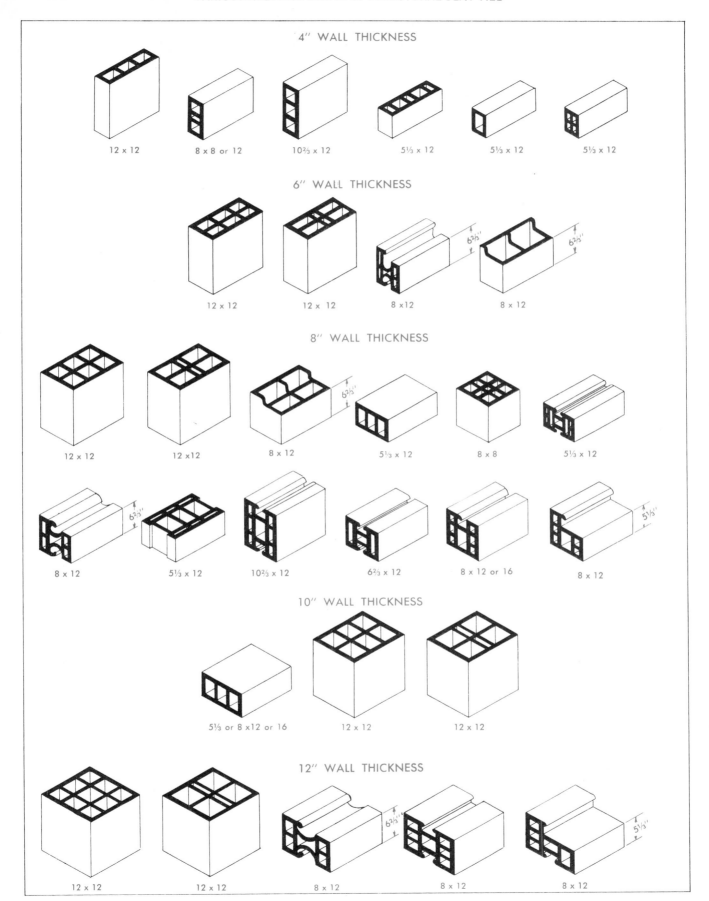

Fig. 1-33. All tile dimensions are nominal. Face dimensions are given height by length.

available. Heights for clay tile units are 5 1/3, 8 and 12 in. nominal.

Tile unit length is generally 12 in. nominal. However, 8 in. and 16 in. units are available in some areas.

Fig. 1-33 shows some of the many shapes and sizes of structural clay tile. Other shapes include header units, jamb units, closures and kerfed units for use in reinforced bond beams or lintels.

The most widely available unit shapes and sizes of structural clay tile are shown in Fig. 1-34. The units shown are all horizontal cell tile. The same sizes are also manufactured in vertical cell tile units.

All structural clay tile are designed to be used with 1/2 in. mortar joints. All are modular sizes. The basic group shown in Fig. 1-34 provides compatability (can be used together) with all facing tile and modular brick sizes.

PROPERTIES OF STRUCTURAL CLAY TILE: Properties include: color, texture, strength and variations in dimensions.

Structural clay tile are produced in a wide range of colors. The greatest influence upon the color will be the chemical compostion of the raw materials and the temperature of the burn. Due to these factors, color alone cannot be used as a measure of quality.

Structural clay tile are produced with surface textures. The two most generally seen are the scored and wire cut. Wire cut is known as "universal finish" in some regions. Both surfaces will readily receive plaster and the wire cut may be painted or left exposed.

The compressive strength of structural clay tile is affected not only by the raw materials and method of manufacture, but also by the design of the unit. ASTM *Standard Specifications for Structural Clay Load-bearing Wall Tile* C34-62 requires that the average compressive strength of five units be not less than the following:

AVERAGE MINIMUM COMPRESSIVE STRENGTH

Grade	End Construction Tile (in psi)	Side Construction Tile (in psi)
LBX	1400	750
LB	1000	750

Most clay tile units produced today will have compressive strengths of two to four time these requirements.

Variations in sizes are also controlled by ASTM. The biggest allowable variation is 3 percent more or less than the dimension called for.

MOST WIDELY AVAILABLE SIZES OF STRUCTURAL CLAY TILE.

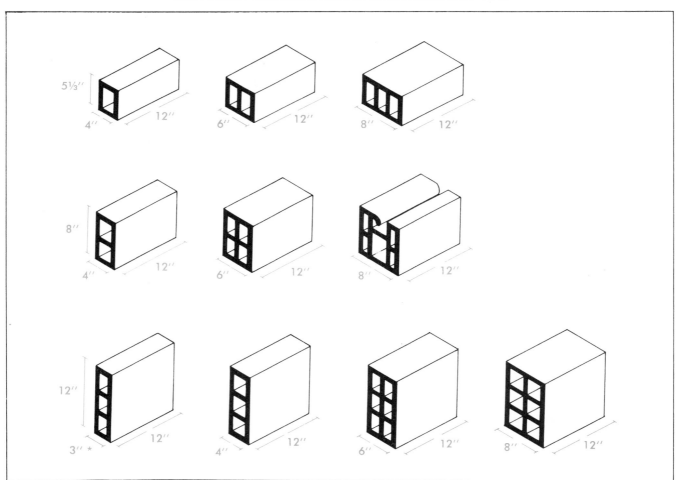

Fig. 1-34. All units shown are horizontal cell tile, but vertical cell tile are also avilable. Dimensions are nominal.

STRUCTURAL FACING TILE

Structural clay facing tile are of two broad types, GLAZED and UNGLAZED, Fig. 1-35. Glazed clay facing tile are produced from high-grade, light-burning fire clay suitable for the application of ceramic glaze. (They may be left unglazed.) There are two grades of glazed facing tile: S and G. Grade S (select) is intended for use with somewhat narrow mortar joints. Grade G (ground edge) is used where face dimension must be very exact. An example of this would be a stacked bond. (Units are placed in the wall so that both the horizontal and vertical joints are continuous.) Both grades may be purchased as single-faced units or two-faced units. (A single-faced unit will have one face exposed. In a two-faced unit, two opposite finished faces will be exposed.)

Unglazed facing tile are made from either light-burning fire clay or shale and other darker-burning clays. These tile are not glazed and have either a smooth or a rough-textured finish. They may be purchased as standard or heavy duty tile. The difference is in the thickness of the face shells.

Another classification for unglazed facing tile is based on factors that control appearance of the finished wall. This classification includes FTX and FTS types:

TYPE FTX UNGLAZED FACING TILE: This tile is suitable for general use in exposed exterior and interior masonry walls

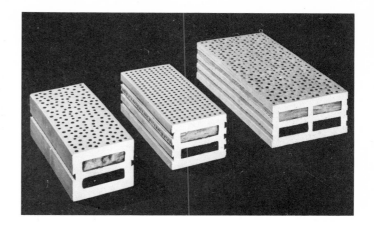

Fig. 1-36. These are typical SCR acoustile units designed to absorb sound. Note use of insulating blanket behind the face.
(Brick Institute of America)

and partitions. It is easily cleaned, resists staining and is low in absorption. This type is also best where these characteristics are important:

1. A high degree of mechanical perfection.
2. Narrow color range.
3. A minimum variation in face dimensions.

TYPE FTS UNGLAZED FACING TILE: Smooth or rough textured, these tile are suitable for:

1. General use in exposed exterior or interior masonry walls and partitions.
2. Special conditions where tile of moderate absorption are required.
3. Usage where moderate variation in face dimensions and medium color range is acceptable.
4. Usage where minor defects in surface finish or small

Fig. 1-35. Structural facing tile is produced with either a glazed or an unglazed face.

STRUCTURAL CLAY FACING TILE
AVAILABLE SIZES

SERIES	NOMINAL FACE DIMENSIONS IN INCHES	NOMINAL THICKNESS IN INCHES
6T	5 1/3 by 12	2, 4, 6, 8
4D	5 1/3 by 8	2, 4, 6, 8
4S	2 2/3 by 8	2, 4
8W	8 by 16	2, 4

NOMINAL MODULAR SIZES*

FACE DIMENSIONS

HEIGHT BY LENGTH IN INCHES	HEIGHT BY LENGTH IN INCHES
4 by 8	8 by 8
4 by 12	8 by 12
5 1/3 by 8	8 by 16
5 1/3 by 12	12 by 12

Thickness: All of the above are in nominal thicknesses of 4, 6 and 8 inches.

*Nominal sizes include the thickness of the standard mortar joint for all dimensions.

Fig. 1-37. The typical sizes of structural facing tile. Joint thickness should be deducted from the dimensions above in determining actual size of the units.

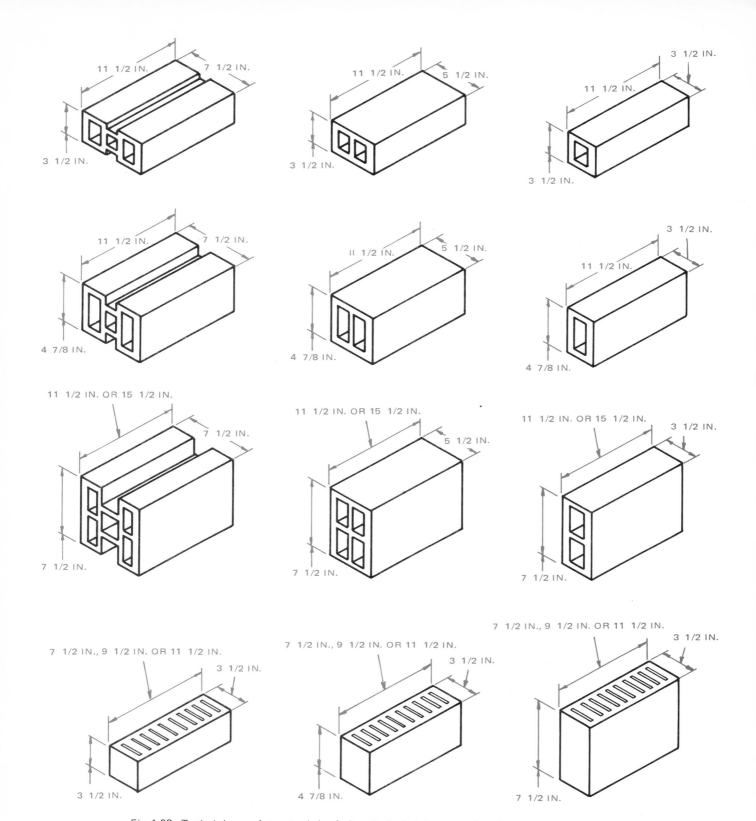

Fig. 1-38. Typical shapes of structural clay facing tile. Included are stretchers, jamb units and corner units.

handling chips are acceptable.

SCR ACOUSTILE: This special type is designed to absorb sound. Typical units are shown in Fig. 1-36. Distinguishing characteristics are the perforations through the faces and the fibrous glass pads behind the faces. The perforations (holes) may be circular or slotted, the same or different in size, and regular or random pattern.

All types of structural clay facing tile, glazed or unglazed, can have an acoustile. These units may be used in wall or ceiling applications.

SIZES AND SHAPES: Some sizes of structural facing tile are listed in Fig. 1-37. Some makers also produce these units with a nominal 4" by 12" face size.

Structural facing tile are designed to be laid with a 1/4 in.

mortar joint. The specified unit dimensions are therefore 1/4 in. less than the nominal dimensions shown in Fig. 1-37 under "Available Sizes."

Nominal modular sizes of structural clay facing tile are shown in the bottom part of Fig. 1-37. Units shown in nominal modular sizes are usually designed to be laid with either 1/4, 3/8 or 1/2 in. joints.

The basic shapes of structural clay facing tile are:
1. Stretchers.
2. Corners and closures.
3. Starters and miters.

A wide variety of supplementary shapes are also available for complex wall layouts. These include:
1. Sills.
2. Caps and lintels.
3. Cove base stretchers and fittings.
4. Coved internal corners.
5. Octagons.
6. Radials.
(Fig. 1-38 shows some of the typical shapes.)

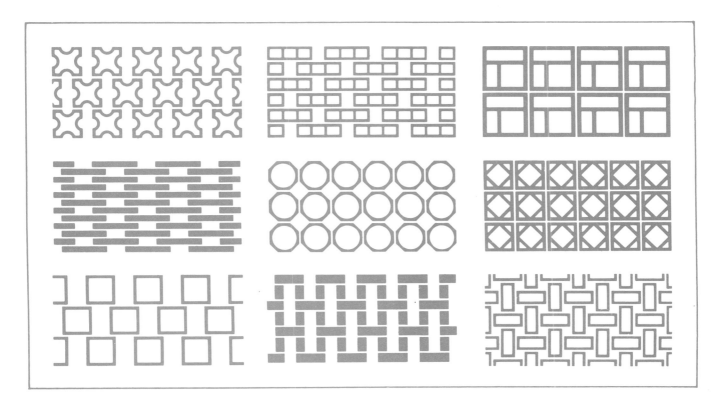

Fig. 1-39. Clay tile in shapes used for decorative screens.

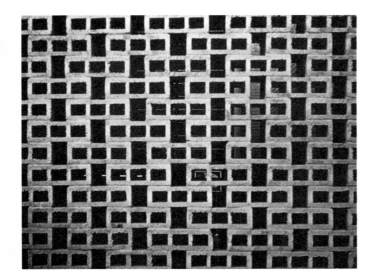

Fig. 1-40. A wall in one of the patterns shown (top center) in Fig. 1-39.

All structural clay facing tile are modular. This means that all units are multiples of 4 in. They will fit together with each other and with all other modular building materials. This reduces cutting. But some units have a dimension of 5 1/3 in. which is one-third of 16 in. rather than a multiple of 4 in.

Clay tile is also produced in various shapes for use as decorative screens. Figs. 1-39 and 1-40 shows some of the different patterns possible with tiles.

ARCHITECTURAL TERRA COTTA

Architectural terra cotta is a custom (made to order) product. It is produced in sizes and shapes to meet specific requirements. An unlimited color range of high-fired, ceramic vitreous (glass-like) glazes is available, Fig. 1-41. The configurations, (arrangement of parts) forms and radius shapes in which architectural terra cotta can be made are almost unlimited. It is not affected by weather.

Fig. 1-41. These are examples of the broad color range in which architectural terra cotta is produced. (Interpace Corp.)

CLASSIFICATION OF ARCHITECTURAL TERRA COTTA

Architectural terra cotta includes several classes of products:

1. Anchored ceramic veneer.
2. Adhesion ceramic veneer.
3. Ornamental or sculptured terra cotta.

ANCHORED CERAMIC VENEER

Anchored ceramic veneer is usually produced in thicker sections held in place by grout and wire anchors connected to its backing. Typical maximum sizes are shown in Fig. 1-42. These are standard die sizes. It is cheaper to use these sizes. Anchored ceramic veneer may be used as facing for exterior and interior walls, column facings, soffits, copings and sills.

ADHESION CERAMIC VENEER

Adhesion ceramic veneer is usually produced in thin sections. It is held in place without metal anchors by a mortar

ANCHORED CERAMIC VENEER
MAXIMUM FINISHED SIZES

A	B	C
1' — 5 3/4''	3' — 0''	7 3/4''
1' — 6 3/4''	3' — 0''	7 3/4''
1' — 7 3/4''	3' — 6''	7 3/4''
1' — 8 3/4''	3' — 6''	7 3/4''
1' — 9 3/4''	3' — 6''	7 3/4''
1' — 11 3/4''	4' — 3''	11 3/4''
2' — 1 3/4''	4' — 3''	11 3/4''
2' — 3 3/4''	4' — 3''	11 3/4''
2' — 5 3/4''	4' — 3''	1' — 2 3/4''
2' — 8 3/4''	4' — 3''	11 3/4''

Fig. 1-42. Anchored ceramic veneer is produced in the maximum finished sizes shown.

backing. Thickness is ordinarily 1 in. Largest lengths and widths are shown in Fig. 1-43.

Adhesion ceramic veneer may be installed over concrete or masonry, metal or wood. It is used primarily as facing for exterior and interior walls and coping units, Fig. 1-44.

ADHESION CERAMIC VENEER
MAXIMUM FINISHED SIZES

A	B	C
5 3/4''	2' — 0''	3 3/4''
7 3/4''	2' — 0''	3 3/4''
11 3/4''	2' — 0''	3 3/4''
1' — 2 3/4''	2' — 2''	3 3/4''
1' — 3 3/4''	2' — 2''	3 3/4''
1' — 5 3/4''	2' — 2''	5 3/4''
1' — 7 3/4''	2' — 6''	4 1/2''

Fig. 1-43. Adhesions ceramic veneer is produced in thse maximum finished sizes.

Fig. 1-44. Adhesion ceramic veneer is available in a variety of colors for use as facing and coping units. (Interpace Corp.)

ORNAMENTAL OR SCULPTURED TERRA COTTA

Ornamental terra cotta may be produced in sculptured patterns or free-standing sculpture. It is used frequently as cornices and column capitals on large buildings.

Ornamental terra cotta may be either anchored or adhesion veneer. It depends on the application. Molded pieces may also be made in larger dimensions than facing veneer.

REVIEW QUESTIONS — CHAPTER 1

1. Structural clay products may be classified into three main types. Name the types.
2. Three methods of manufacturing brick are used. They are the soft mud process, the stiff mud process and the _____ process.
3. Brick remain stationary while being fired in a periodic kiln, but are placed on moving cars in the _____ kiln.
4. Brick are usually classified as facing brick and _____ brick.
5. Name the three grades of brick.
6. When appearance is important as well as durability and strength, _____ brick are used.
7. A brick whose nominal dimensions are based on the 4 in. module is called a _____ brick.
8. The usual mortar joint thickness for brick is _____ or _____ inch.
9. Identify three factors which determine brick color.
10. A masonry unit whose net cross-sectional area in every plane parallel to the bearing surface of 75 percent or more of its gross cross-sectional area measured in the same plane is known as a _____ masonry unit.
11. Efflorescence is caused by _____ salts.
12. Three types of bonds were discussed in the chapter: structural bond, pattern bond and mortar bond. Identify the three methods of producing a structural bond.
13. A _____ of brick is a continuous level row of brick.
14. A brick which is laid across the width of a wall is called a _____.
15. When a structural bond is used to create patterns in the face of the wall, you then have a _____ bond.
16. The simplest of the basic pattern bonds is the _____ bond.
17. Mortar joint finishes are of two types: troweled and _____ joints.
18. Name four types of mortar joints.
19. Hollow clay tile may be classified into two groups: structural clay tile and structural facing tile. Structural clay tile is produced as _____ and _____ types.
20. Tile units are generally a nominal _____ inches in length.
21. All structural clay tile is designed to be used with _____ inch mortar joints.
22. Structural clay facing tile may be divided into two broad types: _____ and _____.
23. Identify the three classification types of architectural terra cotta.

Basic White Sand Mold Brick

Sand Mold Brick in Zinc-flashed Pinks and Cross Sets

Smooth Reds

Smooth Finished in Ebony with Iron Spots

Sand Finished with Soft Pinks and Dusty Rose

Smooth Gray with Manganese Specks

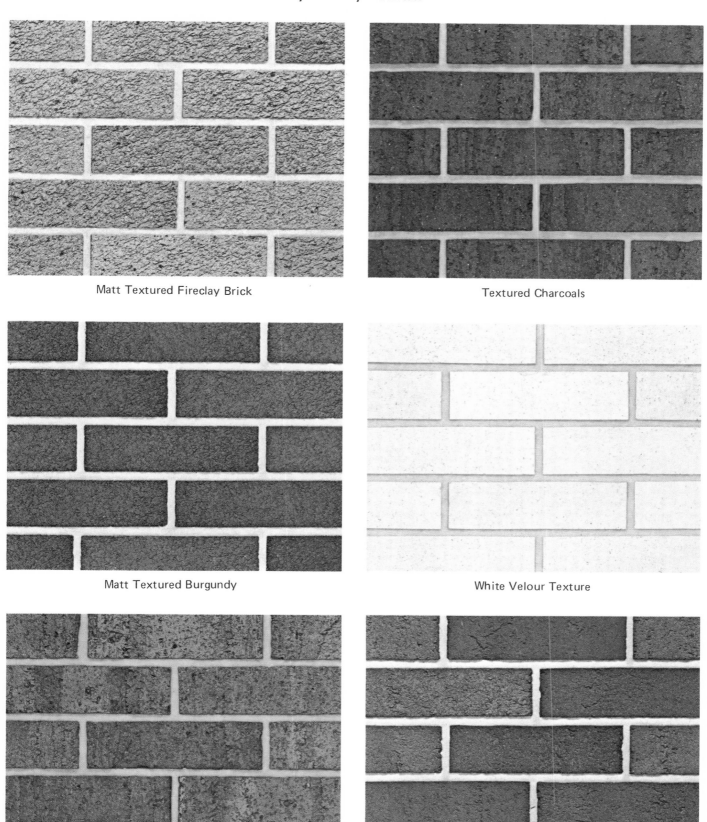

Matt Textured Fireclay Brick

Textured Charcoals

Matt Textured Burgundy

White Velour Texture

Tan Flashed Shades and Cross Sets

Water Struck Brick in Fine Sand-finished Texture

Textured Brick in Tan and Dark Brown

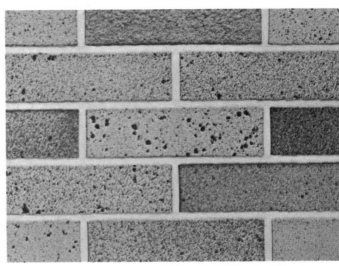

Semismooth Texture in Grays and Flashed Shades

Matt-textured Winter White

Standard Brick, Smooth Finish

Matt Texture in Buff

Semismooth in Reds, Cross Sets, Hearts, Browns and Tans

Clay Masonry Materials

Matt Textured Multicolors

Matt Texture in Pinks and Tans

Sand Finished in Buff Pink, Tan and Flashed Shades

Matt Texture in Gray Shades

Rustic Sand Finish

Antique Finish

Textured Snow White

Sand Finish in Pinks, Reds and Flashed Shades

Sand Texture in Brown

Double-fired, Glazed, Smooth and Textured

Sand Mold Brick in Reds, Tans, Cross Sets and Earth Tones

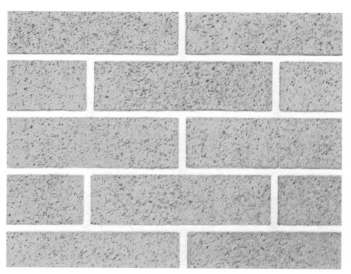

Velour Texture in Pink

Clay Masonry Materials

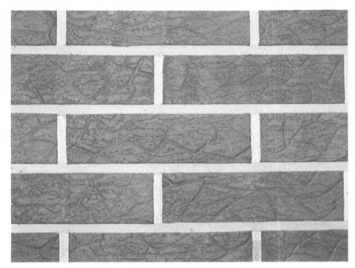

Molded, Textures in Brown

Textured in Grays, Green and Brown

Sand Molded in Grays and Tans

Sand Molded in Earth Tones

Colonial Russet Brown

Standard Striated in Brown

Cast concrete which gives the appearance of stone or masonry units. Florida Southern College at Lakeland, Florida. Designed by Frank Lloyd Wright.

Chapter 2
CONCRETE MASONRY UNITS

The term "concrete masonry" includes all the sizes and kinds of hollow or solid block, brick and concrete building tile, so long as they are made from concrete and are laid by masons.

There is no doubting the usefulness of concrete masonry. We have only to look at the range of sizes and shapes of units being produced today. A modern block manufacturing plant will offer over 100 different sizes and shapes. Across the country, well over 700 different units are made.

Concrete masonry is popular because it is cheap to make and easy to use. It is a good insulation, has good sound reduction properties and stands up well in all kinds of weather. It is fire resistant and can be bought almost anywhere in many sizes, textures and shapes.

HOW CONCRETE MASONRY UNITS ARE MADE

Concrete masonry units are made from a relatively dry mixture of portland cement, aggregates, water and often admixtures. Sometimes other cement-like materials are used in place of portland cement. (The term "aggregate" means sand and gravel or a suitable substitute. "Admixtures" are such things as coloring agents, air-entraining materials, accelerators,

retarders and water repellents.) These materials are mixed. Then the mixture is molded into desired shapes through compaction and vibration. (Compaction is a squeezing action to make a material more dense.) They are then cured under controlled moisture and temperature conditions. After a period of aging they are ready for use.

AGGREGATES

Aggregates normally make up about 90 percent of the block by weight. They have an important effect on the properties of the block. Desirable properties of aggregates include:
1. Uniform amounts of fine and coarse sizes.
2. Toughness, hardness and strength to resist impact abrasion and loading.
3. Ability to resist the forces exerted by freezing, thawing, expansion and contraction.
4. Cleanliness and lack of foreign materials which would reduce strength or cause surface roughness.

There are two classes of aggregates by weight:
1. Normal weight (dense).
2. Lightweight.

Normal weight aggregate includes sand, gravel, crushed

PROPERTIES OF CONCRETE BLOCK MADE OF DIFFERENT AGGREGATES

AGGREGATE (GRADED: 3/8" to 0)		WEIGHT LB./CU. FT. OF CONCRETE	WEIGHT 8" x 8" x 16" UNIT	COMPRESSIVE STRENGTH (GROSS AREA) PSI	WATER ABSORPTION LB./CU. FT. OF CONCRETE
TYPE	DENSITY (AIR-DRY) LB./CU. FT.				
SAND AND GRAVEL	130 — 145	135	40	1200 — 1800	7 — 10
LIMESTONE	120 — 140	135	40	1100 — 1800	8 — 12
AIR-COOLED SLAG	100 — 125	120	35	1100 — 1500	9 — 13
EXPANDED SHALE	75 — 90	85	25	1000 — 1500	12 — 15
EXPANDED SLAG	80 — 105	95	28	700 — 1200	12 — 16
CINDERS	80 — 105	95	28	700 — 1000	12 — 18
PUMICE	60 — 85	75	22	700 — 900	13 — 19
SCORIA	75 — 100	95	28	700 — 1200	12 — 16

Fig. 2-1. This chart shows the effect of aggregate type on weight, strength and other characteristics of concrete block.

limestone and air-cooled slag. Lightweight aggregate is expanded shale or clay, expanded slag, coal cinders, pumice and scoria. (Scoria is the scrap left after the melting of metals.) The effect of aggregate on weight, strength and other characteristics is shown in Fig. 2-1.

CLASSIFICATION OF CONCRETE MASONRY UNITS

Concrete masonry units fall into three main groups. Each group has two or three variations. See Fig. 2-2.

CONCRETE BUILDING BRICK

Concrete building brick, Fig. 2-3, are similar in size and function to clay brick. They are completely solid or have a shallow depression called a FROG. The frog is designed to reduce weight and provide for a better bond when the brick are laid in mortar.

Concrete brick are meant to be laid with a 3/8 in. mortar joint. Their nominal modular dimension is 4" x 8". The thickness of horizontal mortar joints is such that three courses (three brick and three bed joints) equal a height of 8 in. Some manufacturers produce oversized jumbo and double brick units for special applications.

Fig. 2-4. This concrete slump block has the appearance of stone. (Portland Cement Assoc.)

CLASSIFICATIONS OF CONCRETE MASONRY

CONCRETE BRICK	CONCRETE BLOCK		SPECIAL UNITS
BUILDING BRICK	SOLID	HOLLOW	SPLIT-FACE BLOCK
SLUMP BRICK	LOAD-BEARING	LOAD-BEARING	FACED BLOCK
		NON-LOAD-BEARING	DECORATIVE BLOCK

Fig. 2-2. A diagram of 10 types of concrete masonry products.

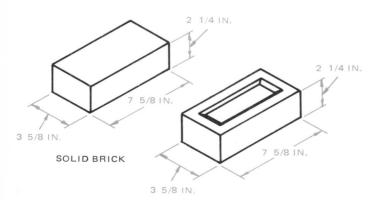

Fig. 2-3. In size, concrete building brick are much like clay brick. They may be solid or frogged.

SLUMP BRICK

Slump brick or block are produced from a mixture wet enough to cause the units to sag or slump when removed from the molds. Resulting faces are irregular. Height, surface texture and appearance vary greatly. See Fig. 2-4. Slump block produce special effects. They resemble stone, Fig. 2-5.

CONCRETE BLOCK

Concrete block units usually are produced in three classes:
1. Solid load-bearing.
2. Hollow load-bearing.
3. Hollow non-load-bearing.

Each of these classes may be purchased with either heavyweight or lightweight aggregates.

Fig. 2-5. Slump block are used here to produce a very attractive masonry planter and wall.   (Portland Cement Assoc.)

A solid concrete block is one in which the hollow parts in a cross section are not more than 25 percent of the total cross-sectional area.

A hollow concrete block is one in which the core or hollow area is greater than 25 percent of its total cross-sectional area. Generally, the core area in hollow units will be from 40 to 50 percent of the gross area.

USES OF CONCRETE BLOCK

Solid load-bearing block are used mostly where great loads are placed on it. Hollow load-bearing block have many uses. They combine high compressive strength with lightweight and flexible design, size and shape. A large percentage of the block produced is hollow load-bearing block. Hollow non-load-bearing block are thin-shelled and lightweight. They are intended to be used mostly in non-load-bearing partitions. If used on non-load-bearing exterior walls, they should be protected from the weather.

CONCRETE BLOCK GRADES
Concrete block units are offered in two grades:
1. Grade N units are designed for areas with freezing and thawing conditions when the block will be in direct contact with moisture. The block might be laid above or below grade on an exterior wall. These units must support at least 800 psi.
2. Grade S units can only be used above grade on outside walls with protective coatings. The walls may not be exposed to weather. These units must support at least 600 psi.

In certain dry areas of the country, concrete masonry units tend to shrink and crack if the block's moisture content is too high. For these areas the ASTM has specified Type I. (The southwest desert areas is an example.)

Type I units have a lower moisture content than Type II units. Type II units have no specified moisture content limitations and are used in most areas of the country.

The usual procedure for specifying grades and types are simply: Grades N-I, N-II, S-I and S-II.

SIZES AND SHAPES
Concrete building units are made in many sizes and shapes to fit different construction needs.

Sizes are usually given in their nominal dimensions. A unit measuring 7 5/8 in. wide, 7 5/8 in. high and 15 5/8 in. long is known as an 8" by 8" by 16" unit. The mortar joint is intended to be 3/8 in. This fits the modular design based on a 4 in. module. Fig. 2-6 shows most of the popular shapes and sizes. All are produced in full and half-length units. Some block units have a two-core design rather than three. This design has some advantages:
1. The shell is thicker at the center web. This increases the strength of the unit and reduces the tendency to crack from shrinkage.
2. Heat conduction is reduced by 3 to 4 percent.
3. Block are about 4 lb. lighter.
4. Hollow portions provide more space for placing conduit or other utilities.

The use of some of the units shown in Fig. 2-6 is explained in the following paragraphs.

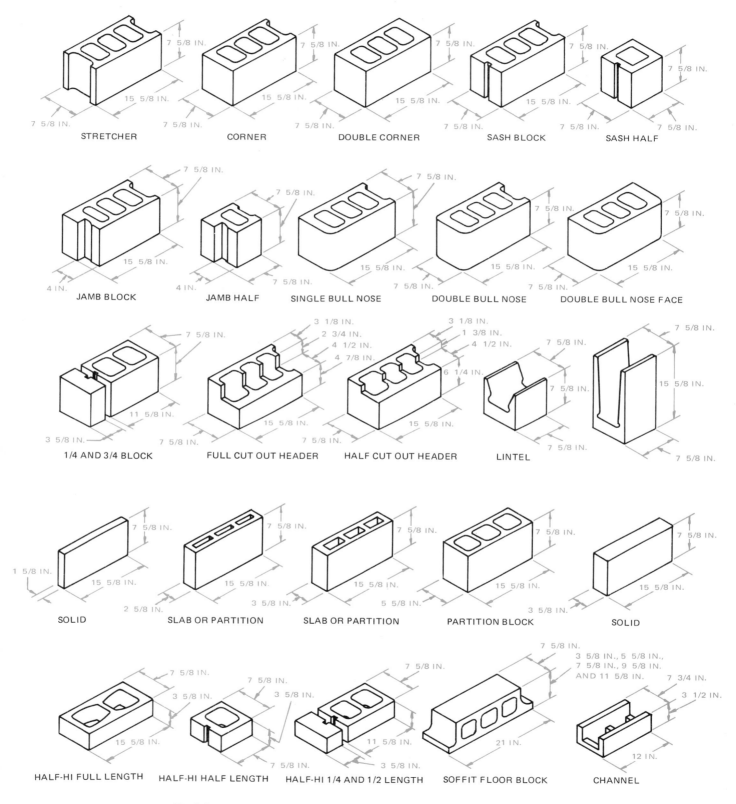

Fig. 2-6. A large selection of the most popular shapes and sizes of concrete block.

Concrete Masonry Units

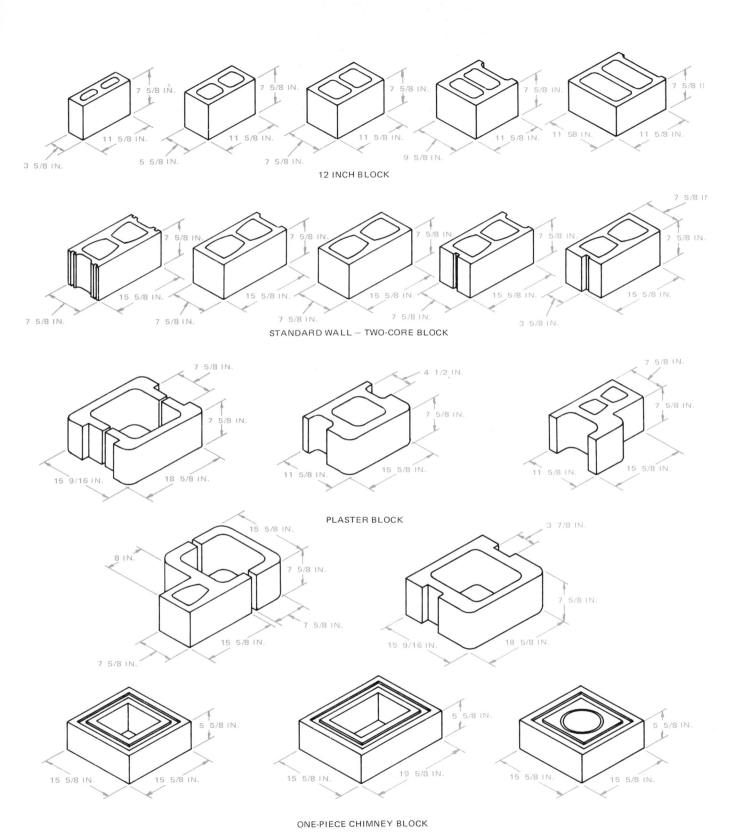

12 INCH BLOCK

STANDARD WALL — TWO-CORE BLOCK

PLASTER BLOCK

ONE-PIECE CHIMNEY BLOCK

Fig. 2-6. Continued from previous page.

Stretcher block are the most common and are used on exterior surfaces.

Corner block have one flush end which may be used for corners, simple window and door openings.

Double corner or pier block are designed for piers or pilasters or any other place where both ends will be exposed.

Bullnose block is used the same way as the corner block, but has a rounded corner.

Jamb or sash block are used to make installation of windows or other openings easier.

Solid top block have a solid top for use as a bearing surface in the finishing course of a foundation wall.

Header block have a recess to receive the header unit in a masonry bonded (brick and block) wall.

Metal sash block are used for window openings in which a metal sash is to be used. The slot anchors the jamb.

Partition block (4 or 6 in.) are used in constructing non-load-bearing partition walls.

Lintel block are used to construct horizontal lintels or beams. They are generally U or W-shaped.

Control joint block are used to construct vertical shear-type control joints.

SURFACE TEXTURE

Surface texture is important in block. It affects sound absorption, appearance and painting characteristics. Coarse textures look nicer and absorb sound better than smooth textures. Fine textures may absorb less sound but also take less paint than coarse textures. Some concrete masonry units have face surfaces ground smooth to produce interesting aggregate colors.

DECORATIVE BLOCKS

Fluted and scored or ribbed units are shown in Figs. 2-7 and 2-8. They provide striations that can be developed into many kinds of patterns. (Striations are a series of parallel ridges or grooves.) The accuracy that is achieved in machine

Fig. 2-8. Attractive wall patterns are formed from concrete block.

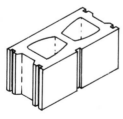

RIBBED UNIT

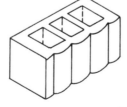

FLUTED UNITS

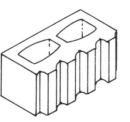

STRI-FACE UNIT

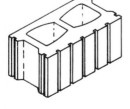

RIBBED UNIT

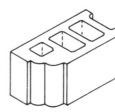

FLUTED UNITS

SPLIT-FLUTED UNIT

Fig. 2-7. Fluted and ribbed concrete block are used to achieve unique textured surfaces.

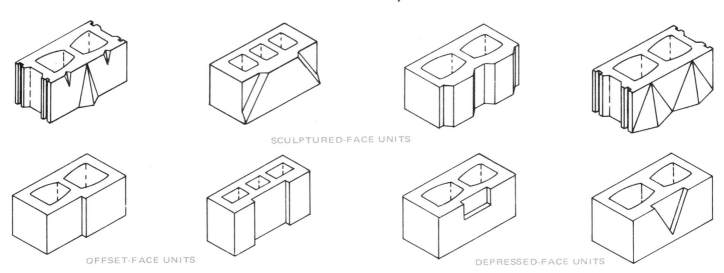

SCULPTURED-FACE UNITS

OFFSET-FACE UNITS

DEPRESSED-FACE UNITS

Fig. 2-9. Concrete block with offset, depressed or sculptured faces.

Fig. 2-10. Split concrete block produce the effect of quarried stone. (Besser Co.)

production makes it possible to produce the effect of long, vertical straight lines. These block can be quite decorative.

Block with recessed corners and raised patterns add endless numbers of effects to concrete masonry wall construction. Fig. 2-9 shows a few recessed and raised designs.

Units that look like rough quarried stone are produced by splitting solid concrete block lenthwise as seen in Fig. 2-10. Several sizes are available. Special hollow block as well as ribbed units may also be split to produce unusual effects.

Still another kind of decorative block is the prefaced block. This block has a ceramic-type glaze on its face, Fig. 2-11. The

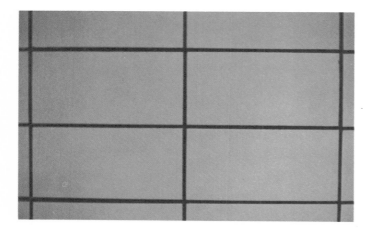

Fig. 2-11. Glazed concrete block are frequently used in schools and hospitals were they provide a long lasting and easily cleaned surface.

surface is colorful, hard, glossy and mar resistant. Surfacing material may be an epoxy or polyester resin. Some contain fine sand or other fillers. Ceramic or porcelainized glazes, mineral glazes and cementitious (cement-like) finishes have also been used. These units are popular in schools, hospitals and kitchens. The surface is easy to clean and lasts long.

SCREEN BLOCK

Screen block are highly decorative, Fig. 2-12. However, they can fill several other useful functions. They can be used,

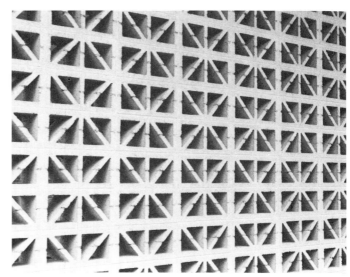

Fig. 2-12. Concrete block screens are highly decorative and may be used to control light or noise.

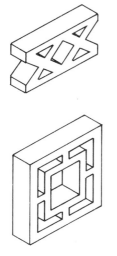
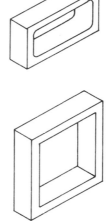
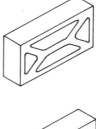

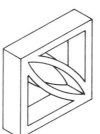
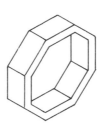
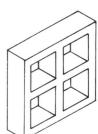

Fig. 2-13. Typical screen block patterns.

for example, to provide only limited, directional vision. Or they can produce a balance between privacy and openness from within or without. They also screen out some light, but produce a degree of shade.

There is an almost unlimited variety of patterns produced. Some are shown in Fig. 2-13. The number of patterns you can find in any given community, however, may be limited.

SOUND BLOCK
A sound block is made to absorb sound. Molded or sawed openings conduct the sound into the cores, Fig. 2-14, and

Fig. 2-14. Sound block absorb sound and may be used where noise is a problem. (The Proudfoot Co.)

absorb it. This design is patented. Sound block are often used in gymnasiums, plants, subways, bowling alleys or any place where noise is a problem. See Fig. 2-15.

BOND AND PATTERNS
Thus far, we have discussed decorative effects that center around characteristics of the individual block-shape, texture,

Fig. 2-16. Projecting concrete blocks produce a dramatic and decorative effect. (Portland Cement Assoc.)

Fig. 2-15. Sound block used in a computer room.

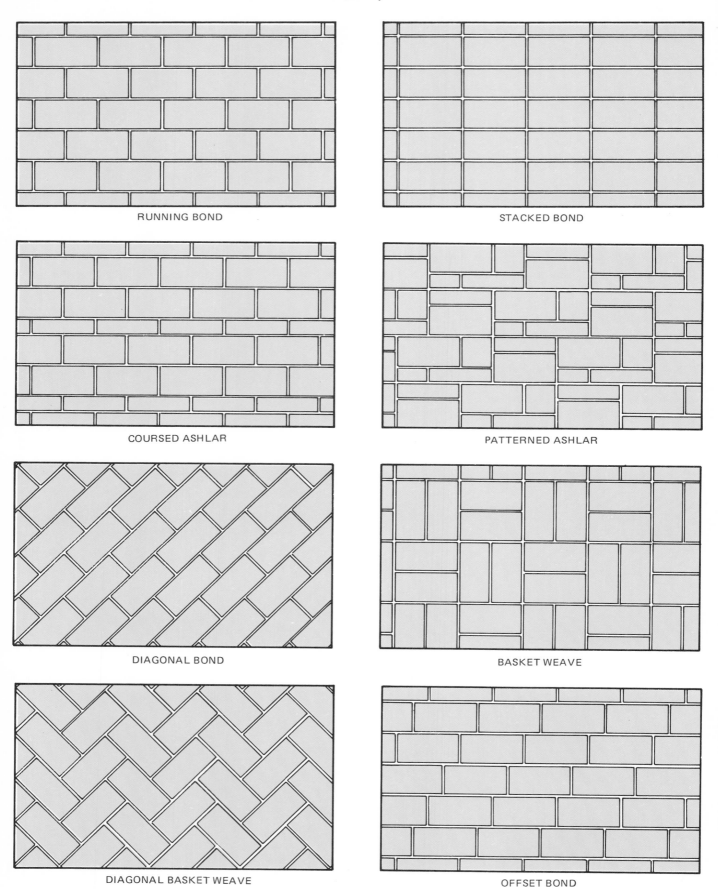

RUNNING BOND

STACKED BOND

COURSED ASHLAR

PATTERNED ASHLAR

DIAGONAL BOND

BASKET WEAVE

DIAGONAL BASKET WEAVE

OFFSET BOND

Fig. 2-17. Common concrete block bond patterns.

size or form of the block. Still other decorative effects can be achieved through bond patterns. See Fig. 2-16. Basically, the patterns belong to several classifications:

1. Running bond.
2. Stacked bond.
3. Coursed or patterned ashlar.
4. Diagonal bond.
5. Basket weave.
6. Diagonal basket weave.

Fig. 2-17 shows each of these patterns. Each classification includes several variations and the number may be enlarged further by using various size blocks.

SAND-LIME BRICK

Sand-lime brick are manufactured from sand and lime. A small amount of cement is sometimes added.

Among the advantages of sand-lime brick are uniform size, shape and weight. These brick are not burned. Therefore, shrinkage, warping and color variation is not as great a problem as in clay and shale brick. To produce various colors the manufacturer adds coloring to the mixture.

Sand-lime brick are used mainly in back-up work. They are generally produced in three grades: SW, MW and NW. Grade SW includes brick that can be exposed to freezing temperatures and water saturation. This grade may be used in foundations, retaining walls or exposed piers. They may be used almost anywhere that clay-burned brick are used.

Grade MW brick may be exposed to freezing conditions, but not to water saturation. Uses of this grade may include wall faces where the frost conditions are not too severe and where the climate is rather dry. Grade NW includes brick that may be exposed to light frost action with very little moisture present. This grade could be used in exterior wall construction in warm, dry climates.

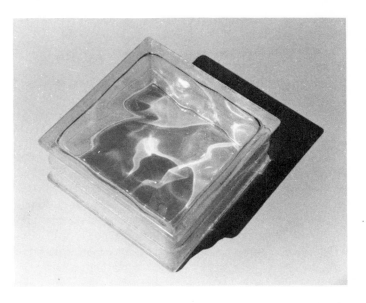

Fig. 2-18. Glass block have rough edges to form a good bond with the mortar.

GLASS BLOCK

Glass block are hollow, partially evacuated units of clear pressed glass, Fig. 2-18. (Partially evacuated means that the air is partly removed from inside the block.) They are produced in two sections and sealed. This provides a dead air space which acts as insulation. Glass blocks are made in square, rectangular, radial and other special shapes.

Mortar-bearing surfaces of glass block are corrugated (a rippled design) with a gritty, alkali and moisture-resistant substance. This coating acts as a bond between the block and the mortar. The construction of the block is such that designs may be imprinted on both sides. These designs distribute and control the direction of light rays. This is one of the most

Fig. 2-19. Glass block are functional as well as decorative. They are produced in several sizes and textures.

important properties of glass block. Some of the other advantages include:

1. Good insulation value.
2. Low cost of maintenance.
3. The elimination of dust, such as comes in through windows.

There are three categories of glass block:

1. Functional.
2. General purpose.
3. Decorative.

The functional type controls both the distribution and diffusion (scattering) of light. See Fig. 2-19. They are used often in schools, hospitals and basement walls. The general purpose type is installed where direct light transmission is required. Bathrooms, windows in homes such as entrance side panels and interior partitions would require this type of glass block. When architectural design is most important, decorative block are used. Examples would include the front entrance of an office building, decorative screens and corridor partitions or stairwell walls.

Glass block are manufactured in three popular sizes: 5 3/4 in. square, 7 3/4 in. square and 11 3/4 in. square. These blocks are all 3 7/8 in. thick. A 1/4 in. mortar joint is required. A glass block layout table is shown in Fig. 2-20. The maximum panel area, using chase construction, (reinforced masonry frame around the glass block panel) is 144 sq. ft. (maximum height is 20 ft.; maximum width, 25 ft.). The maximum panel area using wall anchor construction is 100 sq. ft. (maximum height is 10 ft., maximum width, 10 ft.). Panel and curtain wall sections may be erected up to 250 sq. ft. if properly braced to limit movement and settlement.

REVIEW QUESTIONS – CHAPTER 2

1. Sand and gravel in mortar or concrete is referred to as _____.

2. Aggregates are classified according to their weight as: _____ and _____.

3. Concrete masonry units can be classified in three main groups. Name the groups.

4. Coarse-textured block absorb sound poorly. True or False?

5. Concrete block units are grouped into two grades according to their degree of resistance to frost action, and into two types according to the amount of moisture in each block. Identify the four grade types.

6. Concrete blocks are usually specified by their nominal dimensions. True or False?

7. A concrete block whose nominal dimensions are 8" x 8" x 16" has an actual size of _____.

8. Decorative block with a ceramic type glaze on its face is called _____ block.

9. _____ _____ highly decorative, serves several functions. They provide limited directional vision, screen out some light but produce some shade while giving privacy.

10. A sound block is designed to _____ _____.

11. Name two advantages of sand-lime brick.

12. Name the three grades of sand-lime brick.

13. A unit which is partially evacuated of air and made of clear pressed glass is known as a _____ block.

GLASS BLOCK LAYOUT TABLE

NO. OF BLOCKS	6" 5 3/4 x 5 3/4 x 3 7/8	8" 7 3/4 x 7 3/4 x 3 7/8	12" 11 3/4 x 11 3/4 x 3 7/8
1	0' – 6"	0' – 8"	1' – 0"
2	1' – 0"	1' – 4"	2' – 0"
3	1' – 6"	2' – 0"	3' – 0"
4	2' – 0"	2' – 8"	4' – 0"
5	2' – 6"	3' – 4"	5' – 0"
6	3' – 0"	4' – 0"	6' – 0"
7	3' – 6"	4' – 8"	7' – 0"
8	4' – 0"	5' – 4"	8' – 0"
9	4' – 6"	6' – 0"	9' – 0"
10	5' – 0"	6' – 8"	10' – 0"
11	5' – 6"	7' – 4"	11' – 0"
12	6' – 0"	8' – 0"	12' – 0"
13	6' – 6"	8' – 8"	13' – 0"
14	7' – 0"	9' – 4"	14' – 0"
15	7' – 6"	10' – 0"	15' – 0"
16	8' – 0"	10' – 8"	16' – 0"
17	8' – 6"	11' – 4"	17' – 0"
18	9' – 0"	12' – 0"	18' – 0"
19	9' – 6"	12' – 8"	19' – 0"
20	10' – 0"	13' – 4"	20' – 0"
21	10' – 6"	14' – 0"	21' – 0"
22	11' – 0"	14' – 8"	22' – 0"
23	11' – 6"	15' – 4"	23' – 0"
24	12' – 0"	16' – 0"	24' – 0"
25	12' – 6"	16' – 8"	25' – 0"

Fig. 2-20. This chart shows the course height of glass block in 6 in., 8 in. and 12 in. sizes. For example, 12 courses of 8 in. glass block with a 1/4 in. mortar joint will produce an 8'–0" high wall. An 8'–0" long wall will also require 12 blocks.

Chapter 3
STONE

Stone is one of the oldest building materials. It is used for many things and in many places in construction. Its use as the sole building material for a structure has declined over the years. Today, it is almost always used as a nonstructural material. Generally, it is applied as a facing, a veneer, or for decorative purposes. Other uses such as in sandwich and panel systems are increasing. This trend is expected to continue.

CLASSIFICATION OF STONE

Stone is made up of minerals in various mixtures. Individual samples vary greatly in composition. These variations greatly affect the strength, color, texture and durability of stone. This is especially true of marble, limestone and sandstone.

CLASSIFICATION OF STONE

IGNEOUS	SEDIMENTARY	METAMORPHIC
GRANITE	SANDSTONE	MARBLE
TRAPROCK	LIMESTONE	SLATE
		SCHIST
		GNEISS
		QUARTZITE

Fig. 3-1. This classification diagram identifies the three main types of stone. Division is based on the process of formation.

Stone may be divided into three categories, Fig. 3-1, based on the process of formation:
1. Igneous (volcanic).
2. Sedimentary.
3. Metamorphic.

These processes are many times related and many samples are the result of more than one process.

IGNEOUS STONE

Igneous stone is further classified by its texture, mineral content and origin. It all comes from magmas (molten mixtures of minerals) generally found deep in the earth.

Igneous stone usually contains ferro-magnesian minerals and feldspar or quartz. When composed primarily of light minerals (quartz and potash feldspar) it is called acidic. This type is light in color and weight.

Stone with a greater degree of ferro-magnesian minerals is called basic. It is darker and heavier. In texture, igneous stone ranges from those with large crystals to glassy stone with no crystals at all. Two popular types of igneous stone are GRANITE and TRAPROCK.

GRANITE

Granite is probably the best known of the deeper igneous stone. It is usually light-colored, formed mainly of potash feldspar, quartz and mica. Fine granite has a salt-and-pepper pattern, but feldspar may redden it. Granite is hard and takes a good polish. Fig. 3-2 shows a fine granite quarried in Canada.

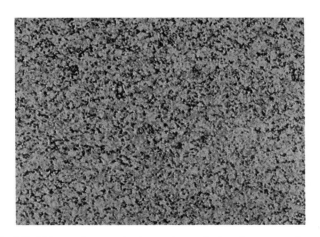

Fig. 3-2. This granite sample shows the typical grain pattern of most granites. It may be purchased in a broad range of colors, including white, gray, pink, green and various combinations.

TRAPROCK

Traprock is the quarryman's terms for diabase, basalt or gabbro. This hard, durable stone still is used in construction, but is limited because of iron minerals which make rusty stains as it weathers. Traprock is excellent as a crushed rock.

SEDIMENTARY STONE

Sedimentary stone differs widely in texture, color and composition, Fig. 3-3. Sedimentary stone is formed in layers.

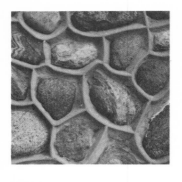

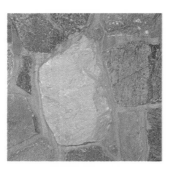

Fig. 3-3. These photos illustrate the varied color, texture and composition of sedimentary stone.

Fig. 3-4. The beauty of limestone is clearly shown in these photos. (Indiana Limestone Institute)

Deposits of shell, disintegrated stone or sand have become cemented together under pressure. Two popular types are SANDSTONE and LIMESTONE.

SANDSTONE

Sandstone is composed mainly of grains of quartz cemented by silica, lime or iron oxide. Silica cement may produce a hard, durable sandstone, but other cements are not as resistant. Porous sandstone may not weather well and special treatment may be required in cold regions. Color and texture are available in such varieties as bluestone, brownstone, silica sandstone, lime sandstone and many mixed varieties.

LIMESTONE

Limestone varies widely in color and texture. It consists mainly of the mineral calcite and is usually marine in origin. Limestone weathers rapidly in humid climates, but very slowly where it is dry. Limestone from Bedford in central Indiana is a well-known building stone. It is widely used in building construction, Fig. 3-4. Bedford limestone is white, even textured and sometimes packed with small fossils. Many other types of limestone are also used for construction around the country. They include: dolomitic, oolitic, crystalline and travertine marble.

METAMORPHIC STONE

Metamorphic stone is formed through reconstitution due to great heat and pressure. Changes may be barely visible, or may be so great that it is impossible to determine what the original stone once was. Metamorphic stone includes: marble, slate, schist, gneiss and quartzite.

MARBLE

Marble is recrystallized limestone. It may be white, yellow, brown, green or black. See Fig. 3-5. Marble does not often

VERMARCO LIGHT CLOUD MARBLE
Source: A Native Vermont marble.
Description: A white marble of very fine grain interspersed with faint clouds of the lightest green tint.

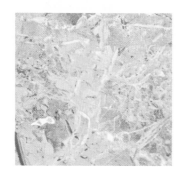

LOREDO CHIARO MARBLE
Source: A product of Italy, quarried in the northern Province of Lombardy.
Description: Characterized by bold strokes of creams and deep garnet against a neutral background which ranges from a soft beige to darker shades. Occasional patches of deep rose and pale green add further interest to the coloring of this marble.

MONTE VERDE MARBLE
Source: Deposits of green marble (serpentines) occur frequently in the Apennine Alps region of the great European Alpine chain.
Description: A beautiful colored marble, displays striking shades of greens. Its background is veined with rich tones of dark green, interspersed with cloud-like patches of white.

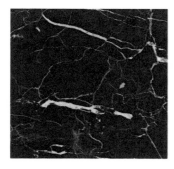

ST. LAURENT MARBLE
Source: Quarried near the village of Laurens in southern France.
Description: A luxurious dark marble with some white and reddish gold veining which is greatly enhanced if highly polished. Its dark overtones and veining of varying widths tend to give it a quality of warmth not often found in other basic dark marbles.

VERMARCO VERDE ANTIQUE MARBLE
Source: Rochester, Vermont.
Description: Has a deep green background and is crossed by veining of various shades of green, including some occasional light, almost white markings.

REGAL WHITE DANBY MARBLE
Source: Imperial Quarry at Danby, Vermont.
Description: A distinctive white coloring and crystalline character with vague, almost indistinct gray and beige markings.

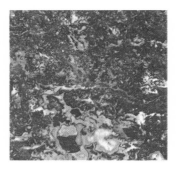

CORAL ROUGE FLEURI MARBLE
Source: Blount County, Tennessee.
Description: Color characteristics vary. In general, the marble is a rich reddish tone, interspersed with random patches of white and occasional areas of blue-gray. Basic color ranges from a relatively solid wine tone with few white areas, to reddish hues broken by large patches of white. Certain blocks have touches of green and occasionally the presence of fossils.

RADIO BLACK MARBLE
Source: The island of Isle LaMotte in Lake Champlain.
Description: A rich, vibrant charcoal-black marble which is essentially monotone in appearance. Upon closer inspection, it becomes evident that the blackness is relieved by a profuse distribution of small fossilized shellfish forms.

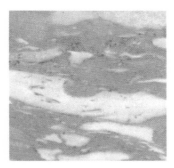

NORWEGIAN ROSE MARBLE
Source: About 150 miles north of the Arctic Circle in Norway.
Description: A warm, rose-colored marble with broad clouds of white and small areas of green. The green varies from light traces to darker accents.

ASBURY PINK MARBLE
Source: One of several light colored marbles produced from the Asbury Quarry east of Knoxville, Tenn.
Description: A buff colored marble, beige in its background color with an overall pink tone. It has a softy variegated appearance. Its warm background is dotted with flecks of rose and cream, creating a harmony in texture.

Fig. 3-5. Marble is noted for its great beauty. These samples show some of the variety, color and patterns available.

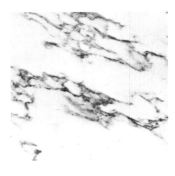

VERMARCO PAVONAZZO MARBLE
Source: A colorful variety obtained from the Maine No. 2 and Covered Kitchen Quarries in West Rutland, Vermont.
Description: Like most West Rutland marbles, the background of Vermarco Pavanazzo is clear white, interspersed with lacy green clouds. Darker gray-green veins are traced through the pastel greens, producing a pattern which is beautiful at a distance as well as up close.

PLATEAU DANBY MARBLE
Source: The Imperial Quarry at Danby, Vermont.
Description: General coloring of Plateau Danby marble is white, but with an added suggestion of very pale gray-green. Medium dark clouds of gray-green sweep sporadically across the marble's surface, disappearing into the background color and combining with random wisps of light golden beige to form a cloudy effect.

PEARL DANBY MARBLE
Source: Imperial Quarry at Danby, Vermont.
Description: Grayish white to light gray in background color, with subtle veins and clouds of darker gray, gray-green and grayish tan.

MONTCLAIR DANBY MARBLE
Source: Imperial Quarry at Danby, Vermont.
Description: A white background, with soft cloud-like markings that are predominately gray-green. Occasional touches of light tan or gold further soften the overall effect of this rich looking marble.

MARIPOSA DANBY CLOUDED MARBLE
Source: Imperial Quarry at Danby, Vermont.
Description: Has a rich, creamy-white background, relieved by relatively bold veins of gray-green. A profusion of pinpoint size crystalline specks enlivens this color effect by reflecting light and making the marble sparkle. Mariposa Danby Clouded is sawed parallel to the normal bedding plane. When sawed perpendicular to the bed, the layer yields Mariposa Danby Veined.

ROYAL DANBY MARBLE
Source: Imperial Quarry at Danby, Vermont.
Description: A white marble with small, soft, blue-gray, cloud-like markings. These markings give the marble an ermine-like quality which is reminiscent of royal robes.

IMPERIAL DANBY MARBLE
Source: Imperial Quarry at Danby, Vermont.
Description: The whitest of the white marbles produced at the Imperial Quarry. It is a fine grained marble with an almost pure white background, marked with faintest clouds of pale tan or gold and gray.

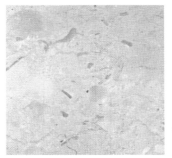

PERLATO D'ITALIA MARBLE
Source: From the hills of northwestern Sicily. It is best known in the United States by the name Perlato d'Italia but is occasionally called Sicilian Perlato or Sicilian Botticino.
Description: Perlato d'Italia marble is a soft pearl beige. It blends light beige background tones with darker overtones. Pearl-like shell formations, fossils, and flecks of white and dark beige are interlaced by distinctive veining.

ROMAN TRAVERTINE MARBLE
Source: The area around Bagni diTivoli, not far from Rome, Italy's capital city.
Description: Texture of Travertine and the voids which dot its surface are the most distinctive features. The basic coloring is a warm beige with some lighter shades of beige appearing throughout. The natural voids in this material may be filled with cements in matching or contrasting colors to produce unusual decorative effects.

HIGHLAND DANBY MARBLE
Source: Imperial Quarry at Danby, Vermont.
Description: Basic coloring of Highland Danby is light gray, but its natural variations range from nearly white to medium gray. Subtle veins of darker gray and light golden brown add interest and variety to this handsome marble.

Fig. 3-5. Marble samples continued.

develop the parallel bonding and mineral arrangement seen in slate and schist. Marble is a classic stone used for the finest work. It is, however, softer and less resistant to weathering than granite.

SLATE

Slate is frequently a blue-gray color, but may be green, red or brown, Fig. 3-6. It may be split into sheets used for roofing or flagstones. Most United States slate comes from Vermont, Maine and Pennsylvania.

SCHIST

Schist is a rather coarse-grained stone with large amounts of mica in it. Several kinds of schist are classified according to the most characteristic mineral present. Examples include: mica schist, hornblende schist, chlorite schist and quartz schist.

GNEISS

Gneiss (pronounced nice) is difficult to define or describe because it is so varied. In general, it is a coarse-textured stone with minerals in parallel streaks or bands. It is relatively rich in feldspar and usually contains mica.

QUARTZITE

Quartzite is sandstone which has been recrystallized. The grain structure in quartzite is not nearly as clear as in sandstone. It is very hard and durable.

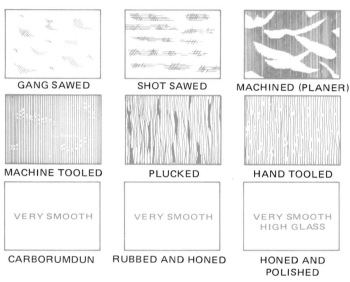

Fig. 3-7. Graphic symbols for typical stone finishes.

GRADES OF STONE

Stone is generally graded into four groups for construction purposes. They are: STATUARY, SELECT, STANDARD and RUSTIC in descending order of quality. Fineness of grain or texture is the basis for this classification.

STONE SURFACE FINISHES

Many surface finishes are possible with stone. Fig. 3-7 shows the graphic symbols for typical stone finishes. Some of the more popular ones are:

1. Gang sawed. This finish has a moderately smooth surface with visible saw marks. It is cheap to produce because it requires no further finishing after leaving the saw. It may be used on all types of stone.
2. Shot sawed. A variable or slightly rough finish according to the amount of shot used and the texture of the stone. It should be used with softer stone.
3. Machined (planer). Finish is smooth with some texture from tool marks. May be used with most types of stone.
4. Machine tooled. Has two to ten grooves per inch. Grooves are parallel and concave in shape. It may be used on all types of stone, but is expensive.

Fig. 3-6. Slate is a favorite floor covering material because it is easy to clean, resists soil and stains and lasts almost forever.

5. Plucked. This surface is obtained by rough planing. Then it is textured by "plucking" out small particles of the stone. Used mainly for limestone.

6. Hand tooled. Finish may be applied in a regular or random pattern. It is very expensive and should be used only in instances that require special accent.

7. Carborundum. A very smooth finish attained by using a carborundum machine rather than a planer. The finish is used primarily for limestone.

8. Rubbed and Honed. A very smooth finish — smoother than the carborundum. It is most often used for interior marble and granite.

9. Honed and Polished. Finish is the smoothest of all finishes. It has a "high glass" sheen. This surface finish is used on marble and granite.

STONE WALL PATTERNS

Stone wall bond patterns fall roughly into three groups:
1. Rubble.

2. Roughly squared stone.
3. Dimensioned or ashlar.

RUBBLE

Stone pattern No. 1, shown in Fig. 3-8, illustrates *uncoursed field stone* or random rubble. Stones of many sizes and shapes are used as they are found in fields and streams. Even though no coursing is possible, stones are usually laid in a horizontal position.

Pattern No. 2 in Fig. 3-8 shows a pattern of *uncoursed cobweb* or polygonal rubble. These stones are dressed with relatively straight line edges and are selected to fit a place in the wall that has a particular shape. Mortar joints are approximately the same width.

ROUGHLY SQUARED STONE

A wall pattern made from *uncoursed and roughly squared stone* is shown as No. 3 in Fig. 3-8. Even though these stones

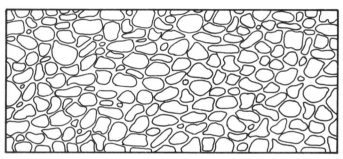

NO. 1 UNCOURSED FIELD STONE
ROUGH OR COMMON RUBBLE

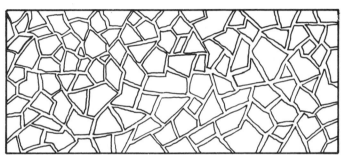

NO. 2 UNCOURSED COBWEB OR POLYGONAL STONE

NO. 3 UNCOURSED AND ROUGHLY SQUARED

NO. 4 COURSED AND ROUGHLY SQUARED STONE

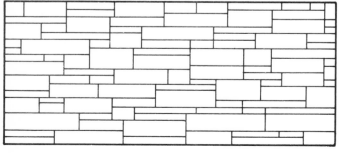

NO. 5 RANDOM, BROKEN COURSE AND RANGE

NO. 6 COURSED BROKEN BOND, BROKEN RANGE STONE

Fig. 3-8. Representative stone patterns used in stone masonry.

have been roughly squared, no coursing is attempted. Both large and small stones are used.

Wall pattern No. 4, shown in Fig. 3-8, is *coursed* and *roughly squared*. Here the horizontal coursing begins to appear. The stones are varied in size but generally are rectangular in shape.

DIMENSIONED OR ASHLAR

Dimensioned or ashlar stone is cut stone of specific dimensions. It is cut, dressed, and finished to precise job requirements at the mill. It is transported to the site as a finished product.

Fig. 3-8, No. 5 shows *random, broken course* and *range* pattern. This pattern makes no attempt to maintain coursing or range. (Range is squared stones laid in horizontal courses of even height.)

A *coursed broken bond, broken range* masonry pattern is shown in No. 6, Fig. 3-8. This pattern keeps the basic coursing without maintaining it fully. Again, the range is broken.

STONE APPLICATIONS

Stone was once widely used for foundations, exterior walls, paving and trim. In recent years, it has been used much less for foundations and solid exterior walls. Stone masonry is slower and more difficult due to the different sizes, and in some styles, odd shapes. Presently, stone is mostly limited to veneer, trim and floor or paving applications.

Limestone is probably the best sill material that can be used on exterior masonry walls, Fig. 3-9. The fact that it is one

Fig. 3-9. One-piece stone window sill protects against water leakage. It is attractive and very durable.

piece insures better protection against water leakage.

Continuous courses of stone just above the foundation, or at other levels, can add to the appearance as well as function of the masonry wall. A grade course of stone prevents moisture from the earth entering the brickwork. It may function as a water table to divert water away from the foundation wall.

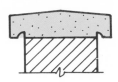

OVERHANGS AND DRIPS BOTH SIDES. SLOPED TO BACK

OVERHANGS AND DRIPS BOTH SIDES. SLOPED BOTH SIDES

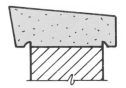

OVERHANG AND DRIP ONE SIDE. SLOPED ONE SIDE

OVERHANGS AND DRIPS BOTH SIDES. SLOPED TO BACK

Fig. 3-10. Stone copings on top of a masonry wall prevents moisture from entering the wall.

Stone is still used extensively for copings (a covering or top for brick walls) on masonry walls. It prevents moisture from entering through the top, Fig. 3-10.

Stone is popular in the form of quoins (large squared stones set at the corners of buildings) for buildings constructed with certain architecural styles.

Paving of patios, courts, entryways and walks using stone is also very popular. See Fig. 3-11. Slate and marble are favorites for paving. They afford a durable, hard surface which requires a minimum of maintenance and provides years of beauty.

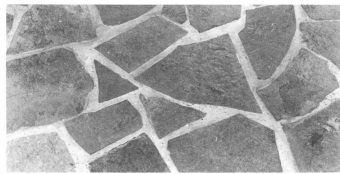

Fig. 3-11. Random slate used as a paving material.

MANUFACTURED STONE

Manufactured stone is a simulated stone veneer made from lightweight concrete. It is colorfast, weatherproof and has the look and feel of natural stone, Fig. 3-12. Produced in a variety of colors and textures, it simulates various types of natural stone. Fig. 3-13 shows some of the popular types.

Manufactured stone may be applied directly to a base coat of stucco, concrete block, brick, concrete or any masonry surface that has not been treated or sealed and which is rough

Fig. 3-12. These are applications of manufactured stone. (Stucco Stone Products)

Fig. 3-13. Manufactured stone is produced in several colors and textures to simulate natural stone. (Stucco Stone Products)

enough to provide a good mechanical bond. Fig. 3-14 shows manufactured stone being applied to concrete block. Wallboard, plaster or wood surfaces must be covered with wire mesh or metal lath before manufactured stone is applied. Painted or sealed stucco and painted or sealed masonry surfaces require either the attachment of wire mesh, sand blasting or a coating of suitable masonry adhesive.

Mortar used for manufactured stone should be a mixture of one part cement to three parts clean sand. To this is added from 1/4 to 1/2 part lime, mortar cream or fire clay. Packaged dry mix mortar that requires the addition of water also works well with manufactured stone.

In any case, the mortar should be mixed to a consistency similar to that of brick mortar. It may be applied either to the wall surface, the back of the stone, or both.

Manufactured stone should be layed out in a convenient location to the work area to insure even distribution of sizes and colors. The stone can be cut with a brick trowel, hatchet or similar tool to achieve sizes and shapes required for fitting stones and keeping mortar joints to a minimum.

Fig. 3-14. Manufactured stone being applied to a concrete block wall.

While the mortar is still soft and pliable the stone should be pressed into place with enough pressure so that the mortar is squeezed out around the edges of the stone. This will insure a good bond between the stone and wall surface.

When the mortar joints become firm (from one to three hours depending on suction of the base and climate) they should be pointed up with a wood or metal striking tool. Excess mortar should be raked out and the stones sealed around the edges to give the finished job the appearance of a natural stone wall.

At the end of each day, the finished wall should be broomed or wire brushed to remove loose mortar and to clean the face of the stone. A wet brush should not be used to treat the mortar joints for at least 24 hours. This is likely to cause staining and discoloration. This is difficult to remove.

REVIEW QUESTIONS — CHAPTER 3

1. Stone may be grouped into three categories based on the process of formation. They are igneous, sedimentary and metamorphic. True or False?
2. Two popular types of _____ stone are granite and traprock.
3. Two popular types of sedimentary stone are _____ and _____.
4. Marble is a metamorphic stone. True or False?
5. Stone is generally classified according to four grades. They are statuary, _____, _____ and rustic.
6. Name three surface finishes that may be applied to stone.
   a. _____.
   b. _____.
   c. _____.
7. When stones are used as they are found in fields and streams, the pattern is known as uncoursed field stone or _____.
8. Dimensioned or ashlar stone is cut stone of specific dimensions. True or False?
9. Large, square stones set at the corners of buildings are called _____.

Brick, placed in the hands of a skilled mason, take on beautiful architectural detail in construction of a building.

Chapter 4
MORTAR

Mortar is the bonding agent that ties masonry units into a strong, well-knit, weathertight structure. It secures each of the units into a wall or other building element.

Mortar serves four functions:

1. It bonds the units together, sealing the spaces between them.
2. It makes up for differences in sizes of the units.
3. It provides bonding for metal ties or other reinforcing.
4. It provides esthetic (nice to look at) qualities by creating pleasing lines and color effects.

MORTAR MATERIALS

The American Society for Testing and Materials has set up standards for materials commonly used in mortars. Fig. 4-1

MATERIAL	ASTM DESIGNATION
PORTLAND CEMENT (TYPES I, II, III)	C 150
AIR ENTRAINING PORTLAND CEMENT (TYPES IA, IIA, IIIA)	C 175
BLENDED CEMENT (TYPES IS, ISA, IP, IPA, S, SA)	C 595
MASONRY CEMENT	C 91
QUICKLIME	C 5
HYDRATED LIME	C 207
AGGREGATE	C 144

Fig. 4-1. The American Society for Testing and Materials standards for materials used in mortars.

shows the standards which apply to these materials.

Mortar is generally made up of cementitious (cement-like) materials together with sand and water.

CEMENTITIOUS MATERIALS

These materials may include *portland cement* (Type I, II or III) and *hydrated lime* (Type S).

Mortar is mostly portland cement. It is a hydraulic material, which means that it hardens under water. ASTM C-150, Standard Specifications for portland cement, covers eight types. But only three are recommended for use in mortar.

They are:

1. Type I — for general use when the special properties of Types II and III are not needed.
2. Type II — for use when moderate sulfate resistance or moderate heat of hydration is wanted.
3. Type III — for use when high early strength is desired.

Types IA, IIA and IIIA are not recommended for masonry mortar because air entrainment reduces the bond between the mortar and masonry units or reinforcement.

Types IS, ISA, IP, IPA, S and SA are not recommended for masonry mortar unless strengths of masonry constructed with mortar containing such cements are first established by appropriate tests.

HYDRATED LIME

Hydrated lime is quicklime which has been slaked (formed into a putty by combining with water) before packaging. This converts the calcium oxide. Hydrated lime can be mixed and used immediately. For this reason, it is more convenient to use than quicklime. (Quicklime, which is essentially calcium oxide, must be mixed with water and stored for as long as two weeks before using.)

Hydrated lime is available in two types, S and N. Only Type S hydrated lime is recommended for masonry mortar because unhydrated oxides and plasticity are not controlled in Type N.

SAND (AGGREGATE)

Aggregate used in masonry mortar is primarily sand. Either natural or manufactured sand may be used. Fig. 4-2 shows recommended sand gradation limits for both natural and manufactured sand. If the sand grains passing through the sieve fall within the percentage range given in Fig. 4-2, it meets the ASTM gradation standards. Well-graded aggregate retards (holds back) separation of material in a plastic mortar mix. This reduces bleeding and improves workability. Sand that has too few fine particles generally produces harsh mortars. Sand with too many fine particles results in weak mortar.

Manufactured sand is made by crushing stone, gravel or air-cooled blast furnace slag. Natural sand has rounder, smoother particles than manufactured sand. Manufactured sand, with its sharp and angular particle shapes, can produce mortars with workability properties different than mortars made with natural sand.

RECOMMENDED SAND GRADATION LIMITS

SIEVE SIZE	PERCENT PASSING	
	NATURAL SAND	MANUFACTURED SAND
No. 4	100	100
No. 8	95 — 100	95 — 100
No. 16	60 — 100	60 — 100
No. 30	35 — 70	35 — 70
No. 50	15 — 35	20 — 40
No. 100	2 — 15	10 — 25
No. 200	—	0 — 10

GRADATION LIMITS ARE GIVEN IN ASTM C 144

Fig. 4-2. Recommended sand gradation limits for both natural and manufactured sand.

WATER

Water for masonry mortar is required by ASTM C-270 to be clean and free of harmful amounts of acids, alkalies or organic materials. Whether or not the water is potable (safe for drinking) is of no concern. However, water from city mains or private wells is generally suitable for mixing mortar.

MORTAR PROPERTIES

Mortars have two distinct sets of properties:
1. Plastic (wet) mortar properties.
   a. Workability.
   b. Water retentivity.
2. Hardened mortar properties.
   a. Compressive strength.
   b. Bond strength.
   c. Durability.

PROPERTIES OF PLASTIC MORTAR

Plastic mortar is mortar that has just been mixed. It must be workable. It is uniform, cohesive, and of a consistency that makes it usable, Fig. 4-3. A usable mortar is easy to spread. Yet, it clings to vertical faces of masonry units, readily extrudes from the mortar joint but does not drop or smear. It will also support the weight of the masonry unit being used. The water content has a great deal to do with the workability of masonry mortar.

Water retention is the ability of the mortar to prevent rapid loss of mixing water to an absorptive masonry unit or to the air on a hot, dry day. Water retention is also related to the ability of mortar to prevent floating of a masonry unit with low absorption. This condition is known as "bleeding." Water retentivity improves with higher lime content and the addition of fine sand. However, the mixture must be within the allowable gradation limits.

PROPERTIES OF HARDENED MORTAR

The compressive strength of mortar increases as the cement content is increased. It decreases as lime is increased. Air entrainment also reduces compressive strength. Structural failures due to compressive loading are rare, but bond strength

is generally more critical.

Bond is that property of hardened mortar which holds the masonry units together. Bond strength is probably the most significant property of hardened mortar. It is also the most difficult to predict. Bond failure results from the masonry structure being subjected to:
1. Eccentric (not exactly vertical) gravity loads.
2. Loads from earth pressure, winds, and other lateral (sideways) loads.

Close contact between mortar and masonry unit surface is essential for good bond. This will more likely be achieved

Fig. 4-3. Mortar mixed to the proper consistency.
(Portland Cement Assoc.)

using mortar with good workability. To insure good bond, mortar should be placed in final position within 2 1/2 hours after the original mixing. This is required by ASTM C-270.

Movement of a masonry unit after the mortar has begun to harden will destroy the bond. The mortar will not be plastic enough to reestablish the bond.

Durability of mortar is measured primarily by its ability to resist repeated cycles of freezing and thawing under natural weather conditions. Mortars of high compressive strength have been found to have good durability.

Air-entrained mortar is highly resistant to freezing and thawing and has good durability. The tiny air bubbles absorb the expansive forces of freezing. Laboratory tests showed that mortars with adequate air entrainment withstood hundreds of freeze-thaw cycles, while other mortars soon failed.

MORTAR PROPORTIONS AND USES

Picking the right type of mortar depends upon the type of masonry and where it will be in the structure. No mortar will produce the highest rating in all properties. Adjustments in the mix to improve one property usually hurts other properties.

MORTAR PROPORTIONS BY VOLUME

MORTAR TYPE	PARTS BY VOLUME OF PORTLAND CEMENT* OR PORTLAND BLAST FURNACE SLAG CEMENT**	PARTS BY VOLUME OF MASONRY CEMENT	PARTS BY VOLUME OF HYDRATED LIME OR LIME PUTTY	AGGREGATE, MEASURED IN A DAMP, LOOSE CONDITION
M	1 1	1 (TYPE II) —	— 1/4	NOT LESS THAN 2 1/2 AND NOT MORE THAN 3 TIMES THE SUM OF THE VOLUMES OF THE CEMENTS AND LIME USED.
S	1/2 1	1 (TYPE II) —	— OVER 1/4 to 1/2	
N	— 1	1 (TYPE II) —	— OVER 1/2 to 1 1/4	
O	— 1	1 (TYPE I OR II) —	— OVER 1 1/4 to 1 1/2	
K	1	—	OVER 2 1/2 to 4	

* TYPES I, II, III, IA, IIA, IIIA
** TYPES IS, ISA
DATA FROM ASTM C-270

Fig. 4-4. Mortar types are shown with the proper proportions of materials by volume.

Fig. 4-4 shows proper proportioning of materials.

The five types of mortars (designated by the letters M, S, N, O and K) recommended for various classes of masonry construction are shown in Fig. 4-5.

TYPE M MORTAR

Type M mortar has high compressive strength and somewhat greater durability than other mortar types. It is very good for unreinforced masonry below grade and in contact with earth. Structures would include foundations, retaining walls, walks, sewers and manholes.

TYPE S MORTAR

Type S mortar is a medium-high-strength mortar. It is intended for use where Type M is recommended but where bond and lateral strength are more important than high compressive strength. Tests indicate that the tensile bond strength between brick and Type S mortar is near the best obtainable with cement-lime mortars. It is recommended for use in:
1. Reinforced masonry.
2. In unreinforced masonry where maximum flexural (bending) strength is required.
3. Where mortar adhesion is the sole bonding agent between facing and backing as it is with ceramic veneers.

TYPE N MORTAR

Type N mortar is a medium-strength mortar suitable for general use in exposed masonry above grade. It is best for parapet walls, chimneys and exterior walls which are subjected to severe exposure. Its principal property is weather resistance.

TYPE O MORTAR

Type O mortar is a low-strength mortar. It is suitable for general interior use in non-load-bearing masonry. It may be used for load-bearing walls of solid masonry where compres-

MORTAR TYPES FOR CLASSES OF CONSTRUCTION

ASTM MORTAR TYPE DESIGNATION	CONSTRUCTION SUITABILITY
M	MASONRY SUBJECTED TO HIGH COMPRESSIVE LOADS, SEVERE FROST ACTION, OR HIGH LATERAL LOADS FROM EARTH PRESSURES, HURRICANE WINDS, OR EARTHQUAKES. STRUCTURES BELOW GRADE, MANHOLES, AND CATCH BASINS.
S	STRUCTURES REQUIRING HIGH FLEXURAL BOND STRENGTH, BUT SUBJECT ONLY TO NORMAL COMPRESSIVE LOADS.
N	GENERAL USE IN ABOVE GRADE MASONRY. RESIDENTIAL BASEMENT CONSTRUCTION, INTERIOR WALLS AND PARTITIONS. CONCRETE MASONRY VENEERS APPLIED TO FRAME CONSTRUCTION.
O	NON-LOAD-BEARING WALLS AND PARTITIONS. SOLID LOAD BEARING MASONRY OF ALLOWABLE COMPRESSIVE STRENGTH NOT EXCEEDING 100 PSI.
K	INTERIOR NON-LOAD-BEARING PARTITIONS WHERE LOW COMPRESSIVE AND BOND STRENGTHS ARE PERMITTED BY BUILDING CODES.

Fig. 4-5. The type of mortar required is shown for a variety of classes of construction.

sive stresses do not exceed 100 psi. In such cases, however, exposures must not be severe. *In general, do not use Type O mortar where it will be subjected to freezing.*

TYPE K MORTAR

Type K mortar is a very low-strength mortar. It is best used in interior non-load-bearing partition walls where high strength is not needed.

SPECIFIC MORTAR USES

The following specific uses are recommended based on properties of the various mortars.

CAVITY WALLS: Where wind velocities will exceed 80 mph, use Type S mortar. For locations where lesser winds are expected, use Type S or Type N mortars.

FACING TILE: Any of the first four mortar types are acceptable. Where 1/4 in. joints are specified, all aggregate (sand) should pass through a No. 16 sieve. For white joints, use white portland cement and white sand in construction mortar. An alternate method is to rake all joints to a depth of 1/2 in. and point or grout with mortar of the desired color.

TUCK-POINTING MORTAR: Use only prehydrated mortars. To prehydrate mortar, all the ingredients (except water) should be mixed thoroughly, then mixed again. Use only enough water to produce a damp unworkable mix which will retain its form when pressed into a ball. After one to two hours, add enough water to bring it to the proper consistency. Prehydrated Type N mortar may be used.

DIRT RESISTANT MORTAR: When a dirt resistant and/or stain resistant mortar is required, add aluminum tristearate, calcium stearate or ammonium stearate to the mortar. The amount added should be 3 percent of the weight of the portland cement.

COLORING MORTAR

Mortar may be colored through the use of aggregates or pigments. Colored aggregates are preferred. White sand, ground granite, marble or other stone usually have permanent color and do not weaken the bond or mortar. For white joints, use white sand, ground limestone or ground marble with white portland cement and lime.

Mortar pigments must be:
1. Capable of giving the desired color when used in permissible quantities.
2. Sufficiently fine to disperse throughout the mix.
3. Non-reacting with other ingredients.

These requirements are usually met by metallic oxides. Iron, manganese and chromium oxides, carbon black and ultramarine blue have been used successfully as mortar colors. *Do not use organic colors, particularly those containing Prussian blue, cadmium lithopone, and zinc and lead chro-* *mates. Paint pigments may not be suitable for mortars either.*

Use the minimum quantity of pigments necessary to produce the desired results. Too much may impair strength and reduce durability. Carbon black, a popular pigment, will impair mortar strength when used in greater amounts than 2 or 3 percent of the cement weight.

For best results, dry mix the color with portland cement in large controlled quantities. This will ensure a more uniform color than when mixing smaller quantities at the job site.

COLD WEATHER MORTAR

The temperature of the mortar should be between 70 and 100 deg. F when used. Higher temperatures may cause fast hardening which will make it more difficult for the mason to produce quality work.

Heating the mixing water is one of the easiest methods of raising the temperature of the mortar. However, the water should not be heated above 160 deg. F. This prevents the danger of "flash" set when the water comes in contact with the cement.

In freezing weather, moisture in the sand will turn to ice. It must be thawed before using. *Never use sand that has been scorched from overheating. Antifreeze materials should not be used to lower the freezing point of mortar.* Such large quantities would be required that it would seriously impair the mortar strength and other properties.

Mortar to which calcium chloride has been added will resist freezing. It should be used in a solution instead of in dry or flake form. To prepare the proper solution dissolve 100 lb. of flake calcium chloride in 25 gal. of water. Not more than one quart should be used with each 94 lb. bag of masonry cement.

The length of protection against freezing should follow requirements of the American Standard Building Code, (A 41.1). It states:

Masonry shall be protected against freezing for at least 48 hours after being laid. Unless adequate precautions against freezing are taken, no masonry shall be built when the temperature is below 32 deg. F on a falling temperature, at the point where the work is in progress. No frozen materials shall be built upon.

REVIEW QUESTIONS — CHAPTER 4

1. Mortar is generally composed of _____, _____ and _____.
2. Mortars have two distinct sets of properties: plastic mortar properties and hardened mortar properties. The main properties of plastic mortar are _____ and _____.
3. List the three properties of hardened mortar.
4. The temperature of mortar should be between _____ and _____ degrees F when used.
5. What is one of the easiest methods of raising the temperature of mortar in cold weather?
6. Why is antifreeze not recommended for use in mortar?
7. Masonry should be protected from freezing for at least _____ hours after being laid.

Mortar

A good workable mortar is soft but has good body. It adheres readily to masonry units and will not drop away while handling.
(Portland Cement Assoc.)

Flat sheets of welded wire fabric are being used as wall reinforcement.   (Wire Reinforcement Institute)

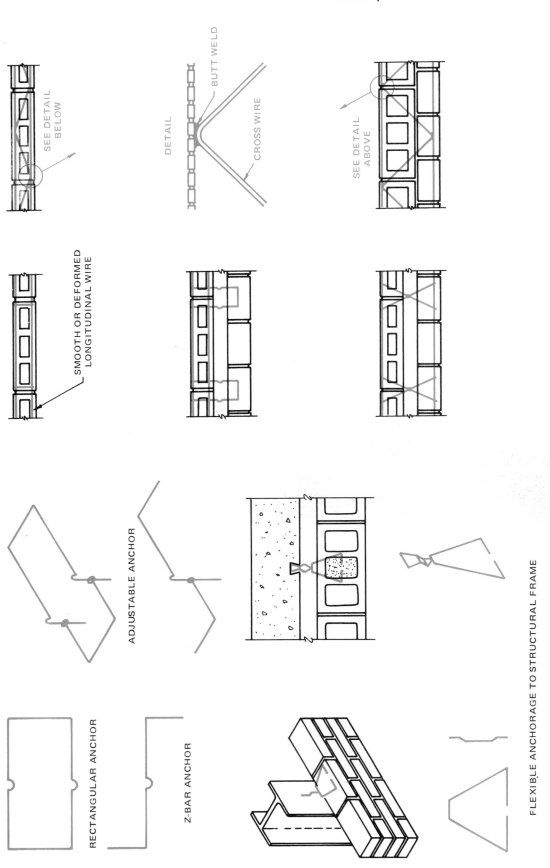

SEE DETAIL BELOW

BUTT WELD

DETAIL

CROSS WIRE

SEE DETAIL ABOVE

HARDWARE CLOTH

SMOOTH OR DEFORMED LONGITUDINAL WIRE

METAL BAR ANCHOR

JOINT REINFORCEMENT

ADJUSTABLE ANCHOR

FLEXIBLE ANCHORAGE TO STRUCTURAL FRAME

RECTANGULAR ANCHOR

Z-BAR ANCHOR

METAL TIES

DOVETAIL ANCHORS

Fig. 5-1. Common masonry anchors, ties and joint reinforcement used in masonry wall construction.

Chapter 5
MASONRY ANCHORS, TIES AND JOINT REINFORCEMENT

Many types of masonry anchors, ties and joint reinforcement are used in masonry construction. Fig. 5-1 shows several of the most common. These materials are placed inside masonry walls by the mason to give it greater strength or to hold it in place. These materials include metal ties, reinforcing rods, anchors and joint reinforcement.

METAL TIES

Structural bonding with metal ties is widely used in three types of construction:
1. Masonry veneer to frame.
2. Solid masonry.
3. Cavity wall, see Fig. 5-2.

Fig. 5-2. Rectangular anchors used in cavity wall construction. (Portland Cement Assoc.)

Some of the popular types include the rectangular, Z-bar, adjustable, flexible and corrugated ties.

Rectangular ties are 3/16 in. diameter, coated steel wire formed in a rectangular shape. They are usually 4" x 6" in size. This tie is often used with horizontal or vertical cell hollow units or with solid units in solid and cavity wall construction.

Z-bar ties are 3/16 in. diameter, coated steel wire bent to look like a "Z" as in Fig. 5-3. They are usually 2" x 6" in

Fig. 5-3. A Z-bar tie may be used in concrete block wall construction. (Portland Cement Assoc.)

size. The Z-bar tie is generally used with horizontal cell units and solid units in solid and cavity wall construction.

Adjustable ties are 3/16 in. diameter, coated steel wire formed into a rectangular or Z-shape. Sizes correspond to the rectangular and Z-bar ties.

Flexible ties are 3/16 in. diameter, coated steel wire or steel band formed into various shapes. They may be used to anchor a masonry wall to a column or beam.

Corrugated ties are galvanized sheet metal corrugated bands 7/8 in. wide and 7 in. long. They are used to attach masonry veneer to a frame structure or masonry backup. Each one serves 2 sq. ft. of wall area.

REINFORCING RODS

Reinforcing rods or bars of various sizes are also used in masonry construction, Fig. 5-4. They are formed from rolled steel and are similar to those used in reinforced concrete. Reinforcing rods may be smooth or may have deformed ridges. They are produced in several standard diameters.

The size is generally specified as a number. The number corresponds to the diameter of the bar or rod in eighths of an

Fig. 5-4. Reinforcing rods are used to strengthen this wall.
(Brick Institute of America)

Fig. 5-5. Anchor bolts may be used to attach roof plates or sills to masonry work.   (Portland Cement Assoc.)

inch. For example, a 1/2 in. rod would be a No. 4.

Reinforcing rods are used in solid reinforced masonry walls, masonry lintels, sills and pilasters. This type of construction is also employed at corners and at window and door openings.

ANCHORS

Anchors are produced in a variety of shapes and sizes. One type is corrugated with a dovetail on one end for veneer construction. These ties are galvanized steel and the dovetail fits into a slot formed in the concrete beam or column. The other end is attached to the veneer.

Another type of anchor is the anchor bolt, Fig. 5-5. Anchor bolts are used to attach roof plates, sills, etc. to masonry work. They are available in various diameters or gauges and lengths. Some are made of band steel and have a bent end for better holding power. Others are round and may have a head or bend on the embedded end.

Bearing walls which intersect are frequently connected with a strap anchor, Fig. 5-6. Anchors are usually 1 1/4" x 1/4" x 30". Three-inch right-angle bends are made at each end for embedding in mortar-filled cores of masonry block.

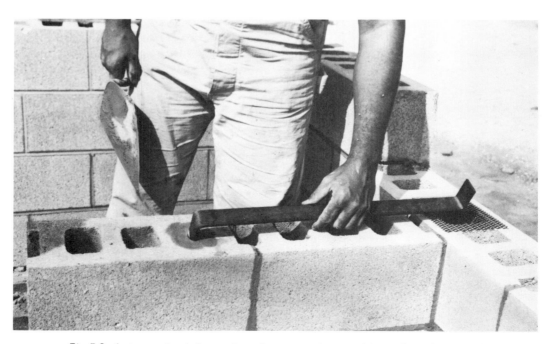

Fig. 5-6. A strap anchor being put into place connecting a partition wall to a bearing wall.

JOINT REINFORCEMENT

Joint reinforcement is produced in many shapes and sizes for various applications. It is especially designed for placement in the horizontal joints of masonry walls to provide greater lateral strength. One popular style of joint reinforcement is made in the shape of a truss or a ladder, Fig. 5-1. Materials used in the fabrication of truss and ladder style reinforcement is smooth or deformed wire or steel rod. Common rod diameters are 1/4 in. and 3/8 in. Wire diameter is generally 1/8 in. (No. 11 gauge) through 3/16 in. (No. 6 gauge).

Fig. 5-7. Hardware cloth may be used to tie intersecting walls together. (Portland Cement Assoc.)

Another type of joint reinforcing is the use of hardware cloth, Fig. 5-7. This material is usually galvanized and may be purchased in various widths and lengths. It is frequently used with a metal tie bar to tie non-load-bearing walls to intersecting walls.

Many other highly specialized anchors and ties are available for unique construction. These products are described in the manufacturer's literature.

REVIEW QUESTIONS — CHAPTER 5

1. Metal ties are placed inside masonry walls to give them greater strength or to hold them in place. True or False?
2. What type of metal tie is generally used to attach masonry veneer to a frame structure or masonry backup?
3. Roof plates and sills are usually attached to masonry work using _____ _____. (type of anchor)
4. Name four types of metal ties.
5. Corrugated ties are used to attach masonry veneer to a _____ structure or masonry veneer. There should be one tie for every _____ sq. ft. of wall area.
6. Reinforcing rods are used in solid reinforced masonry walls, masonry lintels, sills and pilasters. True or False?
7. Why do anchor bolts usually have a bent end?
8. Hardware cloth is used in masonry construction to tie _____ walls to _____ walls.

For safety of concrete workers, safety railings are installed on all catwalks for work in high places.   (Symons Corp.)

Chapter 6
TOOLS, EQUIPMENT AND SAFETY

Tools and equipment and their safe use are a very important part of all trades. Every tool is designed for a specific purpose. Beginning masons must learn how to use each tool skillfully if they are to be good at their trade.

MASONRY TOOLS AND EQUIPMENT

This chapter describes the tools and equipment commonly used in the masonry trade and concrete work. It will show their intended use and how to use them safely.

TROWELS

The tool most used by the mason is the mason's trowel. It has two parts, the blade and handle. The blade is made from a flat piece of forged steel ground to the proper balance, taper and shape. The narrow end of the blade is the point and the wide end nearest the handle is the heel. The blade is connected

Fig. 6-1. Mason trowels are manufactured in several sizes and shapes. Handle may be wood or plastic.

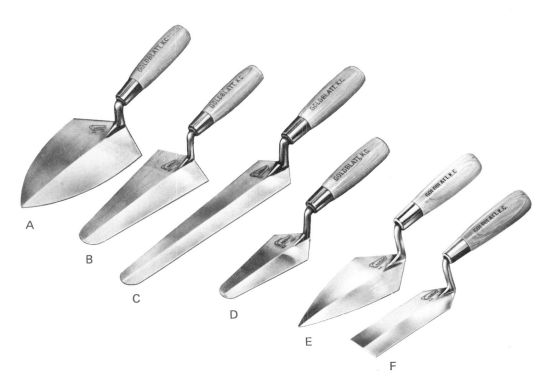

Fig. 6-2. Specialized trowels used by masons. A—Buttering. B—Gauging. C—Duck bill. D—Cross joint. E—Margin. F—Pointing.

69

to the handle by the shank. The handle may be made from wood or plastic. Wooden handles usually have a metal band, called a ferrule, around the shank end. This adds strength and prevents the handle from splitting. Before purchasing a trowel, you will need to consider the following factors: weight, size, materials, construction and angle of the handle to the blade. A fine trowel should have a flexible blade of high grade steel that will withstand long and hard use. A trowel should be light and well balanced.

Trowels are available in lengths from about 9 to 12 inches and in widths from about 4 1/2 to 7 inches. Mason's trowels are produced with wide, sharp heels or narrow, rounded heels, Fig. 6-1.

Special purpose trowels allow the mason to perform some specialized jobs better. These trowels are shown in Fig. 6-2.

A trowel must be clean to perform properly. Old mortar should be removed from the blade and shank. Form the habit of cleaning your tools at the end of each day. They will then be in good condition the next time you use them.

JOINTERS

Jointers are used to finish the surface of mortar joints, Fig. 6-3. They are also called *joint tools* or *finishing tools*. Several types and sizes are commonly used. They are usually either cast or forged rods or stamped sheet metal. The cross-sectional shape of the jointer gives the mortar joint its shape.

Long horizontal joints are finished most easily with a sled runner or joint runner. This tool has a handle and is shown in Fig. 6-3.

Fig. 6-4. Two tools used to produce the rake joint. Above. Plain joint raker. Below. Skate wheel joint raker.

requires the use of this tool. Two styles of joint rakers are shown in Fig. 6-4.

BRICK HAMMERS

The brick hammer is frequently used to drive nails, strike chisels and break or chip masonry materials. The head is flat on one side so that it may be used as a conventional hammer. The other side is drawn out to form a chisel for dressing up

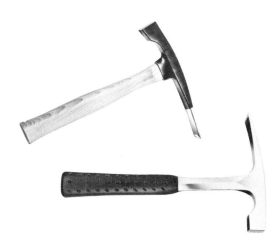

Fig. 6-5. Mason's brick hammer is designed to drive nails, strike chisels and chip masonry units. Above. Wooden handled. Below. Metal handle with plastic grip. (Goldblatt Tool Co.)

cuts. Fig. 6-5 shows two styles of brick hammers. One has a wooden handle and the other is forged from one piece. Most brick hammers weight from 12 to 24 ounces.

RULES

The mason usually has two kinds of rules:

1. A 6 ft. folding rule sometimes with a 6 in. sliding scale on the first section for inside measurements.
2. A 10 ft. retractable steel tape, Fig. 6-6.

The 6 ft. folding rule is a standard tool used by the mason.

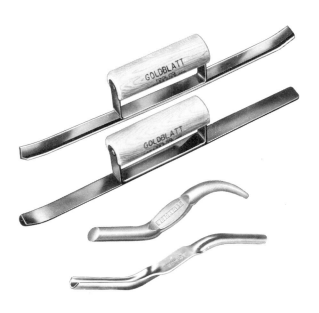

Fig. 6-3. Jointers used by masons to finish mortar joints. Above. Sled runner jointers. Center. Forged heavy duty jointer. Below. Stamped jointer.

JOINT RAKERS

A joint raker is used to remove a portion of the mortar from the joint just before the mortar hardens. The rake joint

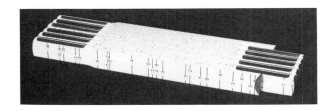

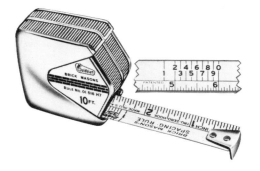

Fig. 6-6. The 6 ft. folding rule and retractable steel tape are standard measuring tools used by the mason. Numbers from 1 to 0 on inset represent height of various brick and mortar joints.

It usually has special markings on the back side representing course heights for various unit sizes and joint thicknesses. A well constructed rule of quality materials will have a long life and gives dependable service.

LEVELS

The mason's level or "plumb rule" is the most delicate piece of equipment the mason uses. Fig. 6-7 shows three popular types. The level is constantly used to check the wall to be sure that it is built absolutely vertical and level.

Several spirit bubble vials are built into the level to permit plumb (vertical) and level (horizontal) use. The vials are generally imbedded in a plaster of paris setting. They are very accurately adjusted.

Masonry levels are usually 42 to 48 in. long. A good level should be lightweight and absolutely straight. It may be made from wood or metal or a combination of the two.

Fig. 6-7. Three common types of levels used in masonry work. Two are wood or plastic with metal edges. The third is made of aluminum. (Goldblatt Tool Co.)

CHISELS

Different chisels are used by masons for different kinds of materials. Fig. 6-8 shows a *brick set*. It is used to cut brick to exact dimensions. Fig. 6-9 pictures a *blocking chisel*. Blocking chisels are used to cut concrete block. They are made in a

Fig. 6-8. Brick set is a type of chisel used to cut brick to specified length. Cutting edge is very blunt.

Fig. 6-9. Blocking chisel is used to cut concrete block.

number of sizes, shapes and weights.

Fig. 6-10 is a display of stone mason's chisels. They are used for scoring or splitting stone.

All chisels must be kept sharp and straight if they are to work properly. Burrs must be removed from the top end to prevent injury during use.

Quality chisels are made from tempered tool steel. They are very hard and must be sharpened by grinding.

Fig. 6-10. Stone mason's chisels are used to split stone. Tool at right is designed to take a handle. (Goldblatt Tool Co.)

LINE AND HOLDERS

The mason's line is usually a strong nylon or dacron cord. The line is used to keep each course level and the wall true and out-of-wind (no bulges or hollows). It is secured at either end with line holders or line pins. Fig. 6-11 shows a pair of adjustable line holders and braided nylon line. Line is available in strengths from 100 to 350 lb. test and in white, yellow and green. It is produced in lengths from 100 to 1000 ft.

BRUSHES

Brushes are produced in a variety of shapes and textures, Fig. 6-12. They are used for the following purposes:

Fig. 6-11. Line and line holders are used to keep each course level and wall true and out-of-wind.

1. To remove mortar from the masonry units after the wall has been constructed.
2. To wash brick surfaces with muriatic acid.
3. For general cleaning.

Fig. 6-12. General purpose masonry brushes have fairly stiff bristles.

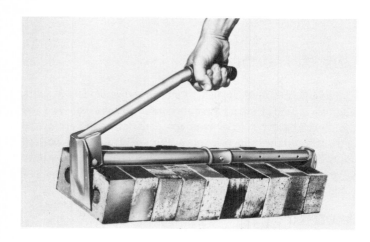

Fig. 6-13. Brick tongs make carrying of brick easier.
(Goldblatt Tool Co.)

BRICK TONGS

Brick tongs, Fig. 6-13, are used to carry brick. They are designed in such a way that they do not chip the brick and are adjustable for various size units.

SCAFFOLDING

Much of a mason's work is done on scaffolding since few projects are so low that they can be reached while standing on the ground. Scaffolding makes masons' work easier by providing a place for their materials and tools at a convenient height.

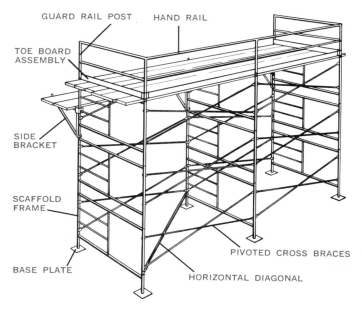

Fig. 6-14. Tubular scaffolding is frequently used by masons to reach high places. Side brackets hold planking to provide deck for workers. Materials are never placed here but on upper deck.
(Goldblatt Tool Co.)

Several types of scaffolding are used by masons. They are the tubular, tower and swing stage scaffolding. Fig. 6-14 shows the tubular type most frequently used.

MORTAR MIXERS

Mortar is usually prepared in a mechanical mixer similar to the one shown in Fig. 6-15. Mixers may be electric or gasoline powered, depending on job conditions or preference. Mixers are produced in several sizes, but a typical size mixes about 4 cu. ft. of mortar at a time. For best results, the dry materials should be mixed first. Then add water.

Mortar should not be allowed to harden in the mixer. This will prevent proper mixing of ingredients because the metal surfaces become mortar clogged.

CEMENT MASONRY TOOLS AND EQUIPMENT

Most tools used in cement masonry are designed for use in finishing horizontal concrete surfaces. They are known as flat-work finishing tools.

Fig. 6-15. Mechanical mortar mixer is driven by gasoline engine.

SCREEDS

A screed is a straightedge or strike-off rod made of any straight piece of wood or metal that has sufficient rigidity (stiffness). It is the first finishing tool used by the cement mason after the concrete is placed. It is used to strike off or screed the concrete surface to the proper level. Fig. 6-16 shows a power screed which is used for large jobs.

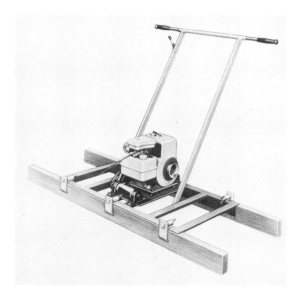

Fig. 6-16. Power screed is used to strike off concrete on larger jobs.

TAMPERS

Hand tampers are produced in several basic designs and are used to compact the concrete into a dense mass. They are used on flatwork construction with low-slump concrete which is

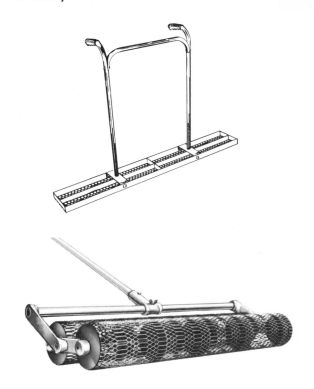

Fig. 6-17. Two styles of hand tampers.   (Marshalltown Trowel Co.)

usually stiff and hard to work. Two styles of tampers are shown in Fig. 6-17.

DARBIES

A darby is a long, flat, rectangular piece of wood, aluminum or magnesium. It is usually 30 to 80 in. long and from 3 to 4 in. wide with a handle on top as shown in Fig. 6-18. It is used to float the surface of the concrete slab immediately after

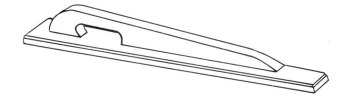

Fig. 6-18. Darby is used to float the surface of concrete immediately after it has been screeded.

it has been screeded. The darby removes any high or low spots left by the screed. It also helps to embed the course aggregate for later floating and troweling.

BULL FLOATS

A bull float is a large, flat, rectangular piece of wood, aluminum or magnesium, Fig. 6-19. It is usually 8 in. wide and 42 to 60 in. long with a long handle. The purpose of the bull float is essentially the same as the darby but it enables the mason to float a much larger area. It is particularly suited for

Fig. 6-19. Lightweight aluminum bull float. Such floats are also made from wood and magnesium.

outdoor use since the long handle (up to 16 ft. long) is difficult to use inside.

EDGERS

Edgers are produced in several sizes and styles, Fig. 6-20. A typical length is 6 in. Widths vary from 1 1/2 to 4 in. Edgers have radii from 1/8 to 1 1/2 in. Edgers are used to produce a radius on the edge of a slab. The radius improves the appearance of the slab and reduces the risk of damage to the edge. The curved-end edger is very popular.

JOINTERS OR GROOVERS

Jointers or "groovers," as they are sometimes called, are usually about 6 in. long and from 2 to 4 1/2 in. wide. Fig. 6-21 shows two styles. The cutting edge or bits are available in depths from 3/16 to 3/4 in. The jointer is used to cut a joint partly through fresh concrete to control the location of any possible cracks.

POWER JOINT CUTTERS

Another method of cutting joints in concrete slabs is to use a power joint cutter, Fig. 6-22. This machine may be either electric or gasoline powered. It has a shatterproof abrasive or diamond blade which produces a narrow joint. The joint is usually cut when the concrete has hardened 4 to 12 hours.

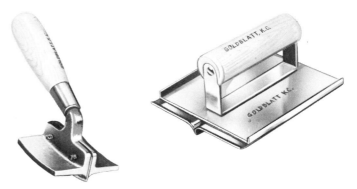

Fig. 6-21. Two popular styles of jointers or groovers. Jointer is used to cut a joint partly through fresh concrete to control location of any possible cracks.

HAND AND POWER FLOATS

Hand floats are made from wood, aluminum, magnesium, cork or molded rubber, Fig. 6-23. They range in size from 10 to 18 in. long and 3 1/2 to 4 1/2 in. wide. Hand floats are used to prepare the concrete surface for troweling.

Some floats are powered by electricity or gasoline engines. They have a rotating disk about 2 ft. in diameter which performs the same task as the hand float.

HAND AND POWER TROWELS

A cement mason's steel hand trowel is the last tool used in the finishing process of a slab of concrete. It is available in many different sizes ranging from 10 to 20 in. long and 3 to

Fig. 6-20. Assortment of edgers used to produce radius on edges of concrete slabs.

Fig. 6-22. Power joint cutter for large jobs must be handled carefully to avoid injury from flying particles.

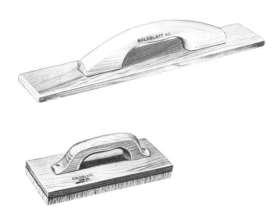

Fig. 6-23. Hand floats are made from wood, aluminum and cork as well as other masons' materials.

4 3/4 in. wide, Fig. 6-24. The first troweling of a slab is generally performed with a wide trowel 16 to 20 in. long. The last few troweling operations are usually done with a "fanning" trowel which is 14 to 16 in. long and 3 to 4 in. wide.

A power trowel, Fig. 6-25, has three or four rotating steel trowel blades. They may be powered by electricity or gasoline

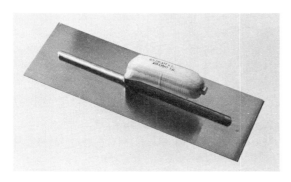

Fig. 6-24. Cement mason's steel hand trowel.

Fig. 6-25. Power trowel used for large jobs has rotating steel blades. (Goldblatt Tool Co.)

engine. The purpose of troweling is to give the surface a smooth, dense finish.

MASONRY SAWS

Cutting masonry units with a power masonry saw is faster and more accurate than using the blocking chisel and hammer. A typical power masonry saw is shown in Fig. 6-26. The blade

Fig. 6-26. Cutting masonry units with a power saw is faster and more accurate than using the blocking chisel.

on a masonry saw is usually 6 or 7 in. in diameter and about 1/8 in. thick. It is made of very hard material such as silicon-carbide or industrial diamonds. This is a dangerous piece of equipment to operate and goggles should always be worn to protect the eyes from flying chips.

SAFE WORK PRACTICES

Most occupations have some hazards and it is important to recognize them and protect oneself against them. A worker's personal safety and the safety of other workers is everyone's concern.

Accidents are quite frequent in the building industry. These accidents often result in lost time on the job, partial or total disability or even loss of life. Accidents can be reduced if each person works safely and uses the precautions that the nature of work requires. Everyone on the job must know where accidents are likely to occur and how to prevent them. Above all, there must be a keen sense of responsibility for other workers. Learning safe work practices is just as much a part of learning a trade as using the tools.

Always follow these general safety precautions:

1. Be alert and spot accidents before they happen.
2. Follow safe work practices. Do not take short cuts or expose yourself to danger unnecessarily.
3. Use the proper safeguards for a given job. Protective clothing, blade guard or goggles could save a hand, an eye or a foot.
4. Keep the work area clean. Many accidents are caused by litter underfoot.
5. Wear proper clothing and keep it in good repair. Shoes with steel toe protection are a must. Never wear shoes with a loose sole. They can hang on a scaffold board and cause a fall. Trouser legs should not be too long because they may get caught on a projecting edge and cause an accident. A hard hat could save your life if a brick were to fall from above and strike your head.
6. Never engage in horseplay on the job.
7. Pick up your tools and store them properly. This should become a habit.
8. Lift heavy objects with your legs rather than your back. Keep the arms and back as straight as possible when lifting. If the object is too heavy, get help.
9. Never place articles on a ledge, ladder or scaffold where they may fall and injure someone below. Check scaffolds and ladders for loose articles before moving them.
10. Get first aid for all injuries immediately. Do not chance getting infection or aggravating a wound.
11. Tell your supervisor about unsafe conditions and violations of safety rules.
12. Do not use faulty tools and equipment. Repair or replace these tools.
13. Do not count on "luck." Learn to work safely.

TOOL AND EQUIPMENT SAFETY

Keeping tools in good repair and using them properly is probably the best way to avoid accidents with them. However, when working near others, remember that a trowel or chisel is sharp. Care must be taken not to injure anyone.

Electrical and mechanical equipment is widely used in the masonry trades. This equipment presents a special hazard because of moving parts.

Equipment should be serviced properly and checked for safe operation. Electricity is the power source for much of the mechanical equipment used on construction jobs. It can be deadly if used improperly. Great care should be taken not to touch any bare wire. *Be sure that all electrical equipment is grounded.*

Never touch a person with live current flowing through him or her. You may be severly injured or killed. Shut off the power, if possible, or use a dry piece of wood to break the contact. Before using any portable equipment, be sure to check for worn or defective insulation, loose or broken connections and a bad ground wire connection.

Keep electrical wires off the ground and never operate electrical equipment in wet locations without proper grounding and safeguards.

SAFE USE OF SCAFFOLDS

Much of a mason's work is performed on scaffolding of one kind of another. The three main sources of injury associated with scaffolding are:

1. Falling from the scaffold.
2. Being struck by tools or materials falling from the scaffold.
3. Faulty scaffolding.

Scaffolding must be erected properly and designed to support the load it is expected to carry. The National Safety Council says that scaffolding should be able to support four times the anticipated load of men and materials.

The Scaffolding and Shoring Institute has developed several safety rules which are directly related to safe use of scaffolding. See Fig. 6-27.

ENCLOSURE SAFETY

In many parts of the country, year-round construction is made possible by enclosing the structure so that work may continue even during freezing weather. Enclosures may create a potentially dangerous condition for workers. The enclosures are generally heated with temporary heaters and may not be properly vented. Salamanders or gas fired heaters give off deadly carbon monoxide gases which are odorless and lethal. *Be sure that all heating devices are properly installed and vented. It may save your life!*

SAFE USE OF CHEMICALS

The first step in using any chemical is to read the manufacturer's directions and follow them closely. More chemicals are being used by masons and some may cause serious burns or loss of sight. If you are not familiar with the use of a chemical, ask someone who has used it how to handle it properly.

Wear the proper protective clothing when using chemicals. Rubber gloves and safety glasses are usually standard items. Have a bucket of water available to wash off any chemical that may come in contact with the skin. Know what to expect from the chemical that you use.

SCAFFOLDING SAFETY RULES
as Recommended by
SCAFFOLDING AND SHORING INSTITUTE
(SEE SEPARATE SHORING SAFETY RULES)

Following are some common sense rules designed to promote safety in the use of steel scaffolding. These rules are illustrative and suggestive only, and are intended to deal only with some of the many practices and conditions encountered in the use of scaffolding. The rules do not purport to be all-inclusive or to supplant or replace other additional safety and precautionary measures to cover usual or unusual conditions. They are not intended to conflict with, or supersede, any state, local, or federal statute or regulation; reference to such specific provisions should be made by the user. (See Rule II.)

I. **POST THESE SCAFFOLDING SAFETY RULES** in a conspicuous place and be sure that all persons who erect, dismantle or use scaffolding are aware of them.

II. **FOLLOW ALL STATE, LOCAL AND FEDERAL CODES, ORDINANCES AND REGULATIONS** pertaining to scaffolding.

III. **INSPECT ALL EQUIPMENT BEFORE USING**—Never use any equipment that is damaged or deteriorated in any way.

IV. **KEEP ALL EQUIPMENT IN GOOD REPAIR.** Avoid using rusted equipment—the strength of rusted equipment is not known.

V. **INSPECT ERECTED SCAFFOLDS REGULARLY** to be sure that they are maintained in safe condition.

VI. **CONSULT YOUR SCAFFOLDING SUPPLIER WHEN IN DOUBT**—scaffolding is his business, **NEVER TAKE CHANCES.**

A. **PROVIDE ADEQUATE SILLS** for scaffold posts and use base plates.

B. **USE ADJUSTING SCREWS** instead of blocking to adjust to uneven grade conditions.

C. **PLUMB AND LEVEL ALL SCAFFOLDS** as the erection proceeds. Do not force braces to fit—level the scaffold until proper fit can be made easily.

D. **FASTEN ALL BRACES SECURELY.**

E. **DO NOT CLIMB CROSS BRACES.** An access (climbing) ladder, access steps, frame designed to be climbed or equivalent safe access to the scaffold shall be used.

F. **ON WALL SCAFFOLDS PLACE AND MAINTAIN ANCHORS** securely between structure and scaffold at least every 30' of length and 25' of height.

G. **WHEN SCAFFOLDS ARE TO BE PARTIALLY OR FULLY ENCLOSED,** specific precautions must be taken to assure frequency and adequacy of ties attaching the scaffolding to the building due to increased load conditions resulting from effects of wind and weather. The scaffolding components to which the ties are attached must also be checked for additional loads.

H. **FREE STANDING SCAFFOLD TOWERS MUST BE RESTRAINED FROM TIPPING** by guying or other means.

I. **EQUIP ALL PLANKED OR STAGED AREAS** with proper guardrails, midrails and toeboards along all open sides and ends of scaffold platforms.

J. **POWER LINES NEAR SCAFFOLDS** are dangerous—use caution and consult the power service company for advice.

K. **DO NOT USE** ladders or makeshift devices on top of scaffolds to increase the height.

L. **DO NOT OVERLOAD SCAFFOLDS.**

M. **PLANKING:**
1. Use only lumber that is properly inspected and graded as scaffold plank.
2. Planking shall have at least 12" of overlap and extend 6" beyond center of support, or be cleated at both ends to prevent sliding off supports.
3. Fabricated scaffold planks and platforms unless cleated or restrained by hooks shall extend over their end supports not less than 6 inches nor more than 12 inches.
4. Secure plank to scaffold when necessary.

N. **FOR ROLLING SCAFFOLD THE FOLLOWING ADDITIONAL RULES APPLY:**
1. **DO NOT RIDE ROLLING SCAFFOLDS.**
2. **SECURE OR REMOVE ALL MATERIAL AND EQUIPMENT** from platform before moving scaffold.
3. **CASTER BRAKES MUST BE APPLIED** at all times when scaffolds are not being moved.
4. **CASTERS WITH PLAIN STEMS** shall be attached to the panel or adjustment screw by pins or other suitable means.
5. **DO NOT ATTEMPT TO MOVE A ROLLING SCAFFOLD WITHOUT SUFFICIENT HELP**—watch out for holes in floor and overhead obstructions.
6. **DO NOT EXTEND ADJUSTING SCREWS ON ROLLING SCAFFOLDS MORE THAN 12".**
7. **USE HORIZONTAL DIAGONAL BRACING** near the bottom and at 20' intervals measured from the rolling surface.
8. **DO NOT USE BRACKETS ON ROLLING SCAFFOLDS** without consideration of overturning effect.
9. **THE WORKING PLATFORM HEIGHT OF A ROLLING SCAFFOLD** must not exceed four times the smallest base dimension unless guyed or otherwise stabilized.

O. For "PUTLOGS" and "TRUSSES" the following additional rules apply.
1. **DO NOT CANTILEVER OR EXTEND PUTLOGS/TRUSSES** as side brackets without thorough consideration for loads to be applied.
2. **PUTLOGS/TRUSSES SHOULD EXTEND AT LEAST 6"** beyond point of support.
3. **PLACE PROPER BRACING BETWEEN PUTLOGS/TRUSSES** when the span of putlog/truss is more than 12'.

P. **ALL BRACKETS** shall be seated correctly with side brackets parallel to the frames and end brackets at 90 degrees to the frames. Brackets shall not be bent or twisted from normal position. Brackets (except mobile brackets designed to carry materials) are to be used as work platforms only and shall not be used for storage of material or equipment.

Q. **ALL SCAFFOLDING ACCESSORIES** shall be used and installed in accordance with the manufacturers recommended procedure. Accessories shall not be altered in the field. Scaffolds, frames and their components, manufactured by different companies shall not be intermixed.

Fig. 6-27. Scaffolding and shoring safety rules. (Scaffolding and Shoring Institute)

REVIEW QUESTIONS — CHAPTER 6

1. The tool most frequently used by the mason is the _____.

2. Name five factors which should be considered when purchasing a mason's trowel.

3. Name three special purpose trowels.

4. A _____ is used by masons to finish the surface of mortar joints.

5. A joint raker is used to make which of the following joints?
   a. Concave joint.
   b. Weathered joint.
   c. Struck joint.
   d. Raked joint.

6. The measuring tool used most frequently by the mason is the _____.

7. A mason uses a _____ to check if a wall is plumb or level.

8. A mason laying conrete block would use a _____ chisel to cut a concrete block.

9. Quality chisels are made from _____ _____ which has been tempered.

10. A mason's _____ is used to keep each course level and the wall true and out-of-wind.

11. _____ helps make a mason's work easier by providing a place for materials and tools at a convenient height.

12. Most tools used in concrete masonry are designed for use in finishing horizontal concrete surfaces. True or False?

13. What is a screed?

14. What is the purpose of a tamper?

15. The tool used to float the surface of a concrete slab immediately after it has been screeded is a _____.

16. An _____ is used to produce a radius on the edge of a slab.

17. A _____ is used to cut a joint partly through fresh concrete to control the location of any possible cracks.

18. The tool which is used for the final finishing process on a concrete slab is usually the _____.

19. Learning safe work practices is just as much a part of learning a trade as using the tools. True or False?

20. Why is horseplay out of place on the job?

21. There is no need to worry about a frayed electrical cord because anyone knows that 120 volts never hurt anyone. True or False?

22. Scaffolding should be able to support _____ times the anticipated load of workers and materials.

23. Gas-fired heaters give off deadly _____ _____ gases.

24. Before using chemical one should _____.

Chapter 7
CONCRETE FUNDAMENTALS

Concrete is one of our most important building materials. It is used in almost every type and size of architectural structure. In footings and foundations, in exterior and interior walls, floor and roof systems, walks and driveways, concrete is a major material. The plastic quality of concrete lends itself to almost unlimited design possibilities in form, pattern and texture. Hardened concrete is durable, needs little maintenance and can be used in many ways.

MATERIALS

Concrete is a mixture of four basic materials in proper amounts (proportions). The basic materials are:
1. Portland cement.
2. Fine aggregate (sand or finely crushed stone).
3. Coarse aggregate (gravel or crushed stone).
4. Water.

Lightweight, manufactured materials are also used to produce lightweight concrete. Sometimes other ingredients are added to the basic concrete mix. Their purpose is to produce certain properties in the finished product. These ingredients are called admixtures.

PORTLAND CEMENT

Portland cement is hydraulic because it sets and hardens by reacting with water. This chemical reaction is called hydration. It forms the cement into a stone-like mass.

Portland cement must contain varying amounts of lime, silica, alumina and iron components. The raw materials are pulverized and combined into a mixture which has the desired chemical makeup.

Manufacturers of portland cement use their own brand names. However, all portland cements are manufactured to meet the American Society for Testing and Materials Standard Specifications, ASTM Designation:

C-150 Portland Cement, Types I through V.
C-175 Air-Entraining Portland Cement.
C-205 Portland Blast Furnace Slag Cement.
C-340 Portland-Pozzolan Cement.

TYPES OF PORTLAND CEMENT

Portland cement is produced to meet various physical and chemical requirements for specific purposes. There are several types.

TYPE I — NORMAL PORTLAND CEMENT: This general purpose cement is suitable for all uses when the special properties of the other types are not required. It is used where it will not be subject to attack by sulfate from the soil or water or to an objectionable temperature rise from the heat generated by hydration. It may be used for pavements, sidewalks, reinforced concrete structures, bridges, tanks, reservoirs, culverts, water pipe and masonry units.

TYPE II — MODIFIED PORTLAND CEMENT: This type of cement produces less heat due to hydration and can resist sulfate attack better. This cement may be used in structures such as piers, heavy abutments and heavy retaining walls.

TYPE III — HIGH EARLY STRENGTH PORTLAND CEMENT: A cement producing high strengths quickly — usually a week or less. It is used when forms are to be removed as soon as possible or when the structure must be used quickly. Its controlled curing period may also be reduced in cold weather.

TYPE IV — LOW-HEAT PORTLAND CEMENT: Use when the heat generated by hydration must be kept low and when a slower developing strength is not objectionable. It is designed for use in massive concrete structures such as dams.

TYPE V — SULFATE-RESISTANT PORTLAND CEMENT: Used only in construction exposed to severe sulfate action such as in some western states with water and soil of high alkali content. It gains strength more slowly than normal cement.

AIR-ENTRAINING PORTLAND CEMENT: Designated as IA, IIA and IIIA. They correspond, in composition, to Types I, II, and III except that small quantities of air-entraining materials are added to the cement during production. These cements improve resistance to freeze-thaw action and scaling caused by chemicals used in snow and ice removal. Concrete made with these cements contains billions of tiny, well-distributed and completely separated air bubbles.

WHITE PORTLAND CEMENT: Manufactured to conform to the specifications of ASTM Designation C-150 and C-175. The manufacturing process is controlled so that the finished product will be white instead of gray. It is used mainly for architectural purposes such as precast wall and facing panels, terrazzo, stucco, cement paint, tile grout and decorative concrete. It is recommended wherever white or colored concrete or mortar is desired.

PORTLAND BLAST-FURNACE SLAG CEMENTS: In these cements, granulated (like sand) blast-furnace slag is ground and

TYPE OF PORTLAND CEMENT		COMPRESSIVE STRENGTH, PERCENT OF STRENGTH OF TYPE I OR NORMAL PORTLAND CEMENT CONCRETE			
ASTM DESIGNATION	CANADIAN STANDARDS ASSOC.	1 DAY	7 DAYS	28 DAYS	3 MONTHS
I	NORMAL	100	100	100	100
II		75	85	90	100
III	HIGH EARLY STRENGTH	190	120	110	100
IV		55	55	75	100
V	SULFATE RESISTING	65	75	85	100

Fig. 7-1. Chart shows relative strengths of concretes made with different cements. These concretes were moist-cured until tested. (Portland Cement Assoc.)

mixed with portland cement. They can be used in general concrete construction when the specific properties of other types are not required.

PORTLAND-POZZOLAN CEMENTS: In these cements, pozzolan is blended with ground portland cement. They are used principally for large hydraulic structures such as bridge piers and dams.

The strength of concretes made with different cements is shown in Fig. 7-1. Type I or normal portland cement concrete is used as the basis for comparison.

AGGREGATES

Sand, gravel, crushed stone and air-cooled blast furnace slag are the most commonly used aggregates in concrete. The aggregates usually make up 60 to 80 percent of the volume of concrete and may be considered to be filler material.

Aggregates must meet certain requirements. They must be clean, hard, strong, durable particles free of chemicals and coatings of clay which could affect the bond. Since aggregates are such a large proportion of the volume of concrete, they greatly affect the quality of the finished product. Requirements for concrete aggregates are specified in ASTM C-33. These specifications place limits on allowable amounts of damaging substances. They also cover requirements as to grading, strength and soundness of particles.

FINE AGGREGATE

Fine aggregate contains sand or other acceptable fine materials. A desirable concrete sand should contain particles varying uniformly in size from very fine up to 1/4 in. in diameter. Fine particles are necessary for good workability and smooth surfaces. These fine particles help fill the spaces between the larger particles. But an excess of fine particles requires more cement-water paste to attain the right strength and will increase the cost.

COARSE AGGREGATE

Coarse aggregate consists of gravel, crushed stone or other acceptable materials larger than 1/4 in. in diameter. Coarse aggregate suitable for use in concrete should be sound, hard and durable. It should not be soft, flaky or wear away rapidly. It should also be clean and free of loam, clay or vegetable matter. These foreign materials prevent the cement paste from properly binding the aggregate particles together and will reduce the strength of the concrete.

The maximum size of coarse aggregate is usually 1 1/2 in. in diameter. This large size aggregate could be used in a thick foundation wall or footing. As a rule of thumb, the largest piece should never be greater than one fifth the thickness of a finished wall section, one third the thickness of a slab or three fourths the width of the narrowest space through which the concrete will be required to pass during placing.

Well-graded, coarse aggregate will range uniformly from 1/4 in. up to the largest size that is satisfactory for the kind of work to be performed. The use of the maximum allowable particle size usually results in less drying shrinkage and lower cost.

MIXING WATER FOR CONCRETE

The primary purpose for using water with cement is to cause hydration (a chemical reaction) of the cement. Water usually constitutes (makes up) from 14 to 21 percent of the total volume of concrete. The amount of water used definitely affects the quality of the concrete.

Concrete mixing water should be clean and free of oil, alkali and acid. Generally, water that is good for drinking is suitable for mixing concrete. However, water with too many sulfates should not be used even though it may be potable (good for drinking).

ADMIXTURES

Admixtures are any materials added to the concrete batch before or during mixing other than portland cement, water and aggregates. Admixtures may be classified as:

1. Air-entraining.
2. Water-reducing.
3. Retarding.
4. Accelerating.
5. Pozzolans.
6. Workability agents.
7. Miscellaneous.

The desirable qualities of concrete can often be attained conveniently and economically through the proper design of the mix and the selection of suitable materials without using admixtures (except air-entraining admixtures).

AIR-ENTRAINING ADMIXTURES

Air-entraining admixtures are used to improve the durability of concrete exposed to moisture during cycles of freezing and thawing. Concrete resists surface scaling caused by ice-removal agents far better with air-entraining admixtures.

The workability of fresh concrete is also improved significantly. Segregation (separating of materials) and bleeding are reduced or stopped. Entrained air may be produced in concrete through the use of an air-entraining cement or through an air-entraining admixture.

WATER-REDUCING ADMIXTURES

Water-reducing admixtures are materials which reduce the quantity of mixing water required to produce concrete of a given consistency. These materials increase the slump of concrete with less water content. (Slump is the relative consistency or stiffness of the plastic concrete.)

Some water-reducing admixtures also retard the setting time of concrete. Others are modified to give varying degrees of retardation and some do not affect the setting time much. Still others may entrain air in the concrete.

An increase in strength is usually gained with water-reducing admixtures if the water content is reduced for a given mix and if the cement content and slump are kept the same. Even though the water content is reduced with these admixtures, increases in drying shrinkage are probable.

RETARDING ADMIXTURES

A retarding admixture is used to retard (slow down) the setting time of concrete. High temperatures of plastic concrete often cause it to harden too fast. This increased hardening rate makes placing and finishing of concrete difficult. One of the most practical ways to combat this difficulty is to reduce the temperature of the concrete by cooling the mixing water and/or aggregates. Retarders will not lower the initial temperature of the concrete.

Retarders may be used when it is desirable to offset rapid setting up (hardening) of concrete due to hot weather. It is also used to delay the initial set of concrete when difficult conditions of placement are encountered.

Retarders usually reduce the strength of concrete for the first one to three days. Effect on other properties such as shrinkage may not be predictable. It is advisable to test the retarders under actual job conditions to be sure of results.

ACCELERATING ADMIXTURES

An accelerating admixture will speed up the setting and the strength development of concrete. However, most of the commonly used accelerators cause an increase in the drying shrinkage of concrete. The development of strength of concrete can also be increased by:
1. Using Type III portland cement.
2. Lowering the water-cement ratio.
3. Increasing the cement content or curing at higher temperatures.

Calcium chloride is the most commonly used accelerating admixture. The amount added should not be more than 2 percent of the cement weight. This admixture should not be used in the following cases:
1. In prestressed concrete.
2. In concrete containing embedded aluminum.
3. Where galvanized steel will remain in contact with the concrete.
4. In concrete subjected to alkali or sulfates.

POZZOLANS

A pozzolan is defined by ASTM C-219 as "a siliceous or siliceous and aluminous material, which in itself, possesses little or no cementitious value but will, in finely divided form and in the presence of moisture, chemically react with calcium hydroxide at ordinary temperatures to form compounds possessing cementitious properties."

These materials are sometimes used in concrete to help control high internal temperatures. These temperature build-ups can occur from slow loss of heat generated by hydration in massive structures. However, these high temperatures can often be controlled by using Type II cement and/or by lowering the temperature of the mixing water and aggregates.

The use of pozzolans can substantially reduce the early strength of concrete during the first 28 days and the effect on concrete varies considerably. Tests should be performed before using Pozzolans in construction.

WORKABILITY AGENTS

Plastic concrete is often harsh due to aggregate characteristics such as particle shape and poor grading. This condition may be improved by increasing the amount of fine aggregate or entraining air. Entrained air is considered the best and acts like a lubricant. Excessive amounts of these fine materials will most likely reduce the strength and increase shrinkage of the concrete.

CONCRETE MIXTURES

The proportioning of concrete mixtures will determine the qualities of the plastic and hardened concrete. Desirable qualities of plastic concrete are consistency, workability and finishability. Fig. 7-2 shows concrete with the proper con-

Fig. 7-2. This concrete has the proper consistency.
(Portland Cement Assoc.)

sistency. Desirable qualities of hardened concrete are strength, durability, water tightness, wear resistance and economy.

In concrete which is properly proportioned each particle of aggregate, no matter how large or small, will be completely surrounded by the cement-water paste. The spaces between aggregate particles will also be filled completely with paste. To lower costs, the proportioning should minimize the amount of cement required without reducing concrete quality.

CONCRETE MIX CHARACTERISTICS

Before a concrete mixture can be proportioned you must know:
1. The size and shape of concrete members.
2. The concrete strength required.
3. The conditions of exposure.

Most of the desired properties of hardened concrete depend upon the quality of the cement paste. The first step then in determining the mix proportions, is to select the right water-cement ratio.

WATER-CEMENT RATIO

The proportion of water to cement, stated in gallons of water per bag (94 lb.) of cement, is called the water-cement ratio. The binding properties of the cement paste are due to chemical reactions between the cement and water.

The reactions require time and favorable conditions of temperature and moisture. They proceed very rapidly at first and then much more slowly for a long period of time if conditions are satisfactory. Only a small amount of water (about 3 1/2 gal. per bag of cement) is necessary to complete the chemical reactions, but more water (4 or 9 gal. per bag of cement) is used to increase workability and aid placement. Also, more water increases the amount of aggregate which can be used. This makes the mix even cheaper.

Good cement paste requires a proper proportion of water to cement. If too much water is used, the paste will be thin and weak when it hardens. Too little water in the mix will make placing and finishing difficult.

Cement paste made with the proper amount of water will be strong, watertight and durable. The concrete will be strong and durable if the aggregates and cement paste are strong and durable. The quality of the concrete is therefore largely determined by the water-cement ratio which is so important a factor in the cement paste.

Strength of concrete is directly related to the water-cement ratio. Fig. 7-3 shows the results of many tests performed on cured concrete. The results were similar for other types of cements. More water resulted in less strength and less water resulted in greater strength.

After concrete is in place, the strength continues to increase as long as moisture and satisfactory temperatures are present. Highest curing temperature is 73 deg. F (23 deg. C). When concrete is permitted to dry, the chemical reactions cease and lower strength results. Concrete should therefore be kept constantly moist as long as possible after placing.

There is also a direct relationship between the durability of concrete and the relative quantities of water and cement in the

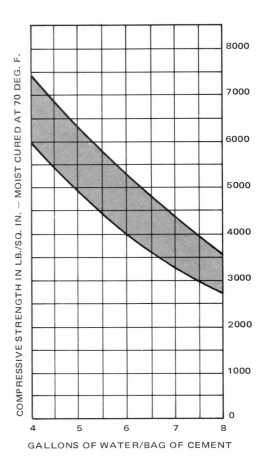

Fig. 7-3. The area within the curves represents the age-compressive-strength relationship for Type I portland cement after 28 days moist curing at 70 F (21 C).

mixture. Freezing and thawing are the most destructive natural forces of weathering on wet or moist concrete. During this process, water expands as it changes to ice. If the water-cement ratio was high during mixing, the concrete is more porous and is more subject to damage by freezing and thawing.

Watertightness, likewise, depends on the water-cement ratio in the mix. The more water used in the mix, the less likely the concrete is to be watertight.

SELECTING CONCRETE MIXTURES

The right concrete mixture for any job is the one which is most economical and workable while giving the desired properties. Fig. 7-4 shows concrete mixtures for various construction applications. More specifically:
1. Column 1 specifies the construction application.
2. Column 2 recommends the approximate maximum water-cement ratio.
3. Column 3 shows the appropriate consistency as measured by inches of slump.
4. Column 4 indicates the maximum size of aggregate.
5. Column 5 specifies the cement content.
6. Column 6 shows the probable approximate compressive strength after 28 days.

Rule of thumb approximation of volumes such as 1:2:3 of

CONSTRUCTION APPLICATION	WATER/CEMENT RATIO GAL. PER BAG	CONSISTENCY (AMOUNT OF SLUMP)	MAXIMUM SIZE OF AGGREGATE	APPROPRIATE CEMENT CONTENT BAGS PER YD.	PROBABLE 28TH DAY STRENGTH (PSI)
FOOTINGS	7	4" TO 6"	1 1/2"	5.0	2800
8 INCH BASEMENT WALL MODERATE GROUND WATER	7	4" TO 6"	1 1/2"	5.0	2800
8 INCH BASEMENT WALL SEVERE GROUND WATER	6	3" TO 5"	1 1/2"	5.8	3500
10 INCH BASEMENT WALL MODERATE GROUND WATER	7	4" TO 6"	2"	4.7	2800
10 INCH BASEMENT WALL SEVERE GROUND WATER	6	3" TO 5"	2"	5.5	3500
BASEMENT FLOOR 4 INCH THICKNESS	6	2" TO 4"	1"	6.2	3500
FLOOR SLAB ON GRADE	6	2" TO 4"	1"	6.2	3500
STAIRS AND STEPS	6	1" TO 4"	1"	6.2	3500
TOPPING OVER CONCRETE FLOOR	5	1" TO 2"	3/8"	8.0	4350
SIDEWALKS, PATIOS DRIVEWAYS, PORCHES	6	2" TO 4"	1"	6.2	3500

Fig. 7-4. Recommended concrete mixtures for various construction applications.

cement to sand to gravel are not recommended for quality concrete mixtures.

Most fine aggregates contain some water. Therefore, allowance should be made for this factor in determining the amount of water to be added to the mix. A simple test will determine whether sand is "damp," "wet" or "very wet." Press some sand in your hand. If it falls apart after you open your hand, it is damp; if it forms a ball which holds its shape, it is wet; if it sparkles and wets your hand, it is very wet.

Fig. 7-5 shows adjustments to the water-cement ratio which should be made when the fine aggregate is damp, wet or very wet. The amount of water specified in the chart plus the water in the fine aggregate will equal the amount of water required for the mix. If the sand is dry, then use the full amount indicated in Fig. 7-4.

IF THE MIX CALLS FOR A WATER-CEMENT RATIO OF:	USE THESE AMOUNTS OF WATER (GALS.) WHEN THE SAND IS:		
	DAMP	WET	VERY WET
5 GAL./BAG	4 3/4	4 1/2	4 1/4
6 GAL./BAG	5 1/2	5	4 1/2
7 GAL./BAG	6 1/4	5 1/2	4 3/4

Fig. 7-5. Recommended adjustment to the water-cement ratio when the fine aggregate is damp, wet or very wet.

MIX CONSISTENCY

The mix consistency or degree of stiffness of plastic concrete is called slump. Slump is measured in inches.

Very fluid (wet) mixes are called high slump concrete while stiff (dry) mixes are called low slump concrete. Slump is related primarily to water-cement ratio. A high water content causes a high slump. Generally, a low slump concrete will produce a better concrete product.

A slump test may be performed in the field by using a sheet metal slump cone, Fig. 7-6. The cone is 8 in. in diameter at the bottom, 4 in. in diameter at the top and 12 in. high. It is placed on a clean dry surface and filled one third full with the concrete being tested. The concrete is rodded 25 times with a metal rod which is 24 in. long and 5/8 in. in diameter. Concrete is then added until the cone is 2/3 full and is again rodded 25 times with rod penetrating the lower layer. Finally, the cone is filled, the top raked off level and the mixture rodded 25 more times. The cone is then removed by lifting it straight up. The slump is measured immediately. The rod is placed across the top of the cone extending over the concrete sample. The distance from the bottom edge of the rod to the concrete is measured with a rule. This distance is the slump rating for the batch.

The slump test may be used as a rough measure of the consistency of concrete, but should not be considered a measure of workability. It should not be used to compare mixes of entirely different proportions or mixes containing different aggregates. The slump test does, however, help to signal changes in grading or proportions of the aggregate or water content in the mix.

MEASURING MATERIALS

The ingredients of each batch of concrete must be measured accurately if uniform batches of proper proportions and consistency are to be produced. The problem of varying amounts of moisture in the aggregate has already been discussed. This must be taken into account if accurate control is to be obtained.

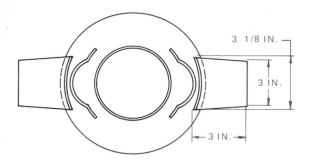

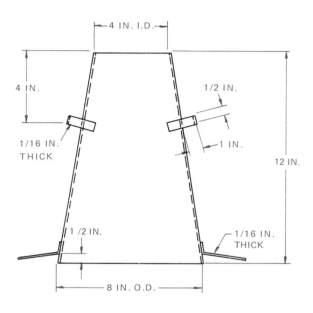

Fig. 7-6. A sheet metal slump cone showing dimensions.

MEASURING CEMENT

If bagged cement is used, the batches should be of such a size that only full bags are used. If this is not possible, then the proper amount should be weighed out each time. Volume is not an accurate method of measuring cement. Bulk cement (not bagged) should always be weighed for each batch.

MEASURING WATER

The effects of the water-cement ratio on the qualities of concrete make it just as necessary to measure accurately the water used as any other materials. Water is measured in gallons and any method which will ensure accuracy is acceptable.

MEASURING AGGREGATES

Measurement of aggregates by weight is the recommended practice and should be required on all jobs that require a high degree of consistency in each batch. Measurement of fine aggregate by volume is not accurate because a small amount of moisture is nearly always present. This moisture causes the fine aggregate to bulk or fluff up. The degree of bulking depends on the amount of moisture present and the fineness of the sand. A fine sand with a 5 percent moisture content will increase in volume about 40 percent above its dry volume.

Another problem with aggregates, especially coarse aggregates, is separation of the various sizes. This problem can be reduced by weighing the coarse aggregate in two or more sizes.

MIXING

All concrete should be mixed thoroughly until it is uniform in appearance and all ingredients are uniformly distributed. Mixing time will depend on several factors:
1. The speed of the machine.
2. Size of the batch.
3. The condition of the mixer.

Generally, the mixing time should be at least one minute for mixtures up to 1 cu. yd. with an increase of 15 seconds for each 1/2 cu. yd. or fraction thereof. Mixing time should be measured from the time all materials are in the mixer.

Mixers should not be loaded above their capacity. They should be run at the speed for which they were designed. Mixers coated with hardened concrete or having badly worn blades will not perform as efficiently as they should. These conditions should be corrected.

Generally, about 10 percent of the mixing water is placed in the mixer before the aggregate and cement are added. Water should then be added uniformly along with the dry materials. The last 10 percent of the water is added after all the dry materials are in the mixer.

READY-MIXED CONCRETE

Ready-mixed concrete is purchased directly from a central plant. It is convenient and usually of high quality. In some ready-mixed operations, the materials are dry-batched at the central plant and then mixed enroute to the site in truck mixers.

Another method is to mix the concrete in a stationary mixer at the central plant just enough to intermingle the ingredients. The mixing is then completed in a truck mixer enroute to the job site. Most truck mixers have a capacity of 1 to 5 cu. yd. and carry their own water supply.

ASTM C-94 requires that the concrete must be delivered and discharged from the truck mixer within 1 1/2 hours after the water has been added to the mix.

REMIXING CONCRETE

Initial set of concrete usually takes place two or three hours after the cement is mixed with water. During this time the concrete may dry out and stiffen if left standing. It may be remixed if it becomes sufficiently plastic and workable. No water may be added to make it more workable (re-tempering). This will reduce the quality of the concrete just as a larger amount of water in the original mix.

PLACING AND FINISHING CONCRETE

Concrete is moved about for placing by many methods. Some of the most popular methods include chutes, push buggies, buckets handled by cranes and pumping through pipes. The method used should not restrict the consistency of the concrete. Consistency should be governed by the placing conditions and the application. In other words, if the conditions permit the use of a stiff mix, the equipment should be designed to handle it.

PREPARATION

Subgrade (ground on which concrete is poured) should be properly prepared before the concrete arrives. Forms should be in place and level. Subgrades should be smooth and moist, Fig. 7-7. (Moistening the subgrade prevents rapid loss of water from the concrete when pavements, floors and similar flatwork are being placed. This is especially important in hot weather.)

Forms should be tight, clean and securely braced. Poorly constructed forms will sag and leak. Forms should be constructed from materials which will give the desired texture to the finished concrete.

Treating the forms with oil or other preparations will make form removal easier after the concrete has hardened. Wooden forms exposed to the sun for long periods should be saturated (soaked) thoroughly with water to tighten the joints.

Reinforcing steel should be clean and free of loose rust and scale. All hardened mortar should be removed from the steel before placing the concrete.

PLACING CONCRETE

Once poured, concrete should be as near as possible to its intended location. For example, it should not be placed in large quantities in one place and allowed to run or be worked over a long distance in the form. Segregation (separating) of ingredients and sloping work planes result from this practice. It should be avoided. Generally, concrete should be placed in horizontal layers having uniform thickness.

If the form is deep, the concrete should be compacted after each layer is placed. Layers are usually 6 to 12 in. thick in reinforced concrete and up to 18 in. for nonreinforced work.

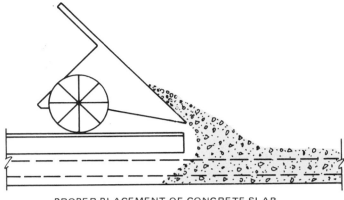

PROPER PLACEMENT OF CONCRETE SLAB

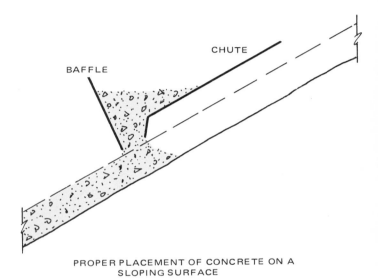

PROPER PLACEMENT OF CONCRETE ON A SLOPING SURFACE

Fig. 7-8. Proper method of dumping concrete when pouring a slab.

Fig. 7-7. This cement worker is using a template to check the subgrade for smoothness and proper height. (Portland Cement Assoc.)

Concrete should not be allowed to drop freely more than 3 or 4 ft. Drop chutes of rubber or metal may be used when placing concrete in thin vertical sections.

SLAB CONSTRUCTION

Placement of concrete in slab construction should be started at the most distant point of the work so that each batch may be dumped against the previously placed concrete, not away from it, Fig. 7-8. Care should be taken to prevent stone pockets (areas of excessive large aggregate) from occuring. If this happens, some of the aggregate may be moved to areas where there is more cement paste to surround them.

Concrete placement in walls should begin at either end and progress toward the center. The same order should be used for each successive layer.

Concrete should be placed around the perimeter (outer edges) first in large flat open areas. Whatever method of placement is used, do not allow water to collect at the ends and corners of forms.

Compacting the concrete is always necessary. Work the mix with a spade or rod to be sure all spaces are filled and air pockets are worked out. (This is called puddling, spading or rodding.) Mechanical vibrators may be used either in the concrete or on the forms.

This process should help to eliminate stone and air pockets and consolidate each layer with that previously placed. It also will bring fine material to the faces and top for proper finishing.

Mechanical vibration does not make the concrete stronger, but it does permit the use of a stiffer mix which will be stronger than a wet mix. Too much vibration will cause segregation of particles. Judgment must be used to determine the proper amount. Indicators of sufficient vibration are the appearance of a line of mortar along the forms and by the sinking of the coarse aggregate into the mortar.

PLACING ON HARDENED CONCRETE

When fresh concrete is placed on hardened concrete, it is important to produce a good bond and a watertight joint. To assure this result, the hardened concrete should be reasonably level, rough, clean and moist. If some coarse aggregate are left exposed, it will aid in bonding. All loose or soft mortar should be removed from the top surface of the hardened concrete before placing the fresh batch.

For floors which require two courses of concrete, the top of the first course (lower level) may be broomed with a steel or stiff bristle broom just as it sets. The surface should be heavily scored and cleaned before the grout coat and top course are placed. The grout coat is a mixture of portland cement and water. It usually has the consistency of thick paint and is scrubbed into the surface of the slab just before the top course is placed.

If old concrete is to receive a new topping, it must be thoroughly roughened and cleaned of dust and loose particles, grease, oil or other materials. The surface might best be chipped with pneumatic tools or sand blasted to expose sound concrete. Hardened concrete must be moistened thoroughly before new concrete is placed on it, but no pools of water should be left standing on the surface of the existing concrete.

Where concrete is to be placed on hardened concrete or rock, a layer of mortar, 1/2 to 1 in. thick, is placed on the hard surface. This mortar provides a cushion for the new concrete and prevents stone pockets and aids in securing a tight joint. The mortar is generally made of the same materials as the concrete, but without the course aggregate. It should have a slump of less than 6 in.

PNEUMATIC APPLICATION OF CONCRETE

Pneumatically applied concrete, frequently called shotcrete, is a mixture of portland cement, aggregate and water shot into place by means of compressed air. Aggregate up to 3/4 in. may be used with some equipment. Shotcrete may be applied by either the wet mix or dry mix process.

In the dry mix process, the cement and aggregates are mixed in a relatively dry condition. This mix is pumped through a hose to a nozzle where water is added. At least 45 psi air pressure and a nozzle velocity of about 400 ft. per second is used to force the dry materials through the hose. Water pressure at the nozzle should be at least 15 psi higher than the air pressure. In the wet mix process, the concrete is premixed before it is applied pneumatically.

Water content may be kept to a minimum using the pneumatic process. Therefore, high-strength, durable concrete may be obtained. The key element in this process is the person controlling the nozzle. In both processes the worker directs the nozzle and controls the thickness of the concrete layer and the angle of application. Pneumatic application may be used for new construction and repair work in difficult locations. It also works well where relatively thin sections and large areas are involved.

FINISHING CONCRETE SLABS

Concrete slabs may be finished several ways. It depends on the effect desired and the use of the product. Some surfaces may be left rough, others broomed, floated or troweled. Still other surfaces may be textured, colored or have exposed aggregate.

SCREEDING

Screeding is usually the first finishing operation after the concrete is placed in the forms. It is performed with a screed. *Screeding is the process of striking off the excess concrete to bring the top surface to the proper grade or elevation, Fig. 7-9.* The edge of the screed may be straight or curved depending on the surface requirements. The screed rests on the top of the form and is moved across the concrete with a sawing motion. It is advanced forward slightly with each movement.

An excess of concrete should be carried along in front of the screed to fill low places as the tool is moved forward. But if too much concrete is allowed to build up in front of the screed, it may tend to leave hollows behind it.

In normal concrete the dry materials used are heavier than water. They will begin to sink or settle to the bottom of a plastic concrete mixture shortly after placement. This settling

Fig. 7-9. This concrete is being screeded to strike off the excess.

action causes excess water to rise to the surface. This condition is called "bleeding." Bleeding does not usually occur with air-entrained concrete.

It is very important that the first operations of placing, screeding and darbying be performed before any bleeding takes place. *If any finishing operation is performed on the surface while the bled water is present, serious scaling, dusting or crazing can result.*

The high and low spots may be eliminated and large aggregate embedded using a darby or bullfloat, Fig. 7-10. This

Fig. 7-10. Bullfloat is used to eliminate high and low spots and to embed large aggregate in the concrete.   (Portland Cement Assoc.)

operation should follow immediately after screeding to prevent bleeding. Some surface finishes may not need any further finishing, but most will require one or more of the following operations.

EDGING AND JOINTING

If edging is necessary, this could be the next operation, Fig. 7-11. Edging provides a rounded edge or radius to prevent chipping or damage to the edge. The edger is run back and forth until the desired finish is obtained. Care should be taken to cover all coarse aggregate and not to leave too deep a depression on the top of the slab. This indentation could be difficult to remove during subsequent finishing operations.

As soon as edging has been completed, the slab is jointed or

Fig. 7-11. Edging produces a radius on the edge of the slab which prevents chipping. In step one, above, trowel is inserted between form and concrete to provide "track" of an edging operation shown in step two, below.

grooved, Fig. 7-12. The bit (cutting edge) of the jointing tool cuts a groove in the slab which is called a control or contraction joint. Any cracking due to shrinkage caused by drying out or temperature change will occur at the joint. These cracks are not noticeable when controlled. The joint weakens the slab and induces cracking at that location rather than some other place.

Fig. 7-12. Jointing a slab helps control any cracking due to shrinkage caused by drying out or temperature change. (Portland Cement Assoc.)

In sidewalk and driveway construction, the tooled joints are generally spaced at intervals equal to the width of the slab, but not more than 20 ft. apart. They should be perpendicular (at right angles) to the edge of the slab. The groove is usually made with a 3/4 in. bit.

Use a straightedge as a guide when making the groove. A 1 x 8 or 1 x 10 board will be ideal. Be sure the board is straight.

Large concrete surfaces may be jointed by cutting grooves with a power saw using an abrasive or diamond blade. When grooves are cut rather than jointed, the operation should be performed 4 to 12 hours after the slab has been finished. The cutting must be done before random shrinkage cracks develop, but after the concrete is hard enough not to be torn or damaged by the blade.

FLOATING

After concrete has been edged and jointed, it should be allowed to harden enough to support a man and leave only a slight foot imprint. Floating should not begin until the water sheen has disappeared. *When all bled water and water sheen has left the surface, the concrete has started to stiffen.* The surface is floated with wood or metal floats or with a finishing machine using float blades. Aluminum or magnesium floats work better especially on air-entrained concrete.

Metal floats reduce the amount of work required by the finisher. Drag is reduced and the float slides more readily over the surface. A wood float tends to stick to the surface and produces a tearing action. The light metal float also forms a

smoother surface texture than the wood float.

There are three reasons why concrete is floated:
1. To embed aggregate particles just beneath the surface.
2. To remove slight imperfections, waves and voids.
3. To compact the concrete at the surface in preparation for other finishing operations.

Be sure not to overwork the concrete. This will bring excess water and fine aggregate material to the surface which will result in surface defects.

When floating is done to provide a coarse texture as the final finish, a second floating may be necessary after the concrete has partially hardened.

TROWELING

When a smooth, dense surface is desired, steel troweling is performed after floating. Frequently the cement mason will float and trowel an area before moving his knee boards.

Troweling produces a smooth, hard surface, Fig. 7-13.

Fig. 7-13. Troweling produces a smooth, hard surface on the concrete. It may be done by hand or with a troweling machine.

During the first troweling, whether by hand or power, the trowel blade must be kept as flat against the surface as possible. If the blade is held at too great an angle, a "washboard" effect will result.

A new trowel is not recommended for the first troweling. An old trowel which is "broken in" can be worked quite flat without the edges digging into the surface.

As the concrete progressively hardens it may be troweled several times to obtain a very smooth and hard surface. Usually smaller size trowels are used for successive applications so that sufficient pressure can be exerted for proper finishing.

If necessary, tooled joints and edges may be rerun after troweling to maintain uniformity.

BROOMING

Steel troweling produces a very smooth surface and is often slippery when wet. Sidewalks, driveways and other outside flatwork frequently are broomed or brushed to produce a slightly roughened surface. The broomed surface is made by drawing a soft-bristled push broom over the surface of the slab after steel troweling, Fig. 7-14. The concrete must be hard

Fig. 7-14. Brooming produces a slightly roughened surface which reduces the danger of slipping on a smooth troweled surface.

enough so that brooming will not damage the edges of tooled joints. Coarser textures for steep slopes may be produced by using a stiffer bristled broom. Brooming is usually perpendicular (at right angles) to the traffic to provide most resistance.

FORM REMOVAL

Generally, it is best to leave the forms in place as long as possible for better curing. Sometimes, however, it is desirable to remove the forms as soon as possible. In either case, leave them in place until the concrete is strong enough to support its own weight and any other loads which may immediately be placed upon it. The concrete should be hard enough to resist damage from form removal.

Usually, the side forms of relatively thick sections may be removed in 12 to 24 hours after placing. Testing the concrete to determine hardness is better than relying on an arbitrary age for form removal. The age-strength relationship should be determined from representative samples of concrete used in the structure and cured under job conditions. Fig. 7-15 shows the times required to attain certain strength under average conditions for air-entrained concrete. It should be stressed that these are averages and strength is affected by materials used, temperature and other conditions.

STRENGTH PSI	AGE	
	TYPE I OR NORMAL CEMENT	TYPE III OR HIGH-EARLY-STRENGTH CEMENT
500	24 HOURS	12 HOURS
750	1 1/2 DAYS	18 HOURS
1500	3 1/2 DAYS	1 1/2 DAYS
2000	5 1/2 DAYS	2 1/2 DAYS

Fig. 7-15. This chart shows the age-strength relationship of air-entrained concrete which must be considered when removing forms.

If the forms are tight and require wedging only wooden wedges should be used. Do not place a pinch bar or other metal tool against the concrete to wedge forms loose. Start removing forms some distance from a projection. This relieves pressure against projecting corners and reduces the likelihood of breaking off the edges.

FINISHING AIR-ENTRAINED CONCRETE

Air-entrained concrete has a slightly different consistency than regular concrete. This requires a little change in finishing operations. Since air-entrained concrete contains many tiny air bubbles that hold all the materials in concrete in suspension, it requires less mixing water than non-air-entrained concrete. It also bleeds less and is the reason for different finishing operations.

There is no waiting for the evaporation of free water from the surface before floating and troweling. If floating is done by hand, an aluminum or magnesium float is essential. A wooden float will drag and increase the work necessary to finish the surface. If floating is done with a power finishing machine, there is practically no different in the finishing procedure for air-entrained and non-air-entrained concrete except that finishing may begin sooner with the air-entrained concrete.

Since most all horizontal surface defects and failures are due to finishing operations performed while bled water or excess surface moisture is present, better results are usually obtained with air-entrained concrete.

CURING CONCRETE

Curing a concrete slab is one of the most important operations in producing quality work. It is also one of the most often neglected operations. Even if the concrete is

Fig. 7-16. Polyethylene film may be used to prevent newly placed concrete from drying out too fast. (Portland Cement Assoc.)

mixed, placed and finished properly, poor quality work will result if proper curing operations are not followed.

Little or no moisture loss should be allowed during the early stages of hardening. If necessary, the concrete should be protected during this critical period. Newly placed concrete should be protected from the sun and not allowed to dry out too fast. This may be accomplished with damp burlap, canvas or polyethylene film coverings, Fig. 7-16. The covering may be applied as soon as the surface is hard enough that it will not be marred. Keep the covering moist for at least three days.

Another method of curing is called ponding. This is done by keeping an inch or so of water on the concrete surface usually by earth dikes around the edges of the slab.

Membrane curing compounds, Fig. 7-17, sprayed on the

Fig. 7-17. Membrane curing compounds may be sprayed on the concrete surface immediately after the concrete has had its final finishing operation. (Portland Cement Assoc.)

surfaces of the concrete are frequently used. Uniform coverage is necessary and often two coats are required to provide adequate protection.

CURING TEMPERATURES

The rate of chemical reaction between cement and water is affected by the temperature. Therefore, the temperature affects the rate of hardening, strength and other qualities of the concrete.

COLD WEATHER CONSTRUCTION: In cold weather, concrete placement may require heated materials, a covering for the fresh concrete or heated enclosure. The hydration of the cement generates some heat, but this may not be enough. The temperature of concrete at the time of placing should generally be 50 to 70 deg. F (10 to 21 deg. C). The materials should never be heated to the point that the fresh concrete is above 70 deg. F. This will reduce the strength.

Concrete should never be placed on a frozen subgrade. When subgrades thaw, uneven settling and cracking of the slab usually results. Forms, reinforcing steel and embedded fixtures should be free of ice when the concrete is placed. A thin layer of warm concrete should be placed on cold, hardened concrete when an upper layer is to be poured. The thick upper layer will shrink as it cools, and the lower layer will expand as it warms. Failure of the bond will result if care is not taken.

In cold weather, moisture for curing is still very important. Keep the concrete moist, especially near heating units. Maintain the temperature of normal concrete at 70 deg. F. (21 deg. C) for three days or at 50 deg. F (10 deg. C) for five days. Keep the temperature of high early-strength concrete at 70 deg. F for two days or at 50 deg. F for three days. Do not allow the concrete to freeze for the next four days.

HOT WEATHER CONSTRUCTION: Avoid high temperatures in fresh concrete in extremely hot weather to prevent rapid drying. Temperature of the mixing materials may be cooled by using chilled water or ice. The ice should be melted by the time the concrete leaves the mixer.

Concrete Fundamentals

Subgrades should be saturated sometime in advance and sprinkled just ahead of placing the concrete. Wood forms should be treated or wetted thoroughly. Placing should not be delayed and it should be screeded and darbied immediately after placing. Covers, such as burlap, which are kept constantly wet, should be placed over the concrete as soon as it is darbied. When the surface is ready for final finishing, a small section should be uncovered immediately ahead of the finishers and recovered as soon as possible.

Air-entrained concrete develops a rubberlike surface in hot weather if finishing is delayed. This concrete is then very hard to surface smoothly.

Keep the concrete moist seven days for proper curing.

JOINTS IN CONCRETE

Three basic types of joints are frequently used in concrete construction. They are:
1. Isolation joints.
2. Control joints.
3. Construction joints.

Isolation joints (sometimes called expansion joints) are used to separate different parts of a structure to permit both vertical and horizontal movement. This type of joint is used

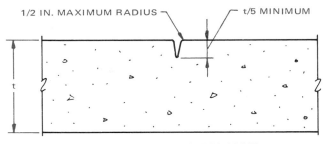

HAND TOOLED CONTROL JOINT

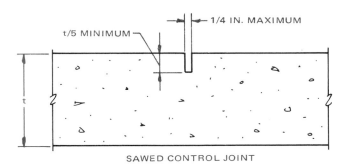

SAWED CONTROL JOINT

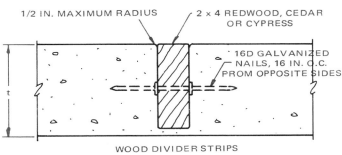

WOOD DIVIDER STRIPS

Fig. 7-19. These types of control joints are used in sidewalks, drives and patios.

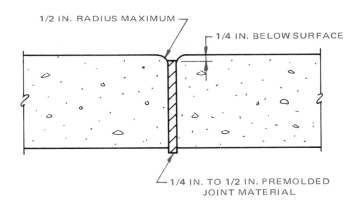

Fig. 7-18. Detail of an isolation joint. The joint material may be flush in areas where no safety hazard from tripping exists. An example is a floating slab against a building wall.

around the perimeter of a floating slab floor and around columns and machine foundations, Fig. 7-18.

Control joints provide for movement in the same plane as the slab or wall is positioned. They are used to compensate for

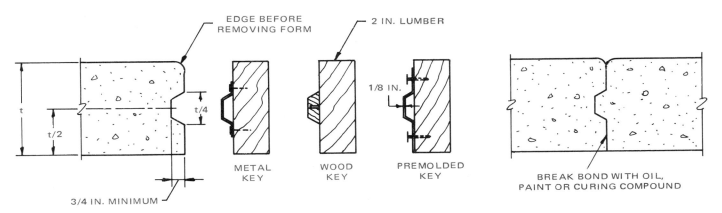

Fig. 7-20. Details of a construction joint. The wood key may be made from a beveled 1″ x 2″ strip and is used for slabs 4 to 6 in. thick. The premolded key may be left in the slab permanently by tacking it lightly to the bulkhead.

contraction caused by drying shrinkage. Control joints should be constructed in such a way that they permit the transfer of loads perpendicular to the plane of the slab or wall. Fig. 7-19 shows three types of control joints.

If control joints are not used in slabs or lightly reinforced walls, random cracks will occur due to drying shrinkage. Control joints are sometimes called contraction or dummy joints.

Construction joints or bonded joints provide for no movement across the joint. They are only stopping places in the process of casting. Construction joints, however, may be made to perform as control joints. Fig. 7-20 shows the details of a construction joint.

DECORATIVE AND SPECIAL FINISHES

A variety of patterns, textures and decorative finishes can be built into concrete during construction:
1. Color may be added to the concrete.
2. Aggregates may be exposed.
3. Textured forms may be used.
4. Concrete may be ground to produce a polished appearance.
5. Geometric patterns can be scored or stamped into the concrete to resemble stone, brick or tile.
6. Divider strips can be used to form interesting patterns.
   The possibilities are unlimited.

COLORED CONCRETE

Concrete can be colored using any of three methods: one-course, two-course and dry-shake, see Fig. 7-21.

The first two methods are similar. In both, the concrete mix is colored by adding a mineral oxide pigment prepared especially for use in concrete. White portland cement will produce brighter colors or light pastel shades when used with light-colored sand. Normal gray cement may be used for black or dark gray colors.

All materials must be accurately controlled by weight to attain a uniform color in each batch. The amount of pigment used should never exceed 10 percent of the weight of the cement.

Color pigments should be mixed with the cement while dry. Use clean tools and a separate mixer to prevent streaking. When using the one-course method, uniform moistening of the subgrade is important for good color results.

The only difference between the one and two-course methods is that the two-course method uses a base coat of conventional concrete. The surface of the base coat is left rough to produce a good bond. The top coat may be placed as soon as the surface water disappears. The top coat of colored concrete is generally 1/2 to 1 in. thick.

In the dry-shake method you will apply a commercially prepared dry color material over the concrete surface after floating, edging and grooving. Apply two coats of the dry-shake to the surface of the slab. Perform the finishing operations after each application of dry-shake. The color must be thoroughly worked into the concrete.

EXPOSED AGGREGATE

One of the most popular decorative concrete finishes is the exposed aggregate finish. It provides an unlimited color

Fig. 7-21. This decorative exposed aggregate slab is further enhanced through the use of colored concrete to form a design. (Portland Cement Assoc.)

selection and a broad range of textures, Fig. 7-22. Not only are exposed aggregate finishes attractive, but they can be rugged, slip resistant and immune to weather.

Fig. 7-22. This exposed aggregate patio is striking in appearance and adds a factor of safety.

There are several ways to produce exposed aggregate finishes. One of the most common is called the *seeding* method. The procedure is to place, screed and bullfloat or darby the concrete in the usual manner keeping the level of the surface about 3/8 to 1/2 in. lower than the forms. This space will be filled with the extra aggregate.

When these finishing operations have been completed, spread the aggregate uniformly with a shovel or by hand so that the entire surface is completely covered with a layer of stone. Next, embed the aggregate by tapping with a wood hand float or straightedge. Work the surface with the hand float until the surface is similar to that of a normal slab after floating.

Starting the next operation requires accurate timing. Usually you will wait until the slab can bear the weight of a worker on kneeboards with no indentations. Then brush the slab with a stiff nylon bristle broom to remove the extra mortar over the stones. Next, apply a fine spray of water along with brushing. If the aggregate becomes dislodged, stop the operation for a while. Continue washing and brushing until the water is clear and there is no noticeable cement film left on the aggregate. A surface retarder may be used for better control of the exposing operations, but is not necessary.

Another method of producing an exposed aggregate finish is to expose the stone in conventional concrete. No extra stone is added, but a high proportion of coarse to fine aggregate is necessary. The coarse aggregate should be uniform in size, bright in color, closely packed and uniformly distributed. The slump of this concrete must be between 1 and 3 in.

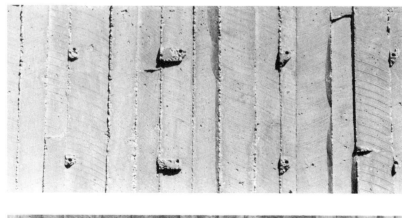

Fig. 7-23. These rustic finishes were produced using rough sawn siding.

A third method is to place a thin topping course of concrete containing special aggregates over a base of regular concrete. This technique is used for terrazzo construction.

Terrazzo toppings on outdoor slabs are generally 1/2 in. thick and contain marble, quartz or grainte chips. Random cracking is eliminated by using brass or plastic divider strips set in a bed of mortar. This type of terrazzo is called *rustic* or *washed terrazzo.*

TEXTURED FINISHES

Interesting decorative textures can be produced on vertical concrete surfaces by using textured form materials. Rough sawn cedar siding was used to produce the rough textured surfaces shown in Fig. 7-23. The variety of textures possible is almost endless. Fig. 7-24 shows cast concrete textured to imitate stone and sculptured concrete block. No special procedure is required for these finishes.

Textured surfaces on slabs may be achieved by brooming, using mortar, dash coat or rock salt, just to name a few.

Fig. 7-24. Cast concrete has been sculptured and used to imitate stone.

Brooming may be executed in a fine or coarse straight line pattern or a wavy texture. The procedure for brooming a surface was discussed earlier.

A travertine or keystone finish, as it is sometimes called, is produced by applying a dash coat of mortar over freshly leveled concrete. The dash coat is mixed to the consistency of thick paint and usually contains a yellow pigment. It is applied in a splotchy manner and with a dash brush. Ridges and depressions are formed by this procedure.

After the coating has hardened slightly, the surface is troweled slightly to flatten the ridges and spread the mortar. The resulting finish is smooth on the high spots and coarse grained in the low areas. This effect looks like travertine marble from which it gets its name.

Another texture can be produced by scattering rock salt over the surface after hand floating or troweling. The salt is pressed into the surface so that the top of each grain is exposed. When the concrete has hardened, the surface is washed and brushed. The salt pellets will be dissolved leaving holes in the surface. Neither the rock salt or travertine finish is recommended for areas which experience freezing weather. Water frozen in the recesses will ruin the surface.

GEOMETRIC PATTERNS

A variety of geometric patterns can be stamped, sawed or scored into a concrete surface to enhance the beauty of walks, drives or patios. Random flagstone or ashlar patterns are popular. They may be produced using a piece of 1/2 or 3/4 in. copper pipe bent into a flat S-shape to score the surface. Scoring must be done before the concrete becomes too hard to push the coarse aggregate aside. The best time is just after darbying or bullfloating. After hand floating, a second scoring will be required to smooth the joints.

Other patterns such as stone, brick or tile can be cut into partially set concrete with special stamping tools. Color may also be added to create varying effects.

Still other patterns may be created using divider strips of wood, plastic, metal or masonry units. These divider strips help to create interest. But they also aid in placing concrete, provide for combinations of various surface finishes and greatly reduce random cracking. Wood divider strips should be made from pressure-treated lumber, redwood or cypress.

NONSLIP AND SPARKLING FINISHES

Nonslip finishes can be applied to surfaces that are frequently wet or that would be especially dangerous if slippery. Abrasive grains may be dry-shaken on the surface and lightly troweled. The two most widely used abrasive grains are silicon carbide and aluminum oxide. The silicon carbide grains are sparkling black in color and are also used to make "sparkling concrete." Aluminum oxide is usually gray, brown or white and does not sparkle. The grains should be spread uniformly over the surface. Use from 1/4 to 1/2 lb. per sq. ft. of slab surface.

COMBINATION FINISHES

Concrete is a very versatile material and may be used in many ways to create a beautiful walk, drive or patio to

Fig. 7-25. Innovative concrete work such as this is interesting and attractive.   (Ideal Cement)

compliment the mood and style of any architectural design.

Striking effects can be attained by combining several colors, textures and patterns in concrete. Alternate areas of exposed aggregate and plain or colored concrete provides an exciting combination, Fig. 7-25. Ribbons and borders of masonry or wood create a dramatic effect with plain concrete or exposed aggregate surfaces.

REINFORCED CONCRETE

Steel reinforcing is frequently used in footings, foundations, slabs and other concrete work to add strength and control stresses. *Reinforced concrete is a composite material which utilizes the concrete in resisting compression forces, and some other material, usually steel bars or wires, to resist the tension forces.*

Concrete has great compressive strength. (This is the ability to support great loads placed directly upon it.) However, it has very little strength to resist stresses or forces that tend to bend or pull it apart. The compressive strength of concrete is about 10 times its tensile strength. The use of steel reinforcing may be used to greatly increase its tensil strength.

KINDS OF REINFORCEMENT

Steel is regarded as the best reinforcing material for concrete. It has almost the same contraction and expansion rate due to temperature changes as concrete and also may be purchased in many sizes and forms. The two main types of steel reinforcing used in concrete are reinforcing bars and welded wire fabric, Fig. 7-26. Bars may be smooth or deformed. Smooth bars are generally smaller in diameter. Deformed bars have ridges along the sides. These increase the bond between the concrete and steel.

Bars are produced in standard sizes. They are designated by a number. Fig. 7-27 shows bars from 1/4 to 1 in. in diameter

with standard number, area and approximate weight per 100 ft. of bar. Larger bars are made for extremely heavy construction. Bars usually come from the mill in 60 ft. lengths, but most local suppliers stock 20 and 40 ft. lengths.

The size of bar required for the job will depend on the amount of tensil strength needed. This is usually calculated by engineers. In commercial construction, professional steelworkers normally make the installation.

Fig. 7-26. Reinforcing bars are used to strengthen the beam being formed and welded wire fabric reinforces the slab.
(Wire Reinforcement Institute)

BAR NUMBER	BAR DIAMETER IN INCHES	BAR AREA IN SQ. IN.	APPROXIMATE WEIGHT OF 100 FT.
2	1/4	.05	17
3	3/8	.11	38
4	1/2	.20	67
5	5/8	.31	104
6	3/4	.44	150
7	7/8	.60	204
8	1	.79	267

Fig. 7-27. Size, area and weight of reinforcing bars from 1/4 to 1 in. in diameter.

Welded wire fabric is manufactured in many types and sizes. It is a prefabricated reinforcement consisting of parallel series of high-strength, cold-drawn wires welded together in square or rectangular grids. Smooth wires, deformed wires or a combination of both may be used in welded wire fabric.

Welded wire fabric provides more uniform stress distribution and more effective crack control in slabs and walls than larger diameter bars more widely spaced. The range of wire sizes and spacings available is broad. Thus, it is possible to furnish almost exactly the cross-sectional steel area required for a specific job.

Cross-sectional area is the basic element used in specifying

WIRE SIZE COMPARISON

W & D SIZE NUMBER		AREA (SQ. IN.)	NOMINAL DIAMETER (IN.)	AMERICAN STEEL & WIRE GAGE NUMBER
SMOOTH	DEFORMED			
W31	D31	0.310	0.628	
W30	D30	.300	.618	
W28	D28	.280	.597	
W26	D26	.260	.575	
W24	D24	.240	.553	
W22	D22	.220	.529	
W20	D20	.200	.504	
		.189	.490	7/0
W18	D18	.180	.478	
		.167	.4615	6/0
W16	D16	.160	.451	
		.146	.4305	5/0
W14	D14	.140	.422	
		.122	.394	4/0
W12	D12	.120	.390	
W11	D11	.110	.374	
W10.5		.105	.366	
		.103	.3625	3/0
W10	D10	.100	.356	
W9.5		.095	.348	
W9	D9	.090	.338	
		.086	.331	2/0
W8.5		.085	.329	
W8	D8	.080	.319	
W7.5		.075	.309	
		.074	.3065	1/0
W7	D7	.070	.298	
W6.5		.065	.288	
		.063	.283	1
W6	D6	.060	.276	
W5.5		.055	.264	
		.054	.2625	2
W5	D5	.050	.252	
		.047	.244	3
W4.5		.045	.240	
W4	D4	.040	.225	4
W3.5		.035	.211	
		.034	.207	5
W3		.030	.195	
W2.9		.029	.192	6
W2.5		.025	.177	7
W2.1		.021	.162	8
W2		.020	.159	
		.017	.148	9
W1.5		.015	.138	
W1.4		.014	.135	10

Fig. 7-28. This chart compares "W" and "D" number wire size with nominal decimal diameters and to the steel wire gage system.
(Wire Reinforcing Institute)

TYPICAL STOCK ITEMS

STYLE DESIGNATION		SPACING IN.		DIAMETER IN.		SECTIONAL AREA BY SQ. IN. PER. FT.		WEIGHT LB. PER. 100 SQ. FT.
BY WIRE GAGE	BY W- OR D-NUMBER	LONGIT.	TRANSV.	LONGIT.	TRANSV.	LONGIT.	TRANSV.	
6x6 — 10x10	6x6 — W1.4xW1.4	6	6	.135	.135	.029	.029	21
6x6 — 8x8	6x6 — W2.1xW2.1	6	6	.162	.162	.041	.041	30
6x6 — 6x6	6x6 — W2.9xW2.9	6	6	.192	.192	.058	.058	42
6x6 — 4x4	6x6 — W4xW4	6	6	.225	.225	.080	.080	58
4x4 — 10x10	4x4 — W1.4xW1.4	4	4	.135	.135	.043	.043	31
4x4 — 6x6	4x4 — W2.9xW2.9	4	4	.192	.192	.087	.087	62
4x4 — 4x4	4x4 — W4xW4	4	4	.225	.225	.120	.120	87

Fig. 7-29. Typical welded wire fabric stock items used in light construction.

wire size. Smooth wire sizes are specified by the letter "W," and deformed wires are specified by the letter "D." This letter is followed by a number indicating the cross-sectional area of a wire in hundredths of a square inch. For example: W10 is a smooth wire with a cross-sectional area of 0.10 sq. in.

This is a new method of designating wire sizes. It is a complete departure from the steel wire gage system previously used. See Fig. 7-28.

Spacings and sizes of wires in welded wire fabric are identified by "style." A typical style designation is: 6 x 12 — W16 x W8. This specifies a welded wire fabric which has a 6 in. spacing for horizontal wires, a 12 in. spacing for transverse

wires, longitudinal wires (W16) of 0.16 sq. in. and transverse wires (W8) of 0.08 sq. in.

A welded deformed wire fabric style would be specified the same way except that D-number wire sizes would be substituted for the W-number wire sizes. The terms "longitudinal" and "transverse" are related to the manufacturing process. They bear no relationship to the position of the wires in a concrete structure. Some of the stock sizes of welded wire fabric used in light construction are shown in Fig. 7-29.

Spacing of wires in welded wire fabric provides a wide variety of combinations. Some of the many spacing combinations possible are indicated on the next page.

TYPE OF CONSTRUCTION	RECOMMENDED STYLE	REMARKS
BASEMENT FLOORS	6 x 6 — W1.4 x W1.4 6 x 6 — W2.1 x W2.1 6 x 6 — W2.9 x W2.9	FOR SMALL AREAS (15 FOOT MAXIMUM SIDE DIMENSION) USE 6 x 6 — W1.4 x W1.4. AS A RULE OF THUMB, THE LARGER THE AREA OR THE POORER THE SUBSOIL, THE HEAVIER THE GAGE.
DRIVEWAYS	6 x 6 — W2.9	CONTINUOUS REINFORCEMENT BETWEEN 25 TO 30 FOOT CONTRACTION JOINTS.
FOUNDATION SLABS (RESIDENTIAL ONLY)	6 x 6 — W1.4 x W1.4	USE HEAVIER GAGE OVER POORLY DRAINED SUBSOIL, OR WHEN MAXIMUM DIMENSION IS GREATER THAN 15 FEET.
GARAGE FLOORS	6 x 6 — W2.9 x W2.9	POSITION AT MIDPOINT OF 5 OR 6 INCH THICK SLAB.
PATIOS AND TERRACES	6 x 6 — W1.4 x W1.4	USE 6 x 6 — W2.1 x W2.1 IF SUBSOIL IS POORLY DRAINED.
PORCH FLOOR A. 6 INCH THICK SLAB UP TO 6 FOOT SPAN B. 6 INCH THICK SLAB UP TO 8 FOOT SPAN	6 x 6 — W2.9 x W2.9 4 x 4 — W4 x W4	POSITION 1 INCH FROM BOTTOM FORM TO RESIST TENSILE STRESSES.
SIDEWALKS	6 x 6 — W1.4 x W1.4 6 x 6 — W2.1 x W2.1	USE HEAVIER GAGE OVER POORLY DRAINED SUBSOIL. CONSTRUCT 25 TO 30 FOOT SLABS AS FOR DRIVEWAYS.
STEPS (FREE SPAN)	6 x 6 — W2.9 x W2.9	USE HEAVIER STYLE IF MORE THAN FIVE RISERS. POSITION FABRIC 1 INCH FROM BOTTOM OF FORM.
STEPS (ON GROUND)	6 x 6 — W2.1 x W2.1	USE 6 x 6 — W2.9 x W2.9 FOR UNSTABLE SUBSOIL.

Fig. 7-30. Recommended styles of welded wire fabric reinforcement for specific concrete applications. (Wire Reinforcement Institute)

Fig. 7-31. A flat sheet of welded wire fabric being placed for the floor of an apartment building.

2″ x 2″	3″ x 3″	4″ x 4″	6″ x 6″
2″ x 4″	3″ x 12″	4″ x 8″	6″ x 12″
2″ x 6″	3″ x 16″	4″ x 12″	
2″ x 12″		4″ x 16″	
2″ x 16″			

Recommended styles of welded wire fabric for some specific applications are shown in Fig. 7-30.

PLACING STEEL REINFORCEMENT

Placing steel reinforcement is a very important operation and is performed at the job site, Fig. 7-31. Reinforcement should be placed so that it is protected by an adequate coverage of concrete, Fig. 7-32.

Reinforcing steel carries the tensile load in concrete. It must, therefore, be lapped at a splice. The overlap distance for deformed bars should be at least 24 bar diameters and never less than 12 in. Smooth bars should be lapped a greater distance than deformed bars, Fig. 7-33. Welded wire fabric should be lapped at least one full stay spacing plus 2 in. An 8″ x 8″ fabric would require a minimum of 10 in. overlap.

CAST-IN-PLACE CONCRETE WALLS

Cast-in-place concrete permits structures of all shapes, sizes and heights. Exterior wall surfaces can be rough or smooth, natural or colored. Some of the popular types of cast-in-place concrete walls include:

1. Rustic.
2. Grid-patterned.
3. Colored aggregate.
4. Window walls.
5. Sculptured.

RECOMMENDED PROTECTION FOR REINFORCEMENT

APPLICATION	MINIMUM CONCRETE PROTECTION
FOOTINGS	3 IN.
CONCRETE SURFACE EXPOSED TO WEATHER	1 1/2 IN. FOR NO. 5 BARS AND SMALLER
SLABS AND WALLS	3/4 IN.
BEAMS AND GIRDERS	1 1/2 IN.
JOISTS	3/4 IN.
COLUMNS	NOT LESS THAN 1 1/2 IN. OR 1 1/2 TIMES THE MAXIMUM SIZE AGGREGATE
CORROSIVE ATMOS- PHERE OR SEVERE EXPOSURES	PROTECTION SHALL BE SUITABLY INCREASED

FROM: "BUILDING CODE REQUIREMENTS FOR REINFORCED CONCRETE," ACI 318

Fig. 7-32. The American Concrete Institute recommends the above minimum concrete protection for reinforcement. It means that rods used for reinforcing must be covered by that much concrete.

RUSTIC CONCRETE WALLS

Rustic concrete walls may be produced by using rough form boards, bushhammering the surface or casting vertical fins in the concrete surface. Forms made from rough-sawn boards have been used for years to produce a textured surface in concrete walls, Fig. 7-34.

Concrete will faithfully reproduce the wood texture and the rough board appearance tends to hide tierod holes and other imperfections. To assure uniformity of surface texture:

1. Use lumber of the same type throughout.
2. Or use a form coating to seal the surfaces of the boards.

Bushhammering produces a coarse concrete texture. This is

Fig. 7-33. An engineer lifts up one of the 2 ft. long overhanging transverse wires on the edge of the sheet of welded wire fabric. (Wire Reinforcement Institute)

Fig. 7-34. This wall texture was the result of using rough sawn form boards. Note that the tie rod holes add to the overall texture effect.

a method of mechanically breaking away the wall surface of hardened concrete to expose coarse aggregate. Interesting color variations and surface textures are produced by bush-hammering.

Vertical fins or ribs, Fig. 7-35, can produce shadow effects on a concrete wall. The fins may be smooth, sandblasted or hammered. Inserts of wood, metal or plastic may be used to create the ribs. Fig. 7-36 shows a ribbed surface which has been hammered.

Fig. 7-35. The fins or ribs on this modern concrete structure add interest to the design.

GRID-PATTERNED CONCRETE WALLS

Well-planned joint patterns provide a low cost architectural treatment for cast-in-place walls or other structural elements. Construction joints can be inconspicuous or hidden by

Fig. 7-36. This concrete ribbed surface has been hammered to expose the aggregate. (American Plywood Assoc.)

rustication strips. These strips are used to produce grooves in the concrete surface which add something to the overall architectural effect and serve as control joints. Removable architectural ties can be used to provide minimum-size, easily patched holes or the tie holes can be accentuated to reduce cost and enhance the appearance.

Fig. 7-37 shows how the tie holes were formed in a grid pattern and left exposed. When form joints and tie holes are left unfinished, they are placed at predetermined locations and the pattern is repeated throughout the structure to create an architectural effect.

Fig. 7-37. Tie holes form a grid pattern in the structure and are left exposed as part of the surface texture.

COLORED AGGREGATE SURFACES

One of the best methods of obtaining color in a cast-in-place concrete wall is through the use of exposed aggregate. A large percentage of coarse aggregate is used in the mix. The surface is sandblasted, bushhammered or chisel-textured to expose the colored aggregate.

Another method of producing exposed aggregate cast-in-place walls is to preplace dry aggregate in the form and then grout under pressure with a cement-sand-water slurry. A hole near the bottom of the form is used for pumping in the grout and an external vibrator is used. The aggregate is exposed by sandblasting two to seven days after grouting.

An alternate, patented method is to wrap wire mesh around the reinforcement and preplace aggregate between the mesh and the outside form. A special concrete mix is cast into the core of the mesh and dispensed to the outside.

CAST-IN-PLACE WINDOW WALLS

Repetition of window openings makes an attractive design in large structures such as the one shown in Fig. 7-38. White portland cement concrete is well suited for window walls because the color is permanent. Any coarse aggregate may be used, but it should be reasonably uniform in color. White aggregate is preferred if maximum whiteness is desired.

Reinforced plastic forms provide a concrete surface that is smooth and hard with few air voids and defects. The plastic forms are reinforced with fiberglass and are usually from 3/16 to 5/8 in. thick. Normal thickness of wall forms is 3/8 in. Column forms are usually 1/2 in. thick.

Fig. 7-38. This attractive design owes its overall effect to the use of white portland cement.   (ASG Corp.)

SCULPTURED CONCRETE WALLS

Sculptured concrete provides esthetic qualities, Fig. 7-39. A wide range of materials may be used to form decorative patterns in concrete walls. A few such materials are: wood, corrugated metal sheets, plastic form liners, fiberglass sheets, formed plastic and tempered fiberboard. Plaster waste molds may be used for fine sculpturing.

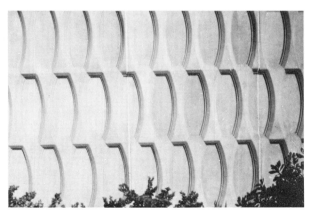

Fig. 7-39. The fine sculptured details of these walls set them apart from the ordinary.

CAST-IN-PLACE CONCRETE ROOF AND FLOOR SYSTEMS

There are four basic cast-in-place concrete roof and floor structural systems. They are:
1. Pan joists.
2. Waffles.
3. Flat plates.
4. Flat slabs.

PAN JOISTS ROOF AND FLOOR SYSTEM

Pan joist construction is a one-way structural system using a ribbed slab formed with pans, Fig. 7-40. This system is

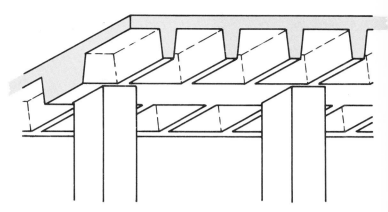

Fig. 7-40. Pan joist roof construction is a one-way structural system using a ribbed slab formed with pans. Spans of up to 50 ft. are common using this construction.

economical because the standard forming pans may be reused. Standard pan forms produce inside dimensions of 20 to 30 in. and depths from 6 to 20 in.

Fig. 7-41. Waffle plate construction is used here to form the roof slab. Spans of up to 60 ft. are possible.

WAFFLE ROOF AND FLOOR SYSTEM

Waffle pans or forming domes are available in standard sizes but may be custom made for a particular job. Like pan joist construction, the forms can be reused. Fig. 7-41 shows a structure using the waffle technique.

Standard 30 by 30 in. square domes have a depth of 8, 10, 12, 14, 16 or 20 in. They have 3 in. flanges which provide for 6 in. wide joist ribs on 36 in. centers. Standard domes 19 x 19 in. square have a depth of 6, 8, 10, 12, or 14 in. and form 5 in. wide joist ribs on 24 in. centers using 2 1/2 in. flanges. Spans from 25 to 60 ft. are possible using waffles.

FLAT PLATE ROOF AND FLOOR SYSTEM

The main features of the flat plate system are minimum depth and architectural simplicity, Fig. 7-42. A flat plate is a two-way reinforced concrete framing system utilizing the very simplest structural shape — a slab of uniform thickness.

The flat ceiling is economical to form and may be used for the finished ceiling without any additional treatment. Canti-

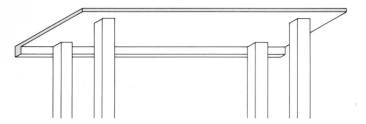

Fig. 7-42. A flat plate roof is a two-way reinforced concrete framing system having a slab of uniform thickness.

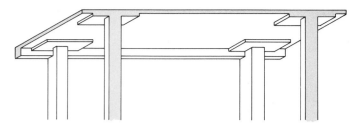

Fig. 7-43. A flat-slab roof is a two-way reinforced structural system that includes either drop panels or column capitals to carry heavier loads.

levers are easily produced as well as other architectural projections. Slabs generally range from 5 to 14 in. thick and provide spans up to 35 ft. Flat plates provide a continuous solid ceiling with complete flexibility for locating partitions and mechanical equipment. Supporting columns need not be in a straight line. This adds flexibility. Electrical ducts and conduits may be embedded in the slab. The flat plate system is well suited for heavy loads such as roof parking.

FLAT SLAB ROOF AND FLOOR SYSTEM

The flat-slab system is designed for heavy roof loads with large open bays below. The difference between the flat-plate system and the flat-slab is that the flat-slab has a supporting panel in the area of each column for added support, Fig. 7-43.

Flat slab thickness is usually 2.5 or 3 percent of the span and a minimum of 4 in thick. The size of the supporting panel is about 33 percent of the span and 25 to 50 percent of the slab thickness. Flat-slabs are designed to span up to 40 ft. with columns an equal distance apart. This system has all the advantages of the flat-plate system and is stronger.

PRECAST/PRESTRESSED CONCRETE SYSTEMS

With expansion of industrial processes in building construction has come greater use of modular layout, planning and cost-conscious construction. Precast units for walls, floors, ceilings and roofs can be mass produced at the factory or job site. Precast concrete units may be cast as tilt-up panels, standard- shaped concrete panels or concrete window walls.

Fig. 7-44. Precast concrete panel units are used to construct these apartment buildings.   (Ideal Cement)

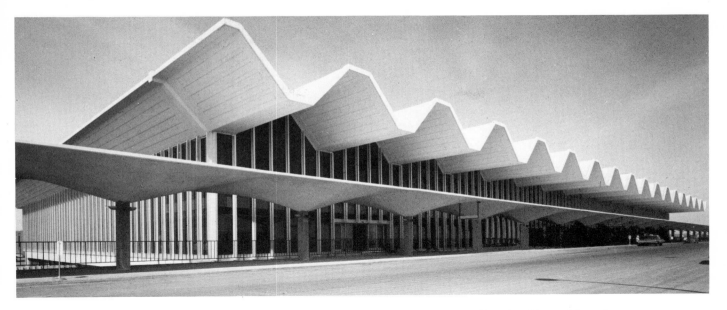

Fig. 7-45. Precast roof units were used in this modern concrete structure. (American Plywood Assoc.)

Tilt-up panels may be cast on the building floor slab or in stacks near the building site, Figs. 7-44 and 7-45. Usually, they may be erected about a week after casting. Panels can be secured by casting a column between them or by bolting or welding them to the columns. Tilt-up panels may be exposed aggregate, plain or colored concrete. They may be either load or non-load-bearing walls.

Standard-shaped concrete panels which are prestressed have brought a new concept to architectural precast concrete. These units are crack-free and highly resistant to deterioration. Some of the most common designs include *double-tee* units, *single-tee* units and *hollow-core panels.* Fig. 7-46 shows a structure using double-tee roof units.

Fig. 7-46. Double-tee concrete units, as shown here, are the most widely used prestressed concrete product for medium range spans. Four foot wide double-tees are commonly used for spans up to 60 ft.

Double-tees are the most widely used prestressed, precast units. Widths are generally 4, 5, 6, 8 and 10 ft. The 4 ft. width is most popular. Depths range from 6 to 36 in. Spans up to 60 feet are possible with the 4 ft. wide unit. Greater spans are

possible with wider units since they usually have a deeper web.

Single-tee units are generally used where long spans, beginning at about 30 ft, are needed. They may be placed flange to flange as in Fig. 7-47, or spread apart with concrete planks or cast-in-place concrete completing the enclosure. Flange widths range from 4 to 10 ft. with depths from 12 to over 36 in. in 4 in. increments. Spans of 30 to 100 ft. are common with single-tee units.

Cored slabs provide a flush ceiling with minimum depth required for the roof or floor system. See Fig. 7-48.

Precast concrete window walls may be cast as curtain walls or load-bearing walls. Forms or molds used to produce complicated designs are made from plastic, wood or steel.

Typically, fiberglass molds are used for smooth concrete with sculptured mullions (wall members supporting a window). Precast window walls may be one-story or multistory. They may even be preglazed (glass installed) at the plant. Fig. 7-49 shows a building with load-bearing wall panels.

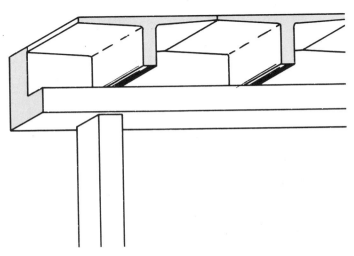

Fig. 7-47. Single-tee precast, prestressed concrete units are generally used for very long spans. Typical spans are from 30 to 100 ft.

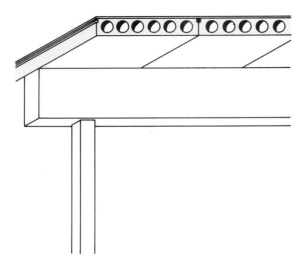

Fig. 7-48. Hollow-core slabs are low cost and widely produced. Depths range from 4 to 12 in. and widths are available from 16 to 48 in. Spans possible with cored slabs are up to 50 ft., but generally under 35 ft.

REVIEW QUESTIONS — CHAPTER 7

1. What are the four basic materials used to make concrete?
2. Portland cement is _____ because it sets and hardens by reacting with water.
3. A general purpose cement suitable for all uses when special properties are not required is:
   a. Type I.
   b. Type II.
   c. Type III.
   d. Type IV.
   e. Type V.
4. Air-entrained cements improve resistance to freeze-thaw action and scaling caused by chemicals used in snow and ice removal. True or False?
5. Aggregates usually constitute about _____ to _____ percentage of the volume of concrete.
6. Aggregates do not affect the quality of finished concrete very much. True or False?
7. Fine aggregate or sand should vary uniformly in size from very fine to _____ inch.
8. The maximum size of coarse aggregate is usually _____ inches.
9. What is the result of using the maximum allowable particle size?
10. Water usually constitutes from _____ to _____ percent of the total volume of concrete.
11. Are any materials added to the concrete batch before or during mixing other than portland cement, water and aggregates?
12. Identify four types of admixtures used in concrete.
13. List three desirable qualities of plastic concrete.
14. List five desirable qualities of hardened concrete.
15. The proportion of water to cement, stated in gallons of water per bag (94 lb.) of cement, is called the _____.
16. The ideal curing temperature of concrete is _____ deg. F.

Fig. 7-49. The precast concrete window walls in this building are load-bearing.   (Kawneer Co., Inc.)

17. Stronger concrete results if it is kept moist for many days as opposed to only a couple. True or False?
18. The mix consistency or degree of stiffness of plastic concrete is called _____.
19. Generally, how long should concrete be mixed?
20. ASTM C-94 requires that concrete must be delivered and discharged from the truck mixer within _____ hours after the water has been added to the mix.
21. When placing concrete, it should not be allowed to drop freely more than _____ feet.
22. Air pockets may be worked out of plastic concrete by _____ and _____.
23. Concrete which is shot into place by means of compressed air is called _____.
24. The first finishing operation performed on concrete is usually _____.
25. A radius may be put on the edge of a concrete slab using an _____.
26. Name three reasons why concrete is floated.
27. Steel troweling produces a very smooth surface that is often slippery when wet. Sidewalks, driveways and other outside flatwork frequently are _____ to produce a slightly roughened surface.
28. When may forms be removed?
29. Why is air-entrained concrete finished slightly different than normal concrete?
30. What is the most important consideration in curing concrete properly?
31. Three basic types of joints are frequently used in concrete construction. What are they?
32. Identify three decorative and special finishes which are used on concrete.
33. What are the two main types of steel reinforcing used in concrete?
34. What are the four basic cast-in-place concrete roof and floor structural systems?

Fig. 8-1. Hold the trowel with the fingers under the handle and the thumb on top of the ferrule as shown.

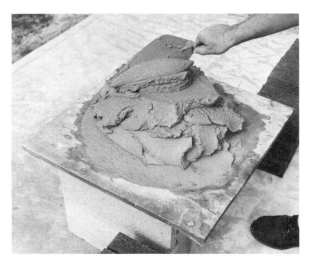

STEP 1: WORK THE MORTAR INTO A PILE IN THE CENTER OF THE MORTAR BOARD.

STEP 2: SMOOTH OFF A PLACE WITH A BACKHAND STROKE.

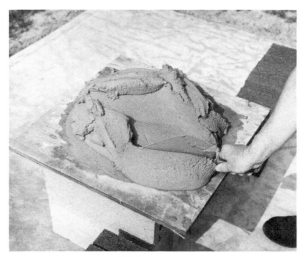

STEP 3: CUT A SMALL AMOUNT FROM THE LARGER PILE WITH A PULLING ACTION.

STEP 4: SCOOP UP THE SMALL PILE WITH A QUICK MOVEMENT OF THE TROWEL.

Fig. 8-2. Steps in loading trowel.

Chapter 8
LAYING MASONRY UNITS

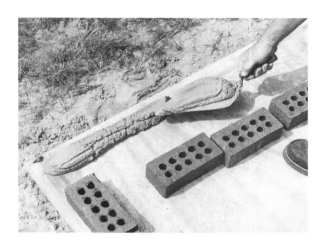 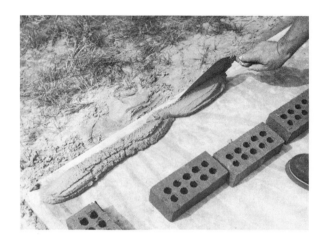

Fig. 8-3. Steps in emptying trowel. As trowel is unloaded, the outside edge is tilted upward with a swift wrist motion (left) while the arm is snapped backward in the direction of the mortar line. As the dump is completed (right) the trowel is nearly vertical.

Laying masonry units (brick, block or stone) involves planning, knowledge of the materials and ability to use the tools properly. Laying masonry is an art and requires a great deal of practice. However, the sense of satisfaction gained from doing a job well is worth the effort.

SPREADING MORTAR

Spreading mortar is part of all masonry construction. It is a simple process, but requires mastery if the mason is to be an expert at it. With practice you will be able to determine just how much mortar to use for a given joint. Placing the unit will be easier if the right amount of mortar has been laid down.

Select a trowel which is best for the job being performed and which "feels good" to you. If it is too small, effort will be wasted in needless movements. If the trowel is too large, you will tire easily. You may want to begin with a trowel about 10 in. long.

Grasp the trowel in the right hand, if you are right handed. Fingers should be under the handle and the thumb on top of the ferrule as shown in Fig. 8-1.

LOADING THE TROWEL

Mortar may be loaded on the trowel in several ways:
1. From the side of the board.

2. From the middle.
3. From the top of the pile.

Fig. 8-2 shows the steps used by one mason to load the trowel. Develop the method which works best for you and perfect it.

Spread the mortar with a quick turn of the wrist toward the body and a backward movement of the arm. As the trowel is near empty, tip the trowel blade even more to help the remaining mortar slide off, Fig. 8-3. The kind of mortar and its

Fig. 8-4. Furrowing the first course to form a uniform bed.

107

consistency most often determines the speed of the stroke.

When solid masonry units are used or the first course of concrete block is being laid, the mortar is spread and furrowed with the point of the trowel as in Fig. 8-4. This helps to form a uniform bed on which to lay the masonry. When mortar is laid down it should be slightly heavier on the outside (face) edge than on the inside. This will force out enough mortar for the next cross joint. Mortar hanging over the sides should be cut off using the front half of the trowel.

MORTAR CONSISTENCY

Mortar on the board should be kept well tempered (sprinkled lightly) with water until it is used. Avoid constant working. This causes the mortar to dry out and become stiff. Stiff mortar does not spread well. Use mortar before it begins to set up.

LAYING BRICK

Any straight wall longer than the plumb rule (4 ft.) is usually laid to a line. The line is a guide that helps the bricklayer build a straight, level and plumb wall. Mason's line is a light, strong cord which can be stretched taut with little or no sag. Pull the line to the same degree of tautness for each course. This is necessary so that all bed joints will be uniform and parallel.

Leads, Fig. 8-5, are usually built first at each corner to establish proper height and provide a place to attach the mason's line. The height of each course is determined by a mason's rule or a story pole with height marks on it.

On long walls, it is impossible to eliminate sagging in the line. The proper course height should be checked with the story pole to insure accuracy.

Laying brick too close to the line (crowding) should be avoided. Stay a line width away so that you do not risk moving the line destroying its usefulness as a reference.

LAYING A 4 IN. COMMON BOND WALL WITH LEADS

Place materials and arrange your work space for efficient work. The mortarboard should be located in the center of the work space about 24 in. from the wall. Brick should be stacked on both sides of the board as in Fig. 8-6.

The following procedure is recommended for laying a 4 in. common bond wall with leads:

1. Establish a wall line, Fig. 8-7. This line can be easily made using a chalk line. The line is coated with colored chalk. When held taut it may be snapped to produce a chalked line on the surface you wish to mark.

   Check the slab or footing to be sure that it is square before snapping the wall line. The diagonal distance from corner to corner should be the same.

   Another method which may be used to check for squareness is to measure 6 ft. along one side and 8 ft.

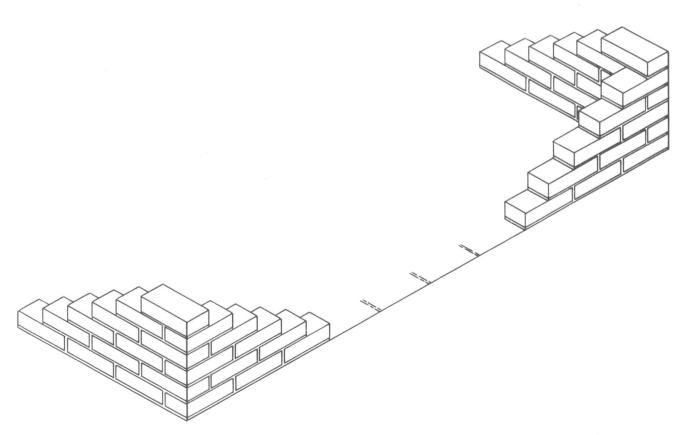

Fig. 8-5. Leads are the built-up sections of wall at each corner. This is done to establish proper height for each course and provide a place to attach the mason's line. It is important that these corners be as near perfect as possible.

Fig. 8-6. Materials arranged for efficient work.

along the intersecting side. The diagonal distance between these points should be 10 ft. This procedure is known as the "6–8–10 rule."

2. After the wall line has been drawn or chalked a dry course of brick should be laid out from corner to corner as in

Fig. 8-7. Establishing wall line with a chalk line.

Fig. 8-8. Arrange them to eliminate as much cutting of units as possible. Be sure to allow for a mortar joint between each brick. Mark the joints and then move the bricks aside, Fig. 8-9.

Fig. 8-9. Marking location of each brick.

Fig. 8-10. Mortar in place and furrowed for first side of lead corner.

3. Spread the mortar on one side of the corner and furrow it as in Fig. 8-10.

4. Lay the corner brick exactly on the point where the

Fig. 8-8. First course is laid out dry to eliminate as much cutting as possible.

Fig. 8-11. The corner brick is laid at the exact corner. Use the trowel to tap it into position.

corner is located, Fig. 8-11. It must be set level and square with the wall line.

5. Lay the remaining four or five bricks of the lead corner. This is called "tailing out" the lead of the corner. After the bricks have been laid, level them with the bricklayer's plumb rule, Fig. 8-12.

Fig. 8-12. Leveling the first five brick of the corner with the mason's level (plumb rule).

6. Plumb the corner brick, Fig. 8-13, and then the tail end, Fig. 8-14. Leveling is done on the outside and top edge of the brick.

7. Line up the bricks between the two plumb points, Fig. 8-15. Here the level is used as a straightedge and the bubbles are disregarded. Always follow this sequence when building a corner: level the unit, then plumb and then line up.

8. After one side of the corner has been laid and trued, start

Fig. 8-13. After leveling, plumb the corner brick to make its edge exactly vertical.

Fig. 8-14. Next plumb the tail end brick. Note how mason braces his arm against his knee to steady the plumb rule. Brick is tapped with the trowel to adjust it to proper position.

Fig. 8-15. First five brick are straightedged with the plumb rule.

the other side of the corner as in Fig. 8-16. Again, spread the mortar and furrow it. Lay three or four bricks. Level, plumb and line them up. Do not tap the level with the trowel or hammer. Use your hand. This time the corner brick does not require plumbing because it has already been plumbed. But, if you think that you might have moved it, check it again.

Fig. 8-18. Completed lead corner will be used as a guide for the rest of the wall.

10. Repeat the sequence until the corner is built, Fig. 8-18.
11. Straightedge the rack of the lead, as shown in Fig. 8-19. This will eliminate any wind, belly or cave-in in the wall.

Fig. 8-16. The other side of the corner is begun. Be careful not to move corner brick. See how mason supports corner brick with fingers.

9. Lay the second course following the same sequence used for the first course. Check for proper height just after the course has been laid, but before the brick have been leveled. See Fig. 8-17. If the brick are too high, tap them down as they are leveled. If they are too low, remove them and add more mortar before leveling.

Fig. 8-19. Mason straightedges the rack of the lead to check for proper spacing of brick, wind, belly or cave-in in the wall. The rack refers to the "steps" formed by the last brick in each course. If the rule rests on each of them the spacing is proper.

12. Lay the second corner of the wall following the same procedure used for the first, Fig. 8-20.
13. Stretch the mason's line between the corners at the top of the first course, Fig. 8-21. The line may be secured with line pins or corner blocks. Pull the line taut enough that it is straight and level. Be sure to use the same amount of tension each time.
14. Begin laying the wall from the lead toward the center, one course at a time, Fig. 8-22. Each brick on the second course should be centered over the cross joint of the first

Fig. 8-17. Checking for proper height. Notice that these brick are being laid on 5's. Hold rule straight up and down for accurate measurement.

Fig. 8-21. Laying the first course to the line. The closer brick is being placed. Note that both ends have been buttered. It must be fully pressed into position.

Fig. 8-20. Above. Beginning the second lead corner using the same procedure as for the first corner. Center. Laying up the corner. Below. Completed corner.

Fig. 8-22. Laying the second course from both ends toward the middle. Above. Mason's line in position. Below. Next to last brick being placed. Note how the fingers extend out over the line without touching it.

course. The cross joints must be of uniform width or the closer brick (the last brick) will not fit correctly. The line should be worked from both leads toward the center. Level, plumb and line up this course as you lay it.

15. Tool or strike the joints when the mortar can be indented with a thumb print without too much pressure. The brick will be smeared if the mortar is too wet. If too dry, the mortar will turn black from metal worn off the jointer. Experience will help determine the proper time to tool the joints.

In striking a wall use the long jointer for the bed joints, Fig. 8-23. Next, use the short jointer for vertical or head joints. See Fig. 8-24. With the edge of the trowel, cut off the tags of mortar, Fig. 8-25, that have been forced out by the jointer.

Fig. 8-25. Remove the tags by moving the blade of the trowel forward along the bed joint.

Fig. 8-23. Bed joints are struck with the long jointer when the mortar is ready.

The proper procedure for removing the tags is to move the trowel horizontally, along the joint. After the head joints have been struck, use the long jointer again to strike the bed joints for best appearance.

16. Brush the wall when the mortar is stiff enough, Fig. 8-26.

Fig. 8-26. The wall is brushed when the mortar is stiff enough.

This may be done at the end of the day or more frequently, if desired. Brushing reduces the amount of cleaning required later.

17. Lay successive courses in the same manner until the wall is the required height, Fig. 8-27.

LAYING AN 8 IN. COMMON BOND WALL WITH LEADS

The common bond is strong and can be laid up fast. It consists of a course of headers between every five to seven courses of stretchers. The heading course serves as a bond between the inside and outside 4 in. tiers. The mortar joint is 1/2 in. thick.

The recommended procedure for constructing an 8 in.

Fig. 8-24. Short jointer is used to strike the head joints.

Fig. 8-27. The completed wall.

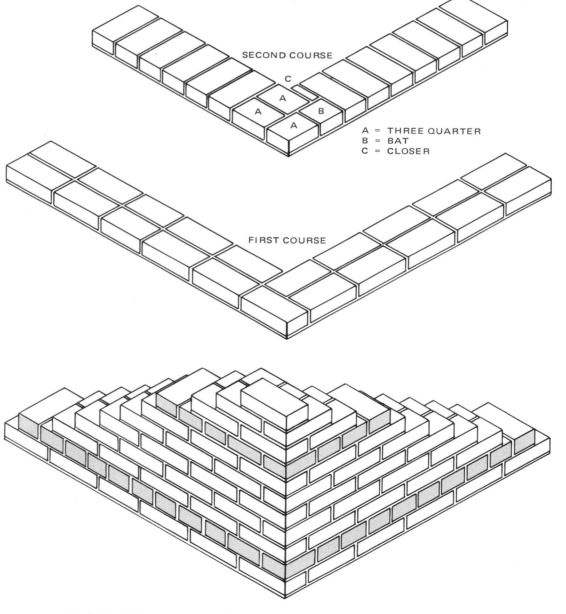

SECOND COURSE

C

A

A   A   B

A   A

A = THREE QUARTER
B = BAT
C = CLOSER

FIRST COURSE

Fig. 8-28. An 8 in. common bond wall lead with first and second courses shown in detail.

American bond wall with leads follows:

1. Lay out the wall location as with the 4 in. wall. In addition, locate a second line (about 8 in.) inside the first wall line. This distance is equal to the average length of the brick being used.
2. Lay out the bond to eliminate excessive cutting for windows and doors. This is an important step!
3. Lay the corner bricks as shown in Fig. 8-28. The header course is shown here starting at the second course, but it could start wherever specified.
4. Lay the first course of stretchers for the corner. Four or five stretchers should be sufficient. Lay the header course as shown in Fig. 8-28.
5. Lay successive courses of stretchers until reaching the next header course. Level, plumb and line up each course as it is laid.
6. Lay the header course breaking the bond in the usual manner.
7. Finish laying the first corner and complete the second in the same manner.
8. Construct the first stretcher course of the wall by laying the outside first and then the inside. These should be laid from the leads to the center. Level, plumb and line up each course before beginning the header course. See Fig. 8-29. A line should be used on all backing courses (the inside tier) to insure a good face.
9. Lay the header course from each lead toward the center. Level, plumb and line up the header course.
10. Lay the outside tier up to the next header course keeping it straight, level and plumb. This is called "header high."
11. Lay the inside tier up to the same height. Be careful to keep it level with the outside tier as it is laid up. This is important as you must have a level surface for the header course.
12. Continue as before until the desired height is reached. Strike the joints and brush the wall at the proper time.

CORNER LAYOUT IN VARIOUS BONDS

Laying up a corner is very similar for most bonds. After a beginner has built several of the more common ones there should be no difficulty raising any corner.

In addition to the common or American bond, the following bonds represent some of the most frequently used patterns.

FLEMISH BOND

The Flemish bond is very popular. It is easy to lay producing an artistic and pleasing wall. It is more costly than the common bond and requires greater care. The bond consists of alternate headers and stretchers in each course. The headers are centered on the stretchers between each course. Bonds may be started at the corner with either a 3/4 bat or a 1/4 closure, Fig. 8-30.

ENGLISH BOND

English bond has alternate courses of stretchers and headers. The headers center on the stretchers and on the joints

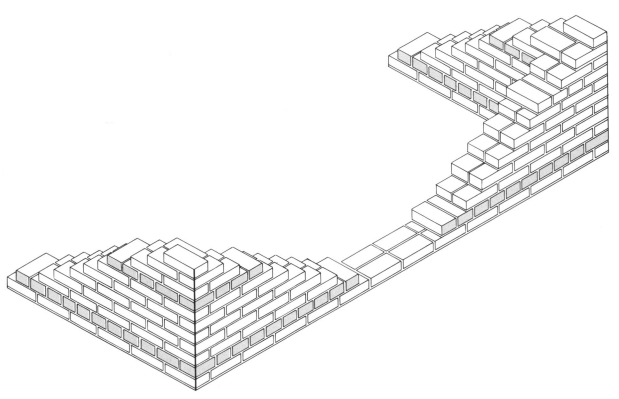

Fig. 8-29. The first stretcher course and leads for the corners are in place in this 8 in. common bond wall. The wall will now be constructed between the leads.

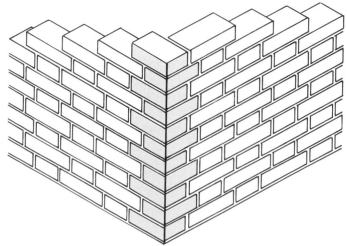

FLEMISH BOND WITH 3/4 CLOSURE AT CORNERS

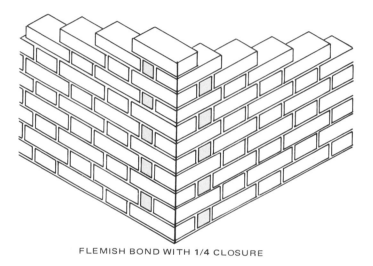

FLEMISH BOND WITH 1/4 CLOSURE

Fig. 8-30. Corner layout in Flemish bond with 3/4 and 1/4 closure.

between the stretchers. The stretchers all line up vertically, one over the other.

Two methods may be used to start the English bond at a corner. Fig. 8-31 shows the English bond with the 1/4 and the 3/4 closure at the corner.

LAYING CONCRETE BLOCK

A well planned concrete block structure will involve mainly stretcher and corner blocks, Fig. 8-32. These blocks are nominally 8" x 8" x 16". Actual size is 7 5/8" x 7 5/8" x 15 5/8". This allows for a 3/8 in. mortar joint which is standard. Blocks of other sizes are sometimes used, but they also use a 3/8 in. mortar joint.

Concrete blocks must be protected from excess moisture before use. If they are wet when placed they will shrink when dry and cause cracks. They should be stacked on platforms and covered with plastic or a tarpaulin to protect them from rain.

Mortar for concrete block masonry should be mixed in

accordance with specifications in Chapter 4. The consistency of mortar for concrete block masonry is just as critical as it is for brick or stone.

LAYING A CONCRETE BLOCK WALL

Good construction requires that adequate reinforcement be used or that the joints be staggered. Therefore, advance planning is necessary for a strong wall with a good appearance. The length of the blocks must be considered as they relate to window and door openings in the wall. It is a good idea to check the building dimensions carefully before beginning the work. Problems can be identified before the wall is begun.

The recommended procedure for laying a concrete block wall follows:

1. The outside wall line should be established. A chalk line may be used, if desired, to provide a straight line for the first course of block. The wall line should be checked for squareness and proper length before proceeding to the next step.

2. String out the block for the first course without mortar to check the layout, Fig. 8-33. Allow 3/8 in. for each mortar

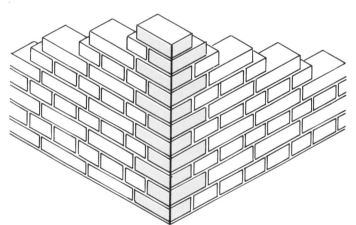

ENGLISH BOND WITH 3/4 CLOSURE AT CORNERS

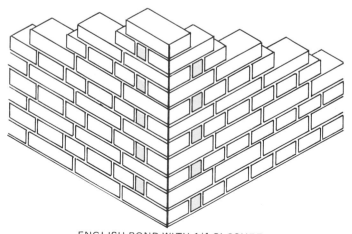

ENGLISH BOND WITH 1/4 CLOSURE

Fig. 8-31. Corner layout in English bond with 3/4 and 1/4 closure.

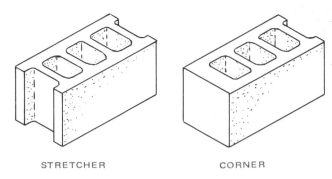

STRETCHER   CORNER

Fig. 8-32. These two concrete blocks are the primary units used to construct most concrete masonry walls. Nominal size is 8" x 18" x 16".

Fig. 8-33. When the corners have been located, the first course of block are strung out to check the spacing. Note that the space between the blocks is equal to the desired mortar joint. (Portland Cement Assoc.)

joint. When you are satisfied with the layout, the block may be set aside.

3. A full mortar bed is spread and furrowed with the trowel, Fig. 8-34. Provide plenty of mortar on which to set the block.

Fig. 8-34. A full mortar bed is being furrowed with the trowel. It is important that ample mortar be used to insure a good bond between the footing and bed course of masonry.

4. Lay the corner block, Fig. 8-35. Position it carefully and accurately. Concrete blocks should be laid with the thicker edge of the face shell up to provide a wider mortar bed.

5. Lay several stretcher blocks along the wall line. Several blocks can be buttered on the end of the face shells if they are stood on end. This speeds the operation. To place them, push them downward into the mortar bed and sideways against the previously laid block.

Fig. 8-35. Corner blocks are laid carefully and accurately. (Portland Cement Assoc.)

6. After three or four blocks have been placed into position, they may be aligned, leveled and plumbed with the mason's level, Figs. 8-36 and 8-37. Tap on the block rather than on the level.

Fig. 8-36. The block are leveled and settled to the proper elevation.

7. After the first course has been laid, the corner lead is built up as in Fig. 8-38. This corner is very important since the remainder of the wall is dependent upon its accuracy. The lead corner is usually laid up four or five courses high

Fig. 8-37. The plumb rule (mason's level) is used to plumb the block. (Portland Cement Assoc.)

Fig. 8-39. The last block is laid and leveled on the lead. (Portland Cement Assoc.)

Fig. 8-38. Alignment is being checked with the level (used as a straightedge) to insure accuracy before completing the corner.

Fig. 8-40. The corner is checked to be sure that it is plumb.

above the center of the wall. Each course is checked to be sure it is aligned level and plumb as it is laid. See Fig. 8-39 and Fig. 8-40.

The faces of the blocks should be in the same plane. Each block is stepped back half a block. The spacing may be checked by placing the level diagonally across the corners of the block as shown in Fig. 8-41. All corners should be lined up on the edge of the level.

8. After the corner leads have been constructed, the blocks are laid between the corners. A mason's line should be stretched from corner to corner at the proper height for each course, Fig. 8-42. Line blocks or line pins may be

Fig. 8-41. Spacing is being checked by lining up the corners with a straightedge. (Portland Cement Assoc.)

Fig. 8-42. The wall being filled in using a mason's line to insure proper height and a straight wall.

used to secure the line. Tension should be the same each time to insure a uniform wall. Work from the corners toward the center of the wall. Do not allow a block to touch the line as this will push the line beyond its normal position causing a curve in the wall. Keep the block a line width away from the line and even with the bottom of the line. If the blocks have been positioned accurately, the closer block will fit properly, Fig. 8-43.

Fig. 8-43. All edges of the opening and the four vertical edges of the closer block are buttered with mortar before it is placed.
(Portland Cement Assoc.)

9. Tool the mortar joints. The most effective joint is one that has been compacted or pressed into place. For this reason, the concave or V-joint is best for exterior work. Joints may be tooled when the mortar has become thumbprint hard. The tool should be slightly larger than the width of the joint so that it will make contact along the edges of the blocks. A 5/8 in. diameter bar is usually used for a 3/8 in. concave mortar joint. A 1/2 in. square bar is used for making a 3/8 in. V-shaped joint.

Tools for tooling horizontal joints in concrete block

construction should be at least 22 in. long, Fig. 8-44. After the horizontal joints are shaped the vertical joints may be tooled with an S-shaped jointer, Fig. 8-45. A trowel should be used to remove any mortar burrs left from the tooling, Fig. 8-46.

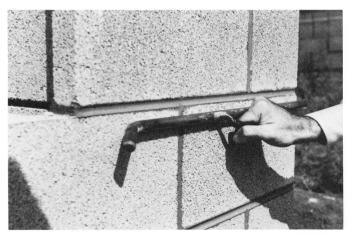

Fig. 8-44. Horizontal joints are tooled before the vertical joints.

Fig. 8-45. Vertical joints are tooled with an S-shaped jointer.
(Portland Cement Assoc.)

Fig. 8-46. Mason's trowel is used to remove any mortar burrs after tooling joints.

10. When the mortar is sufficiently dry, the wall may be brushed or rubbed with a stiff fiber brush or a burlap bag to remove dried particles. See Fig. 8-47.

Fig. 8-47. Use a stiff fiber brush to clean the wall of dried particles of mortar when it is sufficiently dry.

CUTTING BLOCK

Even though concrete blocks are available in half-length units as well as full-length units, it is sometimes necessary to cut block to fit. Block may be cut with a brick hammer and blocking chisel. See Fig. 8-48. Another method is to use a masonry saw, Fig. 8-49. When using the chisel, hold the beveled edge toward you. The piece of block to be cut off should be facing away from you. Score the block on both sides to get cleaner break.

If a neater, cleaner cut is desired, a masonry saw should be used. It is fast and accurate. The block should be dry when cut with the masonry saw.

Fig. 8-49. A masonry saw being used to cut a concrete block.

CONTROL JOINTS IN A CONCRETE BLOCK WALL

Control joints are continuous vertical joints built into concrete masonry walls. They are placed where forces might cause cracks. Instead, movement of the wall occurs along the control joint and is not very visible. Such joints should be the same width as other joints. They should be sealed with caulking compound after the mortar has been raked out to a depth of 3/4 in., Fig. 8-50.

Fig. 8-48. A brick hammer and chisel can be used to cut concrete block. (Portland Cement Assoc.)

Fig. 8-50. Control joint at end of lintel is sealed with caulking. (Portland Cement Assoc.)

Other methods of constructing control joints in concrete block walls include:

1. Placing a Z-tiebar, Fig. 8-51, in the horizontal mortar joint above jamb blocks. This provides lateral (sideways) support for the wall above windows and doors.

Fig. 8-51. Control joint using jamb block and Z-tiebar.

2. Using building paper or roofing felt in the end core of the block to break the bond, Fig. 8-52. The felt extends the entire height of the control joint.
3. Using special tongue-and-groove block which allows movement, Fig. 8-53.

Fig. 8-52. Building paper may be used to break the bond. This method is known as the Michigan control joint. (Portland Cement Assoc.)

WALL INTERSECTIONS

Intersecting walls should not be joined together in a masonry bond as they are at corners. Instead, one wall should end at the face of the other wall with a control joint. One method which provides for movement and also lateral support is to use a metal tiebar. Fig. 8-54 shows such a tie set into a wall. The tiebar is 1/4 in. thick, 1 1/4 in. wide, 28 in. long

Fig. 8-53. These are special tongue-and-groove control joint blocks. They allow up and down movement but stop sideways (lateral) motion.

Fig. 8-54. The bent ends of the metal tiebar are embedded in the cores of the block which are then filled with mortar or concrete. (Portland Cement Assoc.)

Fig. 8-55. Cores being filled with mortar over a metal tiebar.

with 2 in. right angle bends on each end. These tiebars are usually placed 4 ft. apart vertically. The bent ends are placed in cores filled with mortar or concrete. Pieces of metal lath are placed under the cores, to prevent the mortar from falling through. See Fig. 8-55.

Non-load-bearing walls (those which do not carry a structural load) can be tied to other walls using strips of metal lath or hardware cloth. This material is placed across the joint between the two walls, Fig. 8-56. The metal strips are put in every other course.

Fig. 8-57. Anchor bolts placed in the cores of the top two courses of block will be used to attach the plate to the wall.

Fig. 8-56. Metal lath or 1/4 in. mesh galvanized hardware cloth may be used to tie non-load-bearing walls together. (Portland Cement Assoc.)

ANCHORAGE TO MASONRY WALLS

Very often a wood plate must be anchored to the top of a concrete block wall. Anchor bolts are generally used for this purpose. They are usually 1/2 in. in diameter, 18 in. long and are spaced about 4 ft. apart as in Fig. 8-57.

The bolts are placed in the cores of the top two courses of block. The cores should have metal lath under them, so that they can be filled with mortar or concrete. The threaded end of the anchor bolts must extend above the top of the wall so that they pass through the sill and provide enough thread for the washer and nut.

Fig. 8-58. Random rubble stonework with uniform mortar joints.

STONE MASONRY

Stone masonry is the oldest, the most artistic and most expensive of all the types of masonry construction. Over the years it has declined in importance as a solid wall building material. However, it is still being used extensively as veneer over other masonry or frame construction. Stone is also used a great deal for trim on buildings and paving for foyers, patios, drives and walks.

Natural stone is available in a wide variety of sizes, shapes, colors and textures. (See Chapter 3.) These characteristics make possible many wall patterns. Stone masonry is more time-consuming. Due to the different sizes and uneven surfaces of stone units they are more difficult to lay.

There are three main classes of stone masonry construction:
1. Rubble.
2. Ashlar.
3. Trimmings.

RUBBLE MASONRY

Rubble is uncut stone or stone that has not been cut to a rectangular shape. It is generally used in walls where a rustic appearance is desired and precise lines are not so important. See Fig. 8-58. Many sizes and shapes of stone have been used to create the pattern. A liberal amount of mortar must be used because of the stones' irregularities. Rubble stonework is not as strong as other bonds.

ASHLAR

When stone is precut with enough uniformity to allow some regularity in assembly, the wall is generally called *ashlar,* Fig. 8-59. Ashlar stone is easier to lay because the individual units are precut to fit. In this respect, it is much like brick or block.

Fig. 8-59. Ashlar stonework.

When the stone is laid in regular courses, it is referred to as *coursed* ashlar. When it is laid in broken courses without regard to the arrangment of the joints, it is referred to as *broken* or *ranged* ashlar. Stones for ashlar work are produced the following dimensions:

1. Thickness — 2 to 8 in.
2. Heights — 1 to 48 in.
3. Lengths — 1 to 8 ft.

TRIMMINGS

Trimmings are cut on all sides to specific dimensions. They are used for moldings, sills, lintels and ornamental purposes, Fig. 8-60. They may be set in mortar or anchored using a variety of mechanical fasteners.

Fig. 8-60. Stone trimmings to be used as sills, lintels and other purposes. (Georgia Marble Co.)

LAYING A STONE VENEER WALL

To lay a rough stone masonry veneer wall using rubble use the following procedure:

1. Spread out the stone so that each one may be examined for shape, texture or other characteristics, Fig. 8-61. This stone will most likely vary in size from 6 to 18 in.

Fig. 8-61. Rubble stone to be used for a stone veneer wall.

2. Mix enough mortar to last for 1 hour of work, Fig. 8-62. Use it up before it begins to set. Be sure to use a nonstaining mortar. Mix 1 part nonstaining cement, 1 part hydrated lime and 6 parts clean, sharp, washed sand. A comparable manufactured mortar cement can also be used.

3. Select several larger stones for the bed course and place them into position dry. With a piece of chalk, mark them for trimming.

Fig. 8-62. Mortar should be mixed in small amounts so that it will not begin to set before being used.

Stones should be placed in as natural a position as possible. Have a pattern or special effect in mind as you select and place stones together. Do not use too many varieties and keep textures relatively uniform. Stone veneer is generally 4 to 8 in. thick depending on the method of bonding and construction specifications.

Large stones must be split so that they will meet these requirements. A sledge hammer may be used for splitting large stones.

4. Trim each stone as you are ready to place it in the wall, Fig. 8-63. Be careful of flying chips! Each stone should be thoroughly cleaned on all exposed surfaces by washing with a brush and soap powder, followed by a thorough drenching with clear water.

Fig. 8-63. A stone mason's hammer and chisel are used to trim stone to desired shape.

5. Place each trimmed stone in its proper location in the wall and check for proper fit. When you are satisfied with the result, proceed to the next step.

6. Using the trowel, lay a generous full bed of mortar for the trimmed stones and place them into position, Fig. 8-64.

Fig. 8-64. Trimmed stones are placed in a full bed of mortar.

7. Fill in the spaces between the stones with mortar, Fig. 8-65. A narrow caulking trowel, Fig. 8-66, works well for filling narrow spaces and working the mortar into crevices. Joints should be 1/2 to 1 in. wide for rough work and 3/8 to 3/4 in. for ashlar. Sponge the stone free of mortar along the joints as the work progresses.

Fig. 8-66. Narrow caulking trowel works well for filling spaces between stones with mortar.

Fig. 8-65. Spaces between stones are filled with mortar.

9. Remove excess mortar from the stone with the trowel and strike the joint, Fig. 8-67. Joints can be tooled when initial set has occurred. If desired they can be raked out 1 in. deeper and pointed later with mortar.

10. Trim other stones for the deeper bed course and fit them into place using the same procedure.

11. Bend corrugated metal ties into place for successive courses of stone, Fig. 8-68. All ties must be noncorrosive. Use extra ties at all corners and large stones when possible.

Fig. 8-67. Excess mortar may be removed with the trowel.

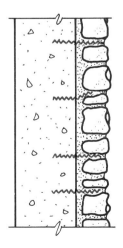

4 — 8 IN. STONE TIED TO CON-CRETE BACKING USING WALL TIES.

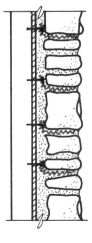

4 — 8 IN. STONE TIED TO FRAME CONSTRUCTION. USE WOOD SHEATHING AND W.P. FELT OR W.P. SHEATHING BOARD.

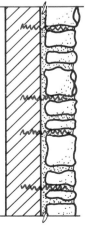

4 — 8 IN. STONE TIED TO BRICK OR BLOCK MASONRY. PROVIDE 1 IN. — 2 IN. AIR SPACE OR SLUSH FILL VOIDS.

Fig. 8-68. Stone veneer sections showing non-corrosive metal ties.

Lead, plastic or wood pads the thickness of mortar joint should be placed under heavy stones to avoid squeezing mortar out, Fig. 8-69. Remove the pads after the mortar has been set and fill the holes. Heavy stones or projecting courses should not be set until mortar in courses below has hardened sufficiently to avoid squeezing.

Fig. 8-69. Wooden pads support heavy stones to avoid squeezing mortar from the joints.

12. Using the above procedure, lay the remaining stones until the wall is the desired height, Fig. 8-70. Work from the corners toward the middle as in laying any masonry wall.

Fig. 8-70. Stone wall being finished off at the proper height.

A mason's line and level may be used to keep the wall straight and plumb. If the structure is to be inclined or tapered as the chimney in Fig. 8-71, then a mason's line should be used to keep the edge straight and moving in the proper direction.

The masonry should be protected at all times from rain and masonry droppings. Adequate protection must be provided during cold weather construction. See Chapter 4.

Fig. 8-71. A beautiful example of stonework in a residential structure.

13. When the wall is completed, it may be scrubbed with a fiber brush and clear water. The stone should be clean and free of mortar.

Strong acid compounds generally should not be used as they will burn and discolor certain types of stone. Limestone is one of these.

Waterproofing may be used. Use a nonstaining asphalt emulsion, vinyl lacquer, cement base masonry waterproofing, stearate, or other approved material.

Fig. 8-72 shows a close-up of rubble stone wall with thin mortar joints of uniform width. Here the stones have been

Fig. 8-72. Uniform mortar joints are a mark of good craftsmanship.

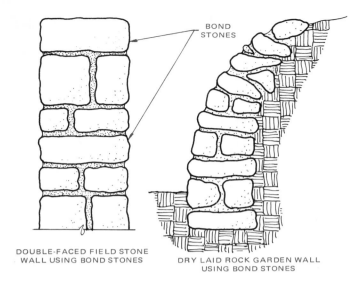

DOUBLE-FACED FIELD STONE WALL USING BOND STONES

DRY LAID ROCK GARDEN WALL USING BOND STONES

Fig. 8-73. Sections of freestanding and retaining stone walls. Note how the bond stones tie the wall together and increase stability.

trimmed with expert accuracy to fit into a specific place in the wall. Practice and patience is required to develop this talent.

BUILDING A SOLID STONE WALL

The procedure for building a solid stone wall using rubble or ashlar is similar to laying a stone veneer wall with a few exceptions:

1. The wall will depend on *bond* stones for strength and stability and tieing the wall together. *A bond stone is one that passes through the wall from side to side, Fig. 8-73.* Bond stones should be placed as frequently as possible. Every 5 to 8 sq. ft. of wall should have one.
2. A solid stone wall is usually thicker than a veneer wall. The backing in a veneer wall provides the structural support and the stone is a protective covering. In a solid stone wall, the stone provides structural support as well as covering. The usual thickness is a minimum of 18 in. Larger stones may be used and they should be placed in their natural position for best appearance.

   Reinforcing may be used in a solid stone wall to give added strength and support. The material used should be noncorrosive.

CLEANING NEW MASONRY

The finished appearance of a masonry wall depends not only on the skill used in laying the units but on the cleaning procedure as well. The appearance of a masonry structure may be ruined by improper cleaning. In many instances, the damage caused by faulty cleaning techniques or the use of the wrong cleaning agent cannot be repaired.

CLEANING BRICK

Cleaning new brick masonry is mainly concerned with removing mortar, mortar stains and any other materials, such as dirt, deposited on the masonry during construction.

Small quantities of various minerals found in some burned clay masonry units will react with some cleaning agents to cause staining. These reactions cannot be predicted in advance and it is therefore recommended that before applying any cleaning agent to a masonry wall, it be applied to a sample section of 10 to 20 sq. ft. to judge its effectiveness. Wait a minimum of one week before inspection.

In the construction of masonry walls, some mortar stains will be present even if the mason is very skilled and careful. Therefore, most specifications require a final washing down of all masonry work.

ACID SOLUTIONS

A solution of hydrochloric acid is used extensively as a cleaning agent for new masonry. The following procedures are recommended as good practice when using an acid or base solution.

CLEANING DARK BRICK: Dark colored brick include red, red flash, brown and black. They are most likely to show light gray, brown or yellow discolorations from failure to rinse off dissolved mortar or dirt. The first procedure requires no acid.

1. When the mortar is thoroughly set and cured, begin the cleaning operation.
2. Remove large particles of mortar with wooden paddles or scrapers before wetting the wall. A chisel or wire brush might be necessary.
3. Saturate the wall with clean water and flush away all loose mortar and dirt.
4. Scrub down the wall with a solution of 1/2 cup of trisodium phosphate and 1/2 cup of household detergent dissolved in one gallon of clean water. Use a stiff fiber brush.
5. Rinse off all cleaning solution and mortar particles using clean water under pressure.

When acid cleaning becomes necessary, the following procedure is recommended:

1. Follow steps 1 through 3 above.
2. Use a clean, stain-free commercial grade of hydrochloric (muriatic) acid mixed one part of acid to nine parts water. Mix in a nonmetallic container. *Pour the acid into the water, not the water into the acid! Use a long handled fiber brush to scrub the wall. Be careful when using this chemical.*
3. Keep the area not being cleaned flushed free of acid and dissolved mortar. This scum, if allowed to dry may be impossible to remove later.
4. Scrub the brick, not the mortar joints. Do not use metal tools. Clean only a small area at a time.
5. Rinse the wall thoroughly with plenty of clean water while it is still wet from scrubbing with the acid.

CLEANING LIGHT COLORED BRICK: Light colored brick include buff, gray, speck and pink. They are more likely to be burned by acid than darker brick. Therefore, do not use acid except in extreme cases.

The first procedure for light colored brick is exactly the same as the first procedure given for dark brick — no acid!

The second procedure for light colored brick is the same as

the second procedure for dark brick except as follows:

1. Use the highest grade acid available. It should be free of any yellow or brown discoloration.
2. Mix 1 part acid with 15 parts water and scrub the wall with a fiber brush. Rinse the wall well with clear water. The acid wash may be neutralized with a solution of potassium or sodium hydroxide, consisting of 1/2 lb. hydroxide to 1 qt. of water (2 lb. per gal.). Allow this to remain on the wall for two or three days before washing again with clear water.

SANDBLASTING

Sandblasting is used extensively in some areas to clean new masonry. It costs about the same as acid cleaning and, with an experienced operator, there is virtually no change in texture of hard brick. However, care must be exercised when sandblasting sand finished brick. This method eliminates the possibility of mortar smear, acid burn and efflorescence which are present in acid cleaning.

For best results with sandblasting, use a very low pressure (from 60 to 120 lb) and a 1/4 in. sandblast nozzle with white urn sand. The secret to successful sandblast cleaning is the distance the cleaner stands from the wall and the manner in which the blast is directed at the brick. Concentrate on hitting the brick rather than the mortar joints.

CLEANING CONCRETE BLOCK

Concrete block walls are not cleaned with acid to remove mortar smears or droppings. Therefore, care must be taken to keep the wall surface clean during construction. *Any mortar droppings that stick to the wall should be allowed to dry and harden.* Large particles of mortar can be removed with a trowel, chisel or putty knife. *If the mortar is removed while it is wet, it will most likely smear into the surface of the block.*

Rubbing the wall with a small piece of block will remove practically all of the mortar. In some instances, a commercial cleaning agent such as a detergent may be used. Be sure to follow the manufacturer's directions and try out the product on a small section of the wall to check the results.

CLEANING STONEWORK

Clean stonework with a stiff fiber brush and clean water. If stains are difficult to remove, use soapy water and then rinse with clear water. If the stonework is kept clean by sponging during construction, the final cleaning will be easy.

Machine cleaning processes should be approved by the supplier before using. Acids, wire brushes and sandblasting are usually not permitted on stonework. Strong acid compounds used for cleaning brick will burn and discolor many types of stone.

REVIEW QUESTIONS — CHAPTER 8

1. Mortar is usually furrowed with a _____.
2. Any straight wall longer than 4 ft. is usually laid to a line. True or False?
3. Why is it important to place the same amount of tension on a mason's line each time it is moved?
4. The corner of a masonry wall which is used to establish proper height for each course is called a _____.
5. Height marks are made on a _____.
6. Brick are laid up tight to the line to insure accuracy. True or False?
7. A dry course of brick is laid out along the wall line when starting a wall to eliminate as much _____ of units as possible.
8. A mason's _____ _____ is used to check whether the wall is level and plumb.
9. A long jointer is used to _____ the bed joints of brick masonry.
10. Tags of mortar are removed from the brick with a _____.
11. Vertical or "head" joints are struck with a _____.
12. A brick wall is usually laid one course at a time from the leads toward the center. True or False?
13. In an 8 in. American bond brick wall, a _____ course is laid every five to seven courses.
14. The _____ bond is composed of alternate courses of stretchers and headers. The headers center on the stretchers and on the joints between the stretchers.
15. The actual size of an 8″ x 8″ x 16″ concrete block is _____.
16. The proper mortar joint thickness for concrete block masonry is _____.
17. How is a corner concrete block different from a stringer?
18. Concrete block may be cut using a hammer and _____ chisel.
19. Vertical joints which are placed in a masonry wall to prevent cracking due to movement are called _____ joints.
20. Intersecting walls should not be joined together in a masonry bond as they are at corners. True or False?
21. _____ bolts are used to attach sills to a masonry wall.
22. Stone masonry is still used extensively for _____ construction.
23. Stone masonry which consists of uncut stones is called _____ masonry.
24. Stone which has been cut on all sides to specific dimensions is called _____.
25. Mortar used for stone masonry should have the following composition: _____ part(s) nonstaining cement, _____ part(s) hydrated lime and _____ part(s) clean, washed sand.
26. The tools used to trim stone are the mason's _____ and _____.
27. Stone veneer is generally _____ to _____ inches thick.
28. Stone veneer may be attached to a masonry or frame backing using _____ _____ ties.
29. How can one prevent heavy stones from squeezing mortar out when they are set into place?
30. Why should strong acids be avoided when cleaning stone?
31. A solid stone wall should be at least _____ thick.

Chapter 9
CONSTRUCTION DETAILS

This chapter includes many of the accepted methods of masonry construction for:
1. Foundation systems.
2. Wall systems.
3. Fireplaces and chimneys.
4. Floors and pavements.
5. Flashing.
6. Steps.
7. Sills and lintels.
8. Garden walls.

This material is meant to be a source of help for the mason in selecting appropriate construction methods for many different jobs.

CONCRETE AND MASONRY FOUNDATION SYSTEMS

One of the first concerns in construction is to prevent the settling of buildings as much as possible. Some settling is to be expected. It cannot be prevented. The worst emeny of buildings is uneven settling. It causes cracks in finished walls and ceilings. It tilts floors, causing doors and windows to bind. It damages woodwork. If severe enough, it can cause total failure of the building.

The foundation is the substructure of a building that resists settling. It supports the weight of the building. Footings transmit (shift) the weight from the foundation walls to the soil, Fig. 9-1.

Footings are usually wider than the foundation wall to provide this extra support that will stop or reduce uneven settling. As a result, the structure remains sound.

There are two general types of foundations:
1. Spread foundations.
2. Pile foundations.

Only spread foundations will be covered. Pile foundations will not concern the work of the mason.

SPREAD FOUNDATIONS

Spread foundations distribute the building loads over a wider area of soil. Spread foundations consist of elements such as walls, pilasters, columns or piers. These rest on a wider base called a footing. Raft and matt foundations are also classified as spread foundations. They are used over soils with low ability to carry weight.

FOOTINGS
All foundation walls built of masonry units bonded with mortar should rest on footings. These footings are usually made from concrete. They are classified as plain, reinforced, continuous, stepped and isolated.

Plain footings carry light loads and are not reinforced with steel. They may be continuous, stepped or isolated.

Steel is embedded in reinforced footings to make them

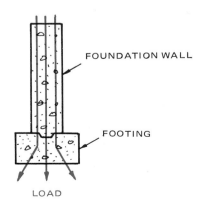

Fig. 9-1. Footings spread the load of the structure over a broader area.

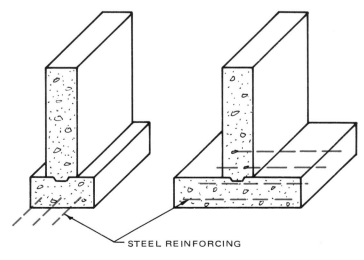

Fig. 9-2. Two methods of reinforcing footings to increase strength.

stronger. See Fig. 9-2. Reinforcing should be used over weak spots in the soil such as where there have been excavations for sewer, gas or other connections.

Continuous footings may be used to support a foundation wall or several columns in a row, Fig. 9-3. When a single footing supports more than one column it is called a combined footing.

Stepped footings are of two types. One widens the base to

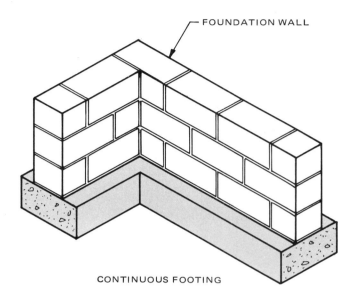

FOUNDATION WALL

CONTINUOUS FOOTING

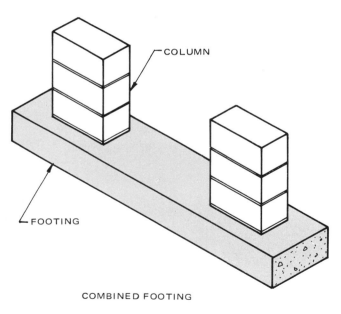

COLUMN

FOOTING

COMBINED FOOTING

Fig. 9-3. Typical applications of a continuous footing.

provide added support. The other changes levels to accomodate a sloping grade. Fig. 9-4 shows both types. However, today, reinforced footings have largely replaced the widened footing.

Isolated footings are not part of the foundation. They receive the loads of free-standing columns or piers, Fig. 9-5.

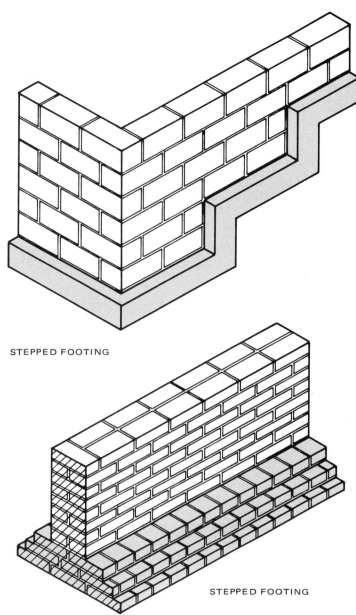

STEPPED FOOTING

STEPPED FOOTING

Fig. 9-4. Stepped footings may be used to change levels on a slopping grade or to widen the base for greater support.

Footing size should be related to soil conditions and the weight of the structure. However, the rule-of-thumb size for residential footings placed on average soil is:
1. The width of the footing should be twice the thickness of the wall.
2. The depth of the footing should be equal to the thickness of the wall or a minimum of 6 in. (Fig. 9-6 shows the proportions.)

RAFT AND MATT FOUNDATIONS

Raft and matt foundations are made of concrete reinforced with steel in such a way that the entire foundations will act as a single unit. Many times they are referred to as *floating* foundations.

A matt foundation is a thickened slab which transmits

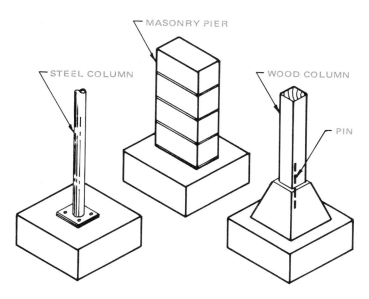

Fig. 9-5. Isolated footings with typical columns and pier.

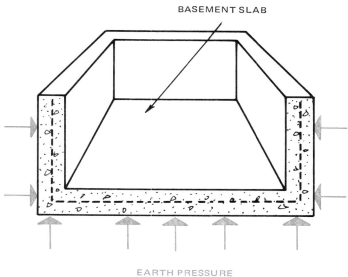

Fig. 9-8. A raft foundation ties walls and floor into one unit.

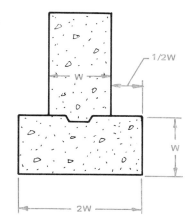

Fig. 9-6. Typical rule-of-thumb footing proportions.

building loads over the entire slab and soil area, Fig. 9-7.

A raft foundation is made with reinforced walls and floor cast into a single unit, Fig. 9-8. Soil is excavated which weighs about the same as the foundation plus the superstructure. This minimizes settling of the structure because the building load on the soil is equal to the weight of the excavated soil.

FOUNDATION WALLS

Like footings, foundation walls are load transferring elements. They must be able to:
1. Support the weight of the building above.
2. Resist pressures from the ground.
3. Provide anchorage for the building superstructure.

They must also be durable and resist moisture penetration.

Foundation walls of hollow concrete block should be capped with a course of solid masonry, Fig. 9-9, to help distribute the loads from floor beams and serve as a termite barrier.

The most common type is the T foundation. The name comes from the shape of the foundation and the footings which look like an inverted T. The footing and foundation wall, in this case, are generally two separate parts, but may be cast as a single unit. Fig. 9-10 shows several styles of foundations.

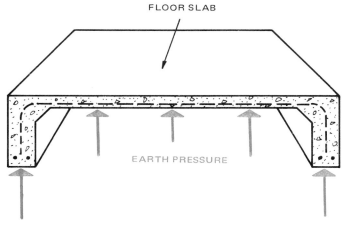

Fig. 9-7. A matt foundation ties entire floor and footings into one unit.

Fig. 9-9. Block masonry foundation wall is capped with a course of solid top block. (Portland Cement Assoc.)

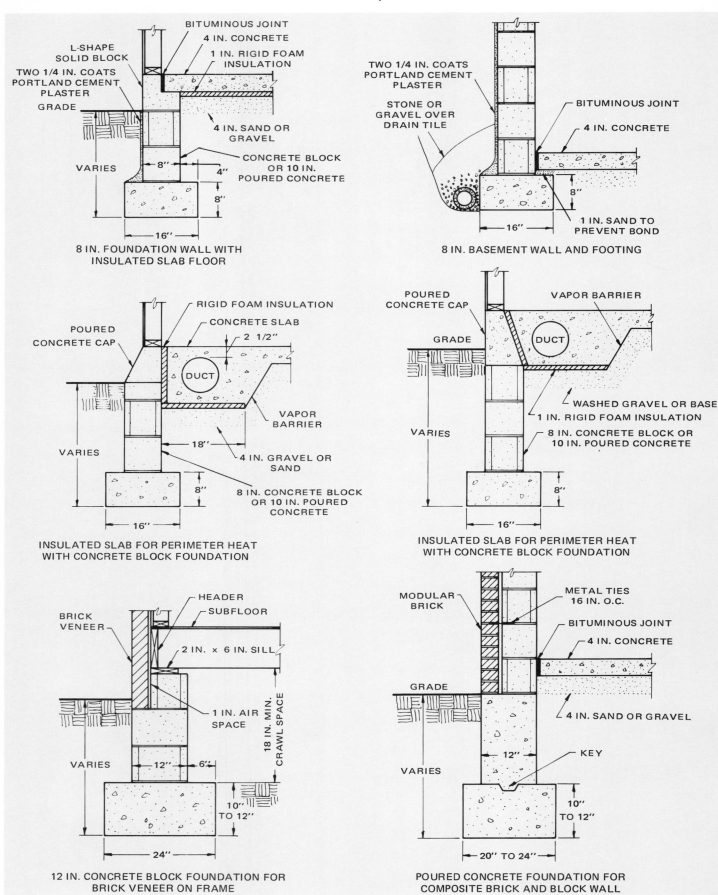

Fig. 9-10. Typical foundation details. (Continued)

Construction Details

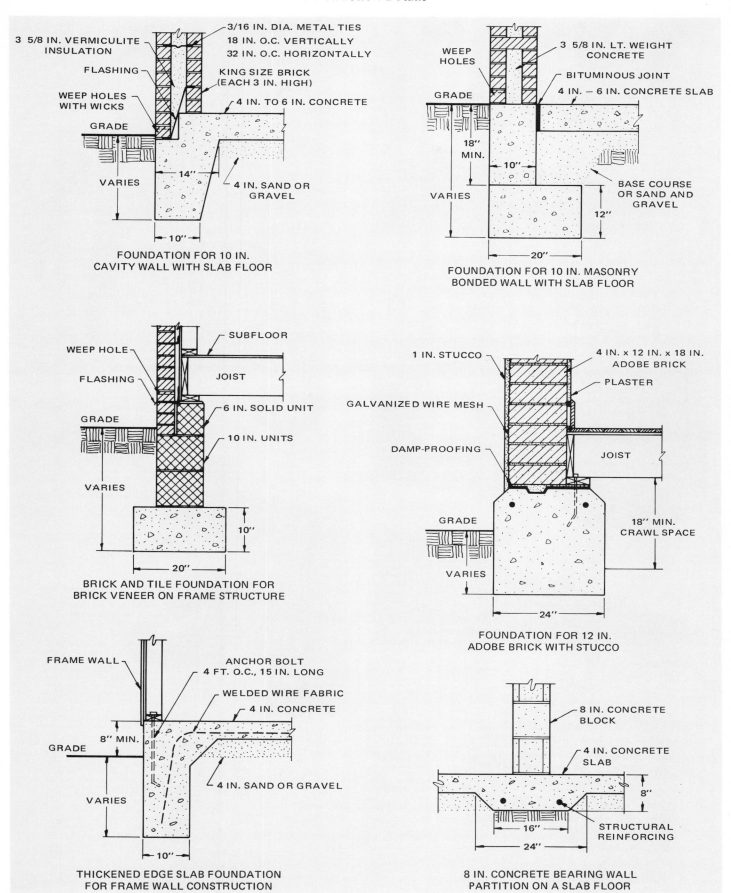

FOUNDATION FOR 10 IN.
CAVITY WALL WITH SLAB FLOOR

FOUNDATION FOR 10 IN. MASONRY
BONDED WALL WITH SLAB FLOOR

BRICK AND TILE FOUNDATION FOR
BRICK VENEER ON FRAME STRUCTURE

FOUNDATION FOR 12 IN.
ADOBE BRICK WITH STUCCO

THICKENED EDGE SLAB FOUNDATION
FOR FRAME WALL CONSTRUCTION

8 IN. CONCRETE BEARING WALL
PARTITION ON A SLAB FLOOR

Fig. 9-10. (Continued) Typical foundation details.

Fig. 9-11. Foundation wall may be capped with solid block 4 in. thick.

Solid blocks 4 in. thick, Fig. 9-11, are available to some areas. If stretcher blocks are used, a strip of metal lath is placed under the cores of the top course. The cores are then filled with concrete or mortar and troweled smooth, Fig. 9-12.

Fig. 9-13. The first coat of parging being troweled over the surface.

Fig. 9-12. When stretcher blocks are used in the final course of a foundation wall, the cores should be filled with concrete or mortar. (Portland Cement Assoc.)

DAMPPROOFING BASEMENT WALLS: The outside of masonry basement walls should be parged with a 1/2 in. thick coat of portland cement plaster or mortar. It should be applied in two coats, each 1/4 in. thick. The wall surface should be clean and damp but not soaked before applying the plaster. The first coat should be troweled firmly over the wall, Fig. 9-13. When this coat has partially hardened, it should be roughened with a scratcher, Fig. 9-14. This operation provides a good bond for the next coat. Keep the first coat moist and allow it to harden for a minimum of 24 hours before applying the second coat shown in Fig. 9-15. Cover the wall from the footing to 6 in. above the finished grade line. Form a cove, Fig. 9-16, over the footing to prevent water from collecting at

Fig. 9-14. A wire scratcher roughens the first plaster coat when it has partially hardened.

this point. The second coat should be moist cured for at least 48 hours.

Some areas of the country and certain soil conditions require additional dampproofing. A heavy coat of tar, two coats of a cement-based paint, or a covering of thin plastic film may be used. This is in addition to the parge coat.

Drain tiles of concrete, clay or plastic are generally placed around footings to carry away ground water. They are connected to a dry well or storm sewer.

Fig. 9-15. The second and final coat of portland cement plaster is troweled over the roughened surface of the first coat.

The concrete and clay tiles are usually 4 in. in diameter and are layed with open joints. The new plastic continuous tile have performed holes which allow the water to enter. All three types are laid in a bed of gravel a foot or more wide and deep. Refer to Fig. 9-10. It shows a drain tile in place along the footing.

COLUMNS, PIERS AND PILASTERS: These elements transmit loads to the footing. Columns and piers are freestanding but pilasters are built into the foundation wall.

Brick or block columns are vertical supports. They are frequently used to hold up wood, steel or masonry beams, Fig. 9-17. The bonding for a 12″ x 12″ column is shown in Fig. 9-18. Columns built with concrete block can be made using one to three blocks in each course, Fig. 9-19.

A pier is an isolated section or column of masonry. It may

Fig. 9-17. This brick column is supporting a reinforced brick masonry beam. (Brick Institute of America)

Fig. 9-16. A cove of mortar is formed over the footing to prevent water from entering. (Portland Cement Assoc.)

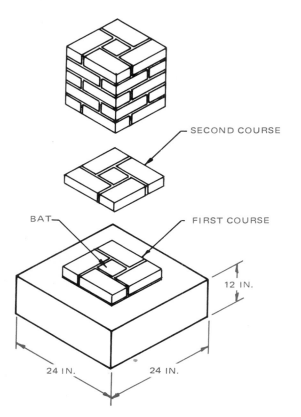

Fig. 9-18. Bonding for a 12" x 12" brick column. The bat is a piece of brick cut to fill.

be part of a bearing wall if it is not bonded to masonry at the sides and its length does not exceed four times its thickness. It may support openings in a wall.

Fig. 9-20 shows several sections of solid masonry piers. The size of piers, unless governed by a local code, is usually a minimum of 8" x 12" for solid units and 8" x 16" for hollow units. The height should not exceed 10 times their smallest cross-sectional dimension. Footings for piers should be at least 8 in. thick.

A *pilaster is a masonry column bonded to a wall.* It has a uniform cross section and serves as a column and/or vertical beam, Fig. 9-21. When pilasters are built with concrete block the hollow cells are sometimes filled with mortar or grout to increase their strength. Fig. 9-22 shows a pilaster being built with concrete blocks.

FOUNDATION ANCHORAGE: The building superstructure should be anchored to the foundation to resist high winds. Sill plates of wood joist floor systems are generally anchored to masonry walls with 1/2 in. bolts extending at least 15 in. into the filled cores of masonry units. Anchor bolts should be spaced not more than 8 ft. apart, with one bolt not more than 12 in. from each end of the sill plate.

Sill plates should be anchored to poured concrete walls with 1/2 in. anchor bolts embedded 6 in. with the same maximum spacing as used for masonry walls. Hardened steel studs, driven by power tools, may be used if permitted by local codes. At most, they should be spaced 4 ft. apart.

MASONRY WALL SYSTEMS

Masonry walls may be classified as:
1. Solid walls.
2. Hollow walls.
3. Veneered walls.
4. Composite walls.
5. Reinforced walls.

Masonry walls are very popular. They provide excellent structural performance, are easily maintained and attractive.

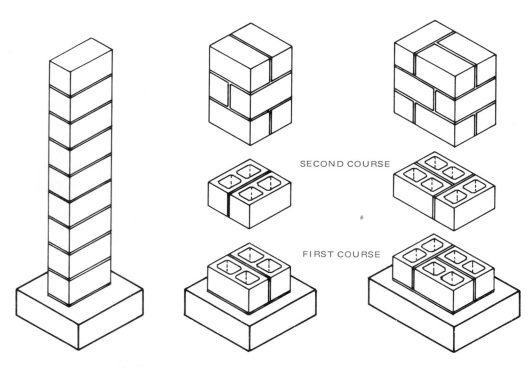

Fig. 9-19. Three popular styles of concrete block columns.

Construction Details

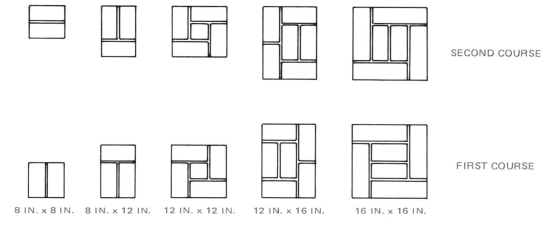

SECOND COURSE

FIRST COURSE

8 IN. x 8 IN.   8 IN. x 12 IN.   12 IN. x 12 IN.   12 IN. x 16 IN.   16 IN. x 16 IN.

Fig. 9-20. Bonding for a variety of sizes of solid brick masonry piers.

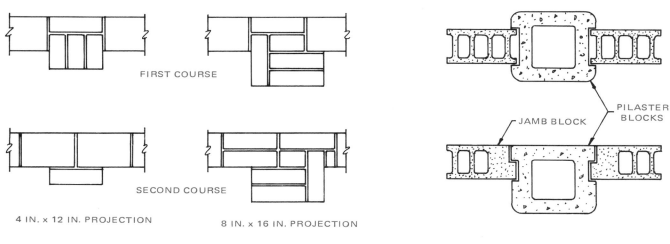

FIRST COURSE

SECOND COURSE

4 IN. x 12 IN. PROJECTION

8 IN. x 16 IN. PROJECTION

SOLID MASONRY PILASTERS

PILASTER BLOCKS

JAMB BLOCK

CONCRETE PILASTER BLOCKS

Fig. 9-21. Typical pilasters made from solid brick masonry and concrete pilaster block. Their purpose is to provide support for a wall. It is usually found at intervals in long wall sections.

Fig. 9-22. A pilaster being constructed from 6 and 8 in. concrete block. The 6 and 8 in. blocks are alternated each course. Left. Pilaster with first course and part of second course in place. Right. Partially completed. Can you see how, by alternating positions of 6 and 8 in. block, the pilaster is tied into the wall?

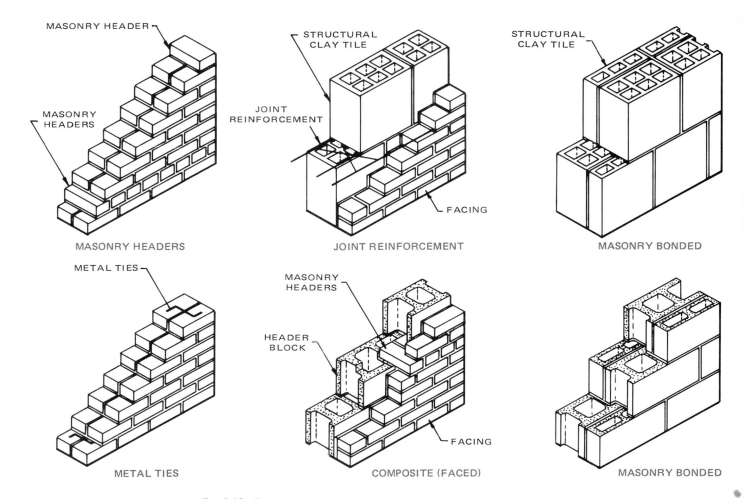

Fig. 9-23. Typical solid masonry walls using brick, concrete block and tile.

SOLID MASONRY WALLS

A solid masonry wall is built up of masonry units laid close together with all joints between them filled with mortar. Fig. 9-23 shows several types. Solid or hollow masonry units or a combination of these materials may be used. Such walls may be either load-bearing or non-load-bearing. The structural bond of these walls may be provided by metal ties, masonry headers or joint reinforcement.

Solid masonry walls may also be classified by the types of units used in their construction. Examples of this classification include:

1. Solid units of brick or concrete brick and block.
2. Hollow units of concrete block or structural clay tile.
3. Composite (faced) walls composed of facing and backup units of different materials bonded in such a way that both facing and backup are load-bearing. Fig. 9-23 shows how this wall is constructed. (See lower center illustration.) Header blocks allow masonry headers to overlap block.

Nationally recognized building codes permit the use of exterior load-bearing 6 in. masonry walls for one-story, single-family dwellings where the wall height does not exceed 9 ft. to the eaves and 15 ft. to the gable peak. Fig. 9-24 shows typical sections of 6 in. walls. The masonry units used for this type of construction are most commonly the "SCR brick." It has a nominal thickness of 2 2/3 in. which produces 16 in. in six courses. It is a nominal 6 in. wide and 12 in. long. This makes it easier to lay in 1/2 in. bond. This brick is also produced in thicknesses of 3 in. (four courses in 12 in.), 3 1/5 in. (five courses in 16 in.), 4 in. (four courses in 16 in.) and 5 1/3 in. (three courses in 16 in.).

Clay tile are also produced in 6 in. sizes which can be used with the SCR brick.

HOLLOW MASONRY WALLS

A hollow masonry wall is built using solid or hollow masonry units. The units are separated to form an inner and an outer wall. They may be either a cavity or masonry bonded type wall.

CAVITY WALLS

A cavity wall is built of masonry units arranged to provide a continuous air space of 2 to 3 in. wide, Fig. 9-25. The facing and backing wythes (a vertical section of masonry one unit thick) or tiers are connected with rigid metal ties. The exterior wythe is usually a nominal 4 in. thick. The interior wythe may be 4, 6 or 8 in., depending on load to be supported and the

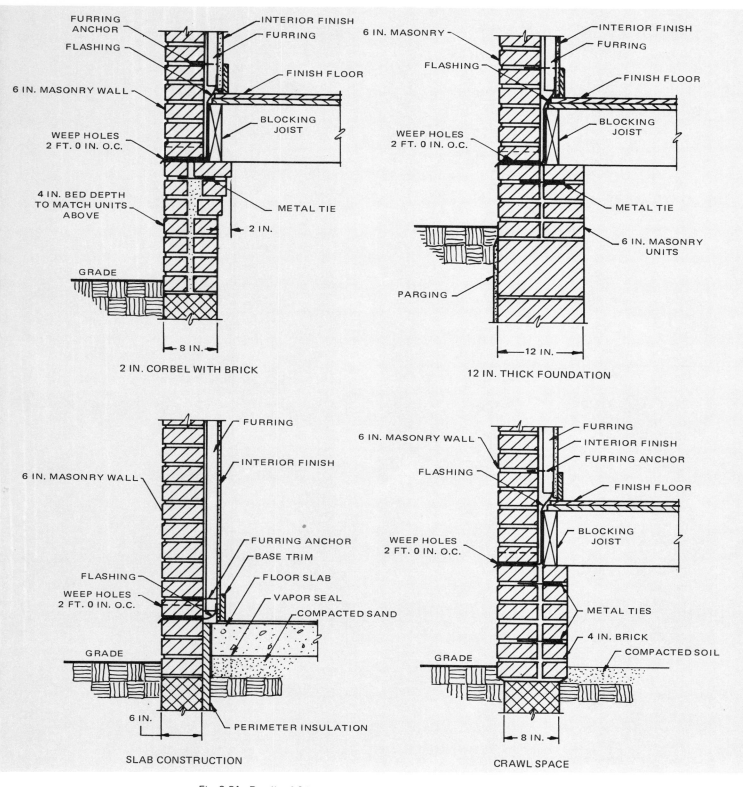

Fig. 9-24. Details of 6 in. masonry wall construction using "SCR brick."

height and length of wall. When a cavity of nominal 2 in. is maintained, the overall thickness will be 10, 12 or 14 in.

The cavity provides two advantages:

1. Air space has insulation value or it may be filled with insulation material for added reduction of heat transfer.
2. It acts as a barrier to moisture.

However, to be effective, the cavity must be kept free of mortar droppings during construction. A board may be used to collect the droppings as shown in Fig. 9-26. When weepholes, Fig. 9-27, are required at the bottom of a cavity wall, flashing should be used. It keeps any moisture which might collect in the cavity away from the inner wall, Fig. 9-28.

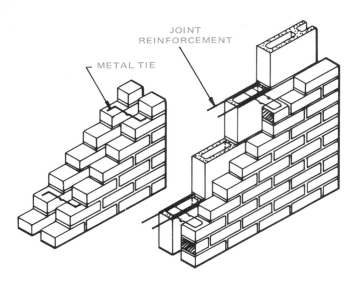

Fig. 9-25. Method of cavity wall construction which uses metal ties and joint reinforcement.

Fig. 9-27. Weep holes provide an outlet for moisture which has collected in the space between wythes of masonry.

Fig. 9-26. A recommended method of keeping the cavity free of mortar droppings is to use a board resting on the joint reinforcement (wire ties). The board is moved up when the next reinforcement level is reached. (Portland Cement Assoc.)

Fig. 9-28. Flashing should be used to prevent moisture from entering the inner wall from the cavity. (Portland Cement Assoc.)

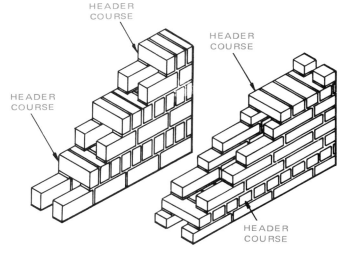

Fig. 9-29. Typical hollow masonry bonded walls.

HOLLOW MASONRY BONDED WALLS

Hollow masonry bonded walls are popular in some parts of the country. They are used for foundation and exterior load-bearing walls. Though economical they are not resistant to high moisture. See Fig. 9-29 for two examples of hollow masonry bonded walls.

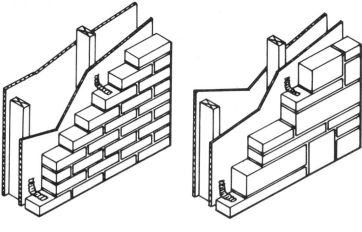

BRICK VENEER ON FRAME   STONE VENEER ON FRAME

Fig. 9-30. Brick and stone veneer attached to frame construction using metal ties. These veneers must rest on an extension of the foundation.

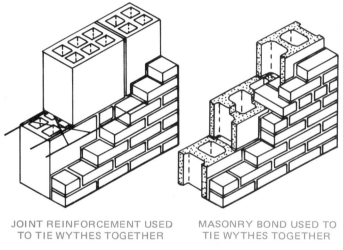

JOINT REINFORCEMENT USED TO TIE WYTHES TOGETHER   MASONRY BOND USED TO TIE WYTHES TOGETHER

Fig. 9-31. Composite walls may be bonded together with masonry or joint reinforcement.

VENEERED WALLS

Masonry units are widely used as a facing veneer, Fig. 9-30. In this application, load-bearing properties of the materials are not used. The veneer is attached to the backing, but does not act structurally with the rest of the wall.

Corrosion-resistant metal ties are used to attach veneer to frame or concrete block construction. Most codes require one tie for every 2 sq. ft. of area covered.

A paperbacked welded wire mesh, which is attached directly to the studding and grouted, is also sometimes used to attach veneer to frame construction. This technique eliminates the need for sheathing.

Veneer wall construction looks much like cavity wall construction in its final form. However construction and function are considerable different. For example, the supporting structure is completed before attaching the veneer

COLLAR JOINT BACK PARGED WITH MORTAR

Fig. 9-32. The outer and inner wythes of masonry are bonded together by the mortar in the collar joint between them. (Portland Cement Assoc.)

which is not load-bearing. Both wythes in a cavity wall are load-bearing and are built at the same time.

COMPOSITE WALLS

A composite wall is two wythes bonded together with masonry or wire ties, Fig. 9-31. The two wythes are joined together in a continuous mass using a vertical collar joint, Fig. 9-32. This joint prevents the passage of water through the wall. (A collar joint is the narrow space between the facing units and the back-up units in a wall.)

When building a 8 in. composite wall, the first course of facing may be either headers as shown in Fig. 9-33 or stretchers. It is important that all facing courses be laid in a

Fig. 9-33. Building a composite wall with six courses of brick between each header course. (Portland Cement Assoc.)

full mortar bed with the head joints filled completely. Mortar which is extruded (squeezed out) on the back side of the facing units, Fig. 9-34, should be cut flush with the trowel

Fig. 9-34. Well placed facing units will have extruded mortar on the back side in a composite wall. This must be cut flush with the trowel.

before it hardens. Parging the back side of the facing is another method of bonding the wythes across the collar joint, Fig. 9-35. Facing headers are laid every seventh course in an 8 in. composite wall.

Fig. 9-35. Bonding a wall across a collar joint by parging the back side of the facing. (Portland Cement Assoc.)

A 12 in. composite wall is constructed in a manner similar to the 8 in. composite wall. Fig. 9-36 shows the facing header course being laid overlapping the header block. The header block may be laid with the recessed notch up or down, depending on construction requirements. Fig. 9-37 shows the seventh course backup wythe being constructed using stretcher block. Header blocks were used for the sixth course bonding.

Fig. 9-36. Header course being laid in composite wall.

Fig. 9-37. Stretcher block being placed as the backup wythe. (Portland Cement Assoc.)

REINFORCED WALLS

Reinforced masonry walls are built with steel reinforcement embedded with the masonry units, Fig. 9-38. The walls are structurally bonded by grout which is poured into the cavity (collar joint) between the wythes of masonry. The grout core seals the space between the wythes of masonry and bonds the

reinforcing steel. Full bed joints are used for reinforced walls.

Grouting techniques vary in different areas of the country, but the *Uniform Building Code* specifies that: "All longitudinal vertical joints shall be grouted and shall be not less than three-fourths inch (3/4 in.) in thickness. In members (walls or columns) of three or more tiers in thickness, interior bricks shall be not less than three-fourths inch (3/4 in.) in

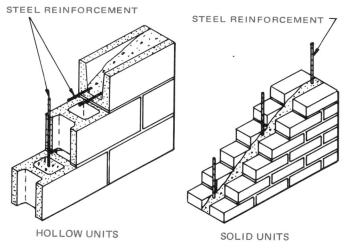

STEEL REINFORCEMENT   STEEL REINFORCEMENT

HOLLOW UNITS   SOLID UNITS

Fig. 9-38. Reinforced masonry walls of brick and concrete block.

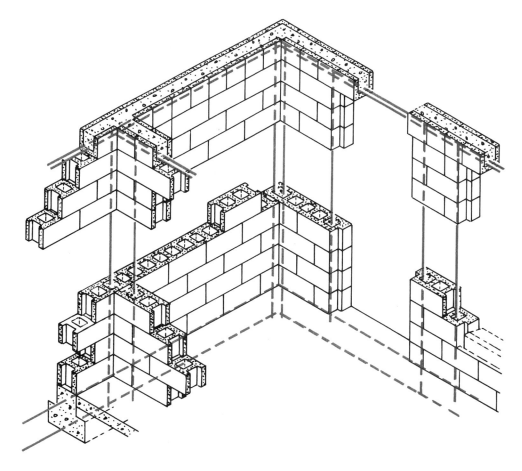

Fig. 9-39. Typical locations of reinforcement for concrete block masonry construction.

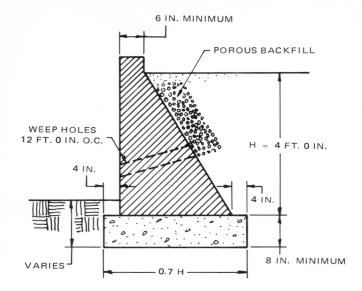

Fig. 9-40. Gravity type retaining wall with specifications. The width of the base must be at least one-half to three-fourths of the wall height. This is expressed in the drawings as "0.7H" or "0.7 of the height."

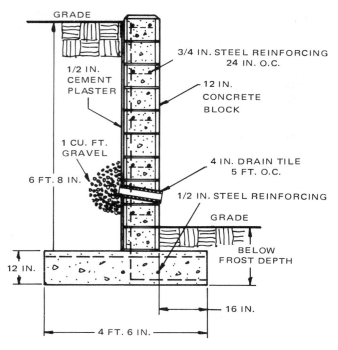

Fig. 9-41. Cantilever type retaining wall constructed of reinforced concrete block.

thickness. In members of three or more tiers in thickness, interior bricks shall be embedded into the grout so that at least three-fourths inch (3/4 in.) of grout surrounds the sides and end of each unit. One exterior tier may be carried up twelve inches (12 in.) before grouting, but the other exterior tier shall be grouted in lifts not to exceed four inches (4 in.) or one unit, whichever is greater. If the work is stopped for one hour or longer, the horizontal construction joints shall be formed by stopping all tiers at the same elevation and with the grout one inch (1 in.) below the top."

Reinforced masonry walls should be reinforced with an area of steel not less than 0.002 times the cross-sectional area of the wall. Not more than 2/3 of this area may be used in either direction. Maximum spacing of principal reinforcement should not exceed 48 in. Horizontal reinforcement should be placed in the top of footings, at the bottom and top of wall openings, at roof and floor levels and at the top of parapet walls.

The primary use of steel reinforcement is in vertical members, (such as columns and walls) and for lintels and bond beams. Typical methods of reinforcing concrete masonry are shown in Fig. 9-39.

RETAINING WALLS

Reinforced masonry is ideal for the construction of retaining walls which must stand lateral earth pressures. By adding reinforcement, the mass may be greatly reduced and the strength maintained.

Two common types of retaining walls are the GRAVITY and CANTILEVER designs. A gravity type, Fig. 9-40, depends

8" WALLS					
HEIGHT OF WALL	WIDTH OF FOOTING	THICKNESS OF FOOTING	DIST. TO FACE OF WALL	SIZE & SPACING OF VERTICAL RODS IN WALL	SIZE & SPACING OF HORIZONTAL RODS IN FOOTING
3'–4"	2'–4"	9"	8"	3/8" @ 32"	3/8" @ 27"
4'–0"	2'–9"	9"	10"	1/2" @ 32"	3/8" @ 27"
4'–8"	3'–3"	10"	12"	5/8" @ 32"	3/8" @ 27"
5'–4"	3'–8"	10"	14"	1/2" @ 16"	1/2" @ 30"
6'–0"	4'–2"	12"	15"	3/4" @ 24"	1/2" @ 25"
12" WALLS					
6'–8"	4'–6"	12"	16"	3/4" @ 24"	1/2" @ 22"
7'–4"	4'–10"	12"	18"	7/8" @ 32"	5/8" @ 26"
8'–0"	5'–4"	12"	20"	7/8" @ 24"	5/8" @ 21"
8'–8"	5'–10"	14"	22"	7/8" @ 16"	3/4" @ 26"
9'–4"	6'–4"	14"	24"	1" @ 8"	3/4" @ 21"

Fig. 9-42. Specifications for reinforced concrete masonry retaining walls up to 9' – 4" high.

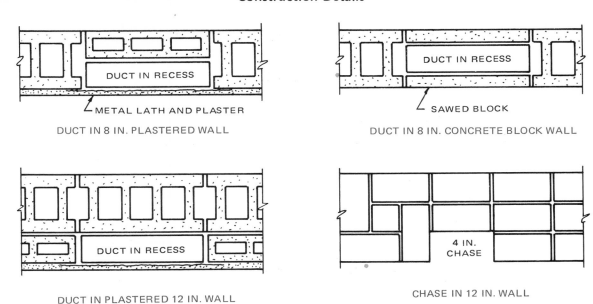

Fig. 9-43. Typical chases and recesses in masonry walls.

primarily on its own weight to hold back the earth pressure.

A cantilever type, Fig. 9-41, uses the weight of the soil together with its own strength to get the same results. For walls over 3 ft. in height, the cantilever type retaining wall provides the best solution. The same principles of good construction should be followed in building a retaining wall as any other reinforced masonry wall.

Fig. 9-42 provides specifications for building reinforced concrete masonry retaining walls up to 9 ft. — 4 in. in height.

WALL OPENINGS

Wall openings may extend through the wall as required for a door or window, or they may take the form of a chase or recess. *Chases and recesses are horizontal or vertical spaces left in a wall for the purpose of containing plumbing, heating ducts, electrical wiring or other equipment.* Fig. 9-43 shows several chases and recesses in masonry walls.

Chases and recesses are formed by the mason as the wall is built. Chases are generally located on the inside of the wall and vary in size from 4 in. to 12 in. in width. Recesses reduce the wall thickness and strength and are usually limited to one-third the thickness of the wall.

LINTELS AND ARCHES

Masonry above a wall opening should be supported by a lintel made from metal, reinforced concrete or masonry or by an arch. Steel lintels should be supported on either side of the opening for a distance of at least 4 in. Reinforced concrete lintels should have a minimum bearing of at least 8 in. at each end. Longer lintels carrying heavier loads should have greater bearing arches at the ends. The lintel should be stiff enough to resist bending in excess of 1/360th of the span. Fig. 9-44 shows several types of lintels used in masonry construction. Specifications for one-piece reinforced concrete lintels are given in Fig. 9-45. Design data for reinforced concrete split

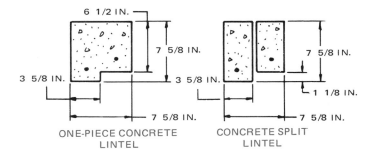

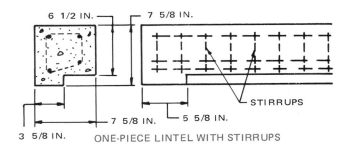

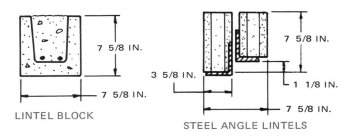

Fig. 9-44. Various types of lintels used in masonry construction.

lintels are shown in Fig. 9-46. Fig. 9-47 gives the size requirements for concrete reinforced lintels with stirrups. These lintels are made for openings supporting wall and floor

| CONCRETE REINFORCED ONE-PIECE LINTELS WITH WALL LOAD ONLY | | | | |
| SIZE OF LINTEL | | | BOTTOM REINFORCEMENT | |
HEIGHT IN.	WIDTH IN.	CLEAR SPAN OF LINTEL FT.	SIZE DESIGNATION OF BARS	SIZE OF BARS
5 3/4	7 5/8	UP TO 7	NO. 2	3/8-IN. ROUND DEFORMED
5 3/4	7 5/8	7 TO 8	NO. 2	5/8-IN. ROUND DEFORMED
7 5/8	7 5/8	UP TO 8	NO. 2	3/8-IN. ROUND DEFORMED
7 5/8	7 5/8	8 TO 9	NO. 2	1/2-IN. ROUND DEFORMED
7 5/8	7 5/8	9 TO 10	NO. 2	5/8-IN. ROUND DEFORMED

Fig. 9-45. Specifications for one-piece reinforced concrete lintels. (Portland Cement Assoc.)

| CONCRETE REINFORCED SPLIT LINTELS WITH WALL LOAD ONLY | | | | |
| SIZE OF LINTEL | | | BOTTOM REINFORCEMENT | |
HEIGHT IN.	WIDTH IN.	CLEAR SPAN OF LINTEL FT.	SIZE DESIGNATION OF BARS	SIZE OF BARS
5 3/4	3 5/8	UP TO 7	NO. 1	3/8-IN. ROUND DEFORMED
5 3/4	3 5/8	7 TO 8	NO. 1	5/8-IN. ROUND DEFORMED
7 5/8	3 5/8	UP TO 8	NO. 1	3/8-IN. ROUND DEFORMED
7 5/8	3 5/8	8 TO 9	NO. 1	1/2-IN. ROUND DEFORMED
7 5/8	3 5/8	9 TO 10	NO. 1	5/8-IN. ROUND DEFORMED

Fig. 9-46. Specifications for reinforced concrete split lintels. (Portland Cement Assoc.)

| CONCRETE REINFORCED LINTELS WITH STIRRUPS FOR WALL AND FLOOR LOADS | | | | | |
| SIZE OF LINTEL | | | REINFORCEMENT | | WEB REINFORCEMENT NO. 6 GAGE WIRE STIRRUPS. SPACINGS FROM END OF LINTEL — BOTH ENDS THE SAME |
HEIGHT IN.	WIDTH IN.	CLEAR SPAN OF LINTEL FT.	TOP	BOTTOM	
7 5/8	7 5/8	3	NONE	2 — 1/2-IN. ROUND	NO STIRRUPS REQUIRED
7 5/8	7 5/8	4	NONE	2 — 3/4-IN. ROUND	3 STIRRUPS, SP.: 2,3,3 IN.
7 5/8	7 5/8	5	2 — 3/8-IN. ROUND	2 — 7/8-IN. ROUND	5 STIRRUPS, SP.: 2,3,3,3,3 IN.
7 5/8	7 5/8	6	2 — 1/2-IN. ROUND	2 — 7/8-IN. ROUND	6 STIRRUPS, SP.: 2,3,3,3,3,3 IN.
7 5/8	7 5/8	7	2 — 1-IN. ROUND	2 — 1-IN. ROUND	9 STIRRUPS, SP.: 2,3,3,3,3,3,3,3,3 IN.

SOURCE: PORTLAND CEMENT ASSOCIATION

Fig. 9-47. Specifications for reinforced concrete lintels with stirrups.

loads. Lintels made of angle steel should meet the following minimum specifications for 4 in. masonry veneer:

STEEL ANGLES TO SUPPORT 4 IN. MASONRY WALLS

SPAN	SIZE OF ANGLE
0' — 5'	3" x 3" x 1/4"
5' — 9'	3 1/2" x 3 1/2" x 5/16"
9' — 10'	4" x 4" x 5/16"
10' — 11'	4" x 4" x 3/8"
11' — 15'	6" x 4" x 3/8"
15' — 16'	6" x 4" x 1/2"

Reinforced masonry lintels (namely brick and tile) are becoming more popular because the steel is completely protected from the elements and the initial cost is less because less steel is required. Fig. 9-48 shows several sizes of reinforced brick masonry lintels.

Slight movement often occurs at the location of lintels. For this reason, control joints are often located at the ends of lintels. A noncorroding metal plate is placed under the ends of

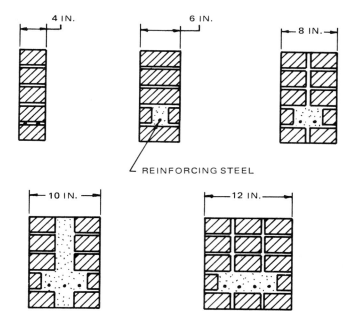

Fig. 9-48. Reinforced brick masonry lintels.

lintels where control joints occur, Fig. 9-49. The metal plate permits the lintel to slip and prevent uncontrolled cracking.

A full bed of mortar should be used over the plate to distribute the lintel load uniformly. When the mortar has hardened sufficiently, it should be raked out to a depth of 3/4 in. and filled with caulking.

Fig. 9-49. A noncorroding metal plate is placed under the ends of lintels where control joints occur. A full bed of mortar will be placed over the plate. (Portland Cement Assoc.)

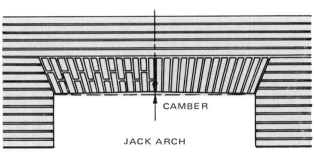

CAMBER

JACK ARCH

CAMBER = 1/8 IN. PER FOOT OF SPAN

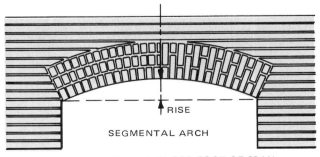

RISE

SEGMENTAL ARCH

MINIMUM RISE = 1 IN. PER FOOT OF SPAN

Fig. 9-50. The jack and segmental are frequently used minor arches.

Arches have been used for centuries to span openings. Some have been built which span distances over 130 ft. These large arches are classed as MAJOR ARCHES. MINOR ARCHES are those whose spans do not exceed 6 ft. with maximum rise-to-span ratio of 0.15. We will discuss only minor arches.

The FLAT or JACK arch and the SEGMENTAL arch are generally used for minor arch construction, Fig. 9-50.

Arches are constructed with the aid of temporary shoring or CENTERING. The centering carries the load of the arch until it has developed sufficient strength to support itself and imposed load. *The centering should not be removed for at least seven days after the arch has been completed.*

Two general methods are used in the construction of clay masonry arches. One involves the use of special shapes which provides for uniform joint thickness. The second method uses regular units of uniform thickness while the joint thickness is varied to get the desired curvature. The arch dimensions and appearance desired will determine the method to be used.

It is important that all mortar joints be completely filled because the arch is a structural member. The crown of the arch, Fig. 9-51, is generally laid in a soldier bond or rowlock header bond and it is frequently difficult to completely fill the joints. Care must be taken.

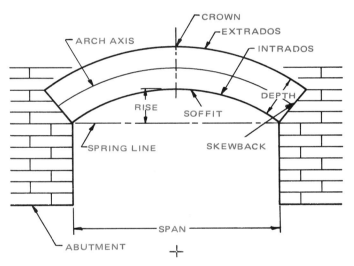

Fig. 9-51. Arch terminology.

JACK ARCH: The jack arch is relatively weak and should be supported by steel if the opening is over 2 ft. wide. The steel must be bent to the camber which is generally 1/8 in. per foot of span. For best appearance, each joint should be the same width the entire length of the joint. Special brick are required or regular brick must be cut to have a perfect job. The brick should be shaped so that the end joints are horizontal rather than perpendicular to the radius of the arch, Fig. 9-52. When constructing any arch, the brick should be laid out to determine the number required, inclination and spacing.

SEGMENTAL ARCH: The rise and span are the most important features to be considered when constructing a segmental arch. The length of the arch must be found in order to determine the number of courses and size of brick needed. The number of courses of brick in an arch is determined by finding the length of the extrados. *The extrados is the upper portion of the arch which is between the crown and the skewback.* See Fig. 9-51. The size of the brick may be

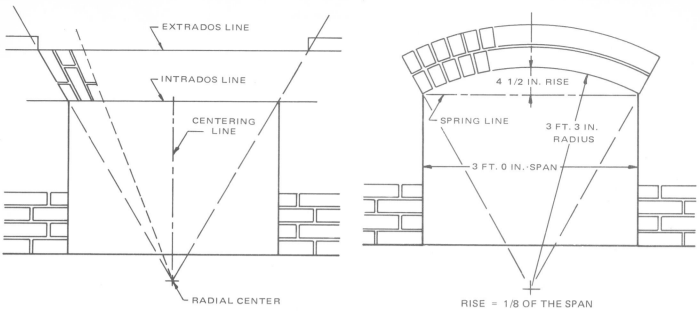

Fig. 9-52. Construction layout for a jack arch.

Fig. 9-53. Construction layout for segmental arch.

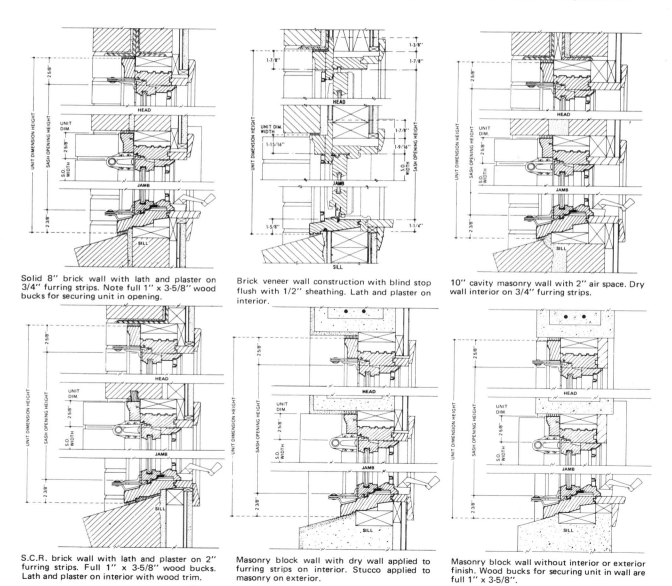

Solid 8″ brick wall with lath and plaster on 3/4″ furring strips. Note full 1″ x 3-5/8″ wood bucks for securing unit in opening.

Brick veneer wall construction with blind stop flush with 1/2″ sheathing. Lath and plaster on interior.

10″ cavity masonry wall with 2″ air space. Dry wall interior on 3/4″ furring strips.

S.C.R. brick wall with lath and plaster on 2″ furring strips. Full 1″ x 3-5/8″ wood bucks. Lath and plaster on interior with wood trim.

Masonry block wall with dry wall applied to furring strips on interior. Stucco applied to masonry on exterior.

Masonry block wall without interior or exterior finish. Wood bucks for securing unit in wall are full 1″ x 3-5/8″.

Fig. 9-54. Window details for masonry construction. (Anderson Corp.)

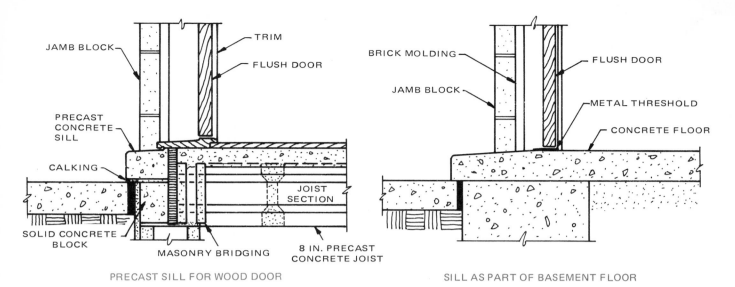

Fig. 9-55. Typical door sill construction in masonry walls.

determined by finding the length of the intrados. *The intrados is the lower surface of the arch.* The rise of a segmental arch should be 1/6, 1/8, 1/10 or 1/12 of the span. Fig. 9-53 shows the key dimensions of one segmental arch example.

WINDOW AND DOOR DETAILS

There are many types of windows and doors which could be used in a structure. Details of the windows and doors being used in a given situation should be provided so the mason can plan ahead. Details shown in Figs. 9-54 and 9-55 are rather typical and illustrate the relationship of the *head, jambs* and *sill* to the masonry wall. *The head is the top of the window or door. The jambs are the vertical sides and the sill is the bottom piece connecting the jambs.*

Sills are usually classified as SLIP SILLS or LUG SILLS, Fig. 9-56. Slip sills are the same length as the window opening, but lug sills extend into the masonry on either side of the opening. The slip sill may be left out when the opening is being laid and set at a later time. The lug sill, however, must be set when the masonry is up to the bottom of the window or door opening. Sills are usually made of cut stone or brick.

FLOORS, PAVEMENTS AND STEPS

Because of color, pattern and high resistance to abrasion, brick, tile and stone are very desirable surfaces for roadways, walks, patios and interior floors. These materials can be used over the ground, over a floor or they can be incorporated within a structural floor system.

Several factors affect the selection of brick and jointing materials. They include traffic, exposure to moisture and freezing and appearance desired. Hard-burned brick are very resistant to wear and a Grade SW will prove satisfactory for residential and nonindustrial floors. For exterior pavements, brick's resistance to freezing and thawing damage is just as important as its resistance to abrasion. Brick having an average saturation coefficient of 0.78 or less are recommended for

severe weathering conditions. (ASTM Specifications C-216 and C-62, Grade SW.)

Finished appearance of the surface depends largely on the color, size of units and pattern in which they are laid. When brick are laid flat with the largest plane surface horizontal,

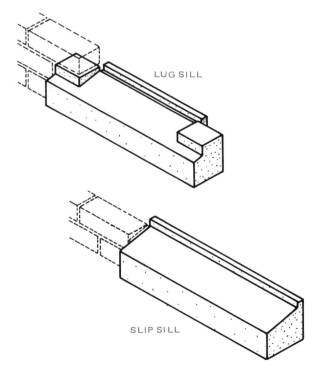

Fig. 9-56. How lug and slip sills of stone or cast concrete fit into a brick wall.

noncored brick are generally used. Cored brick can be used if they are placed on edge. Figs. 9-57 and 9-58 show several pattern bonds which may be used for floors and pavements.

Interior masonry floors are usually laid over concrete slabs,

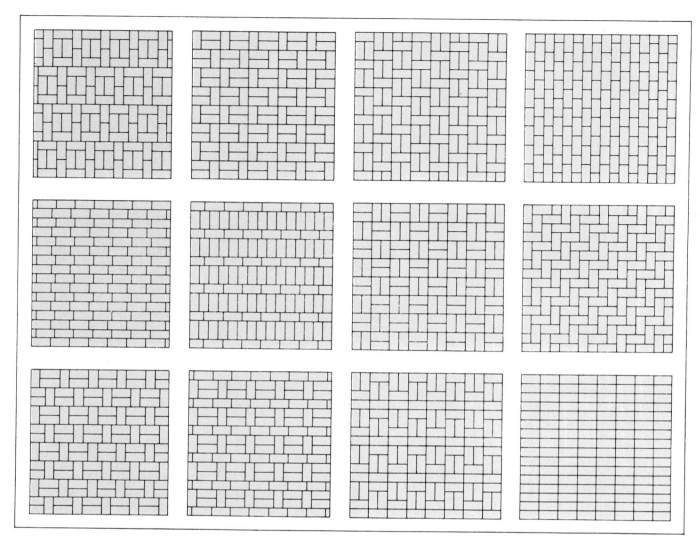

Fig. 9-57. Brick pattern bonds commonly used for floors and pavements.

Fig. 9-59. The primary elements of an unreinforced masonry floor are:

1. The base, which provides the support.
2. The cushion, an intermediate layer which aids in leveling.
3. The wearing surface of brick, tile or stone.

4. The joints.

Floors or pavements may be laid without mortar joints, but most interior applications have mortar joints. The procedure for laying a masonry floor on a slab is simple. First, decide on a pattern and arrange the materials for easy access. Form a

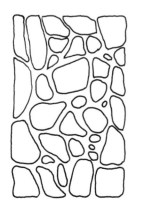
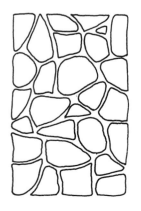

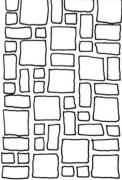

IRREGULAR (NOT FITTED)   IRREGULAR (FITTED)   SEMI-IRREGULAR   RECTANGULAR

Fig. 9-58. Stone pattern bonds frequently used for floors and pavements.

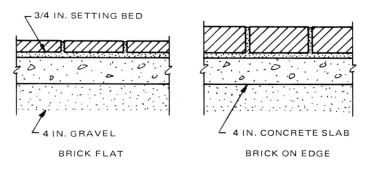

Fig. 9-59. Sections of floors with brick laid over a concrete slab.

cushion of mortar for each unit as it is laid and leveled. Strike the joints and brush away excess mortar with burlap rags as it begins to harden. The joints may be tooled for maximum hardness when the mortar has sufficiently hardened.

Exterior pavings for walks, patios and drives are similar to interior floors. However, more care must be taken in selecting materials which will withstand the weather. Also, drainage must be considered. Walks should be sloped to one or both sides to shed surface water, Fig. 9-60. Slope patios away from

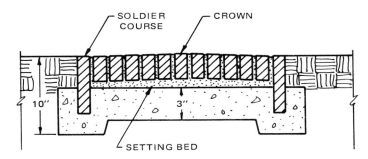

Fig. 9-60. Brick sidewalk on a concrete foundation, sloped to both sides.

buildings, retaining walls or other obstructions which will hold water on the surface. The slope is generally 1/8 to 1/4 in. per foot for walks and patios.

Below-surface drainage may also be required for large paved areas or locations with a high water table. A layer of gravel under the slab, or masonry units if no slab is used, will usually provide the necessary drainage and stop the upward capillary flow of moisture. An expansion joint should be installed between the slab and building walls to provide for expansion.

Pavings tend to spread or shift at the edges under normal traffic and weather conditions. Therefore, an edging is desirable. Fig. 9-61 shows a soldier course of brick used as an edging. The edging is generally placed first and can be used as a guide for elevation and slope.

Several companies produce PAVER brick which are 1 1/4" x 3 5/8" x 7 5/8". They are approximately 1 in. thinner than regular brick and are designed especially for paving.

STEPS

Steps made of brick, block or stone offer a practical solution to a sloping terrain. They also give a particular appearance in an interior setting. A good approach to building

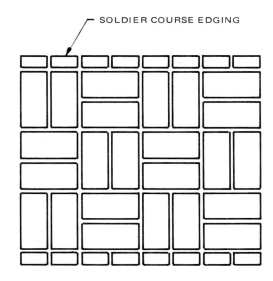

Fig. 9-61. This sidewalk uses a soldier course of brick as an edging.

steps or stairs with masonry units is to pour a concrete foundation for the structure and cap the treads and risers with the material selected. Fig. 9-62 shows a typical set of steps capped with brick.

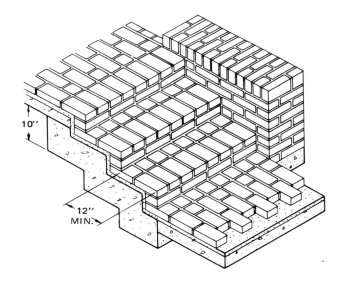

Fig. 9-62. Typical brick steps laid over a concrete foundation.

In designing the steps, use care to insure that each tread is the same size and each riser is identical. Any change in height of riser or depth of tread will invite an accident. Treads for all outside steps are usually 12 inches. A riser height equal to two regular brick plus mortar joints is an appropriate height. Treads should be sloped about 1/4 in. to the front edge to aid in water drainage.

FIREPLACE AND CHIMNEY CONSTRUCTION

The design and construction of fireplaces and chimneys has a long history. Much data has been collected regarding the most efficient designs. It is wise to follow recommended practices and procedures if the finished product is to perform satisfactorily.

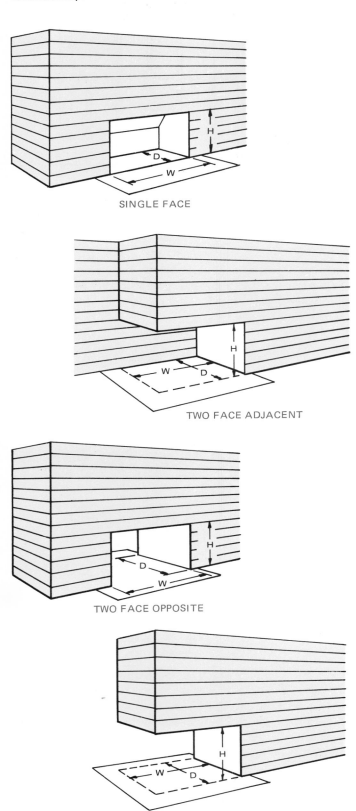

SINGLE FACE

TWO FACE ADJACENT

TWO FACE OPPOSITE

THREE FACE

Fig. 9-63. Basic fireplace designs.

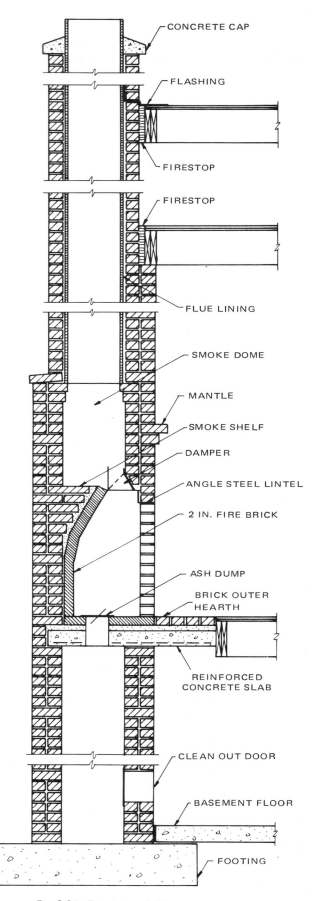

CONCRETE CAP

FLASHING

FIRESTOP

FIRESTOP

FLUE LINING

SMOKE DOME

MANTLE

SMOKE SHELF

DAMPER

ANGLE STEEL LINTEL

2 IN. FIRE BRICK

ASH DUMP

BRICK OUTER HEARTH

REINFORCED CONCRETE SLAB

CLEAN OUT DOOR

BASEMENT FLOOR

FOOTING

Fig. 9-64. Fireplace and chimey components.

FIREPLACE DIMENSIONS			
FIREPLACE TYPE	OPENING HEIGHT H IN IN.	HEARTH SIZE W BY D, IN.	MODULAR FLUE SIZE, IN.
SINGLE FACE	29	30 BY 16	12 BY 12
	29	36 BY 16	12 BY 12
	29	40 BY 16	12 BY 16
	32	48 BY 18	16 BY 16
TWO FACE — ADJACENT	26	32 BY 16	12 BY 16
	29	40 BY 16	16 BY 16
	29	48 BY 20	16 BY 16
TWO FACE — OPPOSITE	29	32 BY 28	16 BY 16
	29	36 BY 28	16 BY 20
	29	40 BY 28	16 BY 20
THREE FACE	27	36 BY 32	20 BY 20
	27	36 BY 36	20 BY 20
	27	44 BY 40	20 BY 20

Fig. 9-65. Specifications for fireplace opening height, hearth size and flue size.

DESIGN AND CONSTRUCTION

There are several basic designs of fireplaces which are currently in use. They include:
1. The single face.
2. Two face opposite.
3. The two face adjacent.
4. Three face.

Fig. 9-63 shows these four fireplaces. Other fireplaces with exotic shapes and designs are also possible, but they require special engineering to function properly and they are not within the scope of this text. All fireplaces, regardless of their design, have basically the same component parts. Fig. 9-64 illustrates these components.

SIZES OF FIREPLACES

The size of a fireplace is important from an aesthetic (how it looks) as well as functional (how it works) point of view. If it is too small, it may work well, but not provide enough heat. If too large, it would be too hot for the room. Also, it would require a large chimney and an abnormal amount of air to supply its needs for combustion (burning). Research has shown that a room with 300 sq. ft. of floor area can be adequately served by a fireplace with an opening 30 to 36 in. wide.

Fireplace opening sizes, hearth dimensions and flue sizes are shown in Fig. 9-65. Fireplace openings are seldom greater than 32 in. The dimensions shown in Fig. 9-65 may be changed slightly to meet brick courses and joints if necessary, but should not be changed significantly.

COMBUSTION CHAMBER

The shape and depth of the combustion chamber affect both the draft and the amount of heat radiated into the room, Fig. 9-65. The slope of the back forces the flame forward and forces the gases with increasing velocity into the throat. The combustion chamber should be lined with fire brick which are laid with thin joints of fire clay mortar. The back and end walls of an average size fireplace are generally 8 in. thick. Large sizes will have thicker walls to support the load above.

STEEL DAMPERS					
WIDTH OF FIREPLACE IN INCHES	DAMPER DIMENSIONS IN INCHES				
	A	B	C	D	E
24 TO 26	28 1/4	26 3/4	13	24	9 1/2
27 TO 30	32 1/4	30 3/4	13	28	9 1/2
31 TO 34	36 1/4	34 3/4	13	32	9 1/2
35 TO 38	40 1/4	38 3/4	13	36	9 1/2
39 TO 42	44 1/4	42 3/4	13	40	9 1/2
43 TO 46	48 1/4	46 3/4	13	44	9 1/2
47 TO 50	52 1/4	50 3/4	13	48	9 1/2
51 TO 54	56 1/4	54 3/4	13	52	9 1/2
57 TO 60	62 1/2	60 3/4	13	58	9 1/2

CAST IRON DAMPERS					
WIDTH OF FIREPLACE IN INCHES	DAMPER DIMENSIONS IN INCHES				
	A	B	C	D	E
24 TO 26	28	21	13 1/2	24	10
27 TO 30	34	26 3/4	13 1/2	30	10
31 TO 34	37	29 3/4	13 1/2	33	10
35 TO 38	40	32 3/4	13 1/2	36	10
39 TO 42	46	38 3/4	13 1/2	48	10
43 TO 46	52	44 3/4	13 1/2	48	10
47 TO 50	57 1/2	50 1/2	13 1/2	54	10
51 TO 54	64	56 1/2	14 1/2	60	11 1/2
57 TO 60	76	58	14 1/2	72	11 1/2

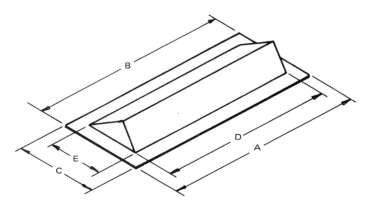

Fig. 9-66. Steel and cast iron damper sizes for fireplace construction. (Donley Brothers Co.)

THROAT

The throat is a critical spot in the fireplace and should not be less than 6 in. and preferably 8 in. above the highest point of the fireplace opening. The sloping back is the same height and supports the back of the damper.

The metal damper is placed in the throat and extends the full width of the fireplace opening. The best designs open upward toward the back and form a barrier to stop downdrafts. Typical damper sizes are shown in Fig. 9-66.

SMOKE SHELF AND CHAMBER

The smoke shelf is located in the throat. This shelf should be directly under the bottom of the flue. It extends the full width of the throat and is constructed horizontally. The smoke chamber is the space above the smoke shelf. The back and front walls of the chamber are vertical with the side walls sloping to the bottom of the flue lining.

Metal lining plates may be purchased for the smoke chamber which give it a proper form with smooth surfaces and simplify the bricklaying.

FIREPLACE FLUE

Each fireplace should have its own clay flue which is free of openings and connections. The first section of lining must start at the center line of the fireplace opening. If the lining is offset to one side, the draft will not be uniform over the full width of the fireplace. The lining should be supported on at least three sides by a ledge of projecting brick which is flush with the inside of the lining. Flue lining sizes are given in Fig. 9-65.

HEARTH

The modern hearth is usually constructed by building a reinforced brick masonry cantilevered slab, Fig. 9-64, or a reinforced concrete slab. The simplest method of constructing this slab is to cover the entire area of the chimney base the width of the hearth. Openings for the ash chute and basement flue (if required) are formed in the hearth.

CHIMNEY CONSTRUCTION

Masonry chimneys should be supported on foundations of masonry or reinforced concrete. The foundation wall should be a minimum of 8 in. thick. The footing for a chimney is generally 12 in. thick and extends 12 in. beyond the outside dimensions of the foundation. Reinforcing in the footing is recommended.

The chimney should not change in size or shape as it passes through the roof. Changes must not be made within 6 in. above or below the roof joists or rafters.

Chimney walls may not be less than 4 in. thick for residential structures and lined with approved fire clay flue liners not less than 5/8 in. thick. The flue liners should be installed ahead of construction of the chimney as it is laid up. The joints must be close fitting and smooth on the inside. In masonry chimneys with walls less than 8 in. thick, liners are separate from the chimney wall. The space between the liner and masonry is not filled. Liners should extend from the throat of the fireplace to the top of the chimney as nearly straight up as possible.

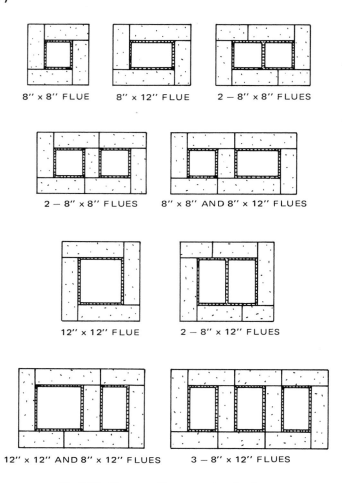

8" x 8" FLUE   8" x 12" FLUE   2 – 8" x 8" FLUES

2 – 8" x 8" FLUES   8" x 8" AND 8" x 12" FLUES

12" x 12" FLUE   2 – 8" x 12" FLUES

12" x 12" AND 8" x 12" FLUES   3 – 8" x 12" FLUES

Fig. 9-67. Masonry chimney section details.

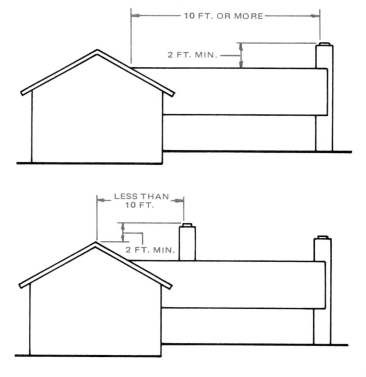

Fig. 9-68. Minimum recommended chimney heights above the roof.

When a chimney has two flues not separated by masonry, the joints of adjacent flue linings should be staggered at least 7 in. If more than two flues are located in the same chimney, masonry wythes at least 4 in. wide should be built between them so that no more than two are grouped together, Fig. 9-67.

The tops of chimneys above the roof must meet certain specifications. The flue lining should project 4 in. above the chimney cap. The cap should be finished with a straight or concave slope to direct air current upward and drain water from the top of the chimney. The top of the flue liner should extend a minimum of 3 ft. above the highest point where the chimney passes through the roof and at least 2 ft. above any part of the building within 10 ft. of the chimney, Fig. 9-68.

At the intersection of the chimney and roof, the connection should be weathertight. Flashing and counter flashing should be used, Fig. 9-69. Copper or other rust-resisting metal may be used.

Fig. 9-70. Prefabricated steel heat circulating fireplace with complete firebox, damper and smoke shelf. (Superior Fireplace Co.)

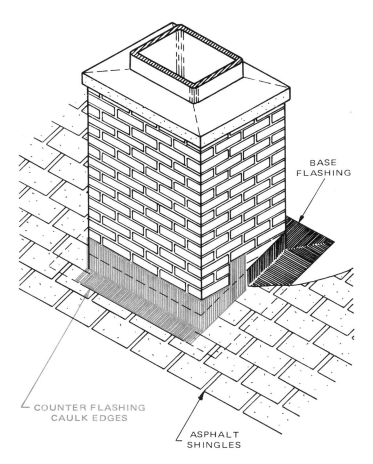

Fig. 9-69. Recommended flashing for chimneys at the roof intersection.

PREFABRICATED STEEL HEAT CIRCULATING FIREPLACES

Prefabricated steel heat circulating fireplaces are becoming very popular in residential construction. They include not only the firebox and heating chamber, but also the throat, damper, smoke shelf and smoke chamber, Fig. 9-70. These units are

very efficient because sides and back are double walled where air is heated and returned through vents. Fig. 9-71 shows another model. Design specifications are given in Fig. 9-72.

Fig. 9-71. A prefabricated steel heat circulating fireplace. (Vega Inc.)

155

PREFABRICATED STEEL HEAT CIRCULATING
FIREPLACE SIZES (IN INCHES)

| OVERALL DIMENSIONS | | | FINISHED OPENING | | FLUE SIZE |
WIDTH	HEIGHT	DEPTH	WIDTH	HEIGHT	(MODULAR)
34 3/4	44	18	28	22	8 x 12
38 1/2	48	19	32	24	8 x 12
42 1/2	51	20	36	25	12 x 16
47 3/4	55	21	40	27	12 x 16
55	60	22	46	29	12 x 16
63	65	23	54	31	16 x 16

SOURCE: THE MAJECTIC CO., INC.

Fig. 9-72. Design specifications for prefabricated fireplace.

STONE QUOINS

Quoins are large squared stones used at corners and around openings of buildings for ornamental purposes, Fig. 9-73. The height of each stone should be 3, 5, 7 or any odd number of

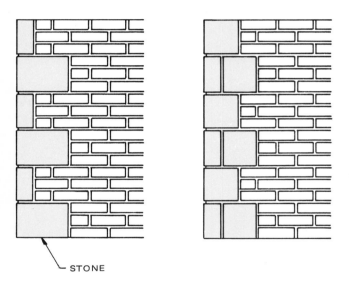

STONE

Fig. 9-73. Quoins used at corners for ornamental purposes.

brick courses. This permits the brick above and below the stone to have a full lap in bond. The length of each stone should be equal to one or more of the masonry units used in the wall.

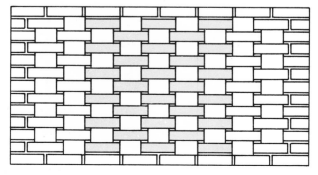

Fig. 9-74. Perforated brick wall with each unit bonding one quarter its length on either end. Patterns are limited only to one's imagination.

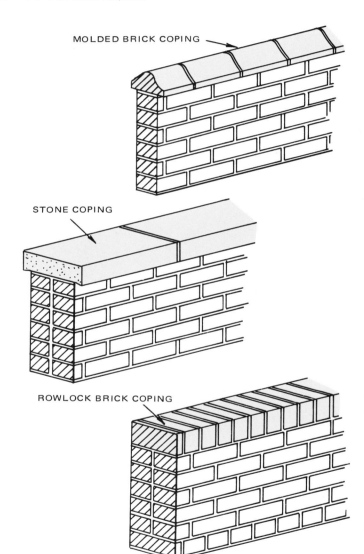

MOLDED BRICK COPING

STONE COPING

ROWLOCK BRICK COPING

Fig. 9-75. Straight garden walls with three commonly used copings.

GARDEN WALLS

There any many types of garden walls (free standing structures) and new variations are constantly being designed. Perforated walls can be used with any of the typical types. Perforated walls are sometimes used rather than solid walls because they admit light, breezes and do not block all vision. Fig. 9-74 shows a typical perforated brick wall.

STRAIGHT WALLS

This form of garden wall depends on the texture and color of the masonry for its character. The thickness must be sufficient to provide lateral stability against wind and impact loads.

It is recommended that for 10 pounds per square foot (psf) wind pressure, the height above grade not exceed 3/4 of the wall thickness squared. For example, an 8 in. thick wall could be a maximum height of 48 in. ($3/4 \times 8^2 = 48$). This formula does not depend on a bond between the foundation and the wall. Therefore, reinforcing will greatly increase the height that may be attained for a given wall thickness.

Fig. 9-75 shows three typical straight walls. Each has a different coping.

PIER AND PANEL WALLS

A pier and panel wall is composed of a series of relatively thin panels 4 in. thick. They are braced by masonry piers, Fig. 9-76. This type of wall is relatively easy to build and is

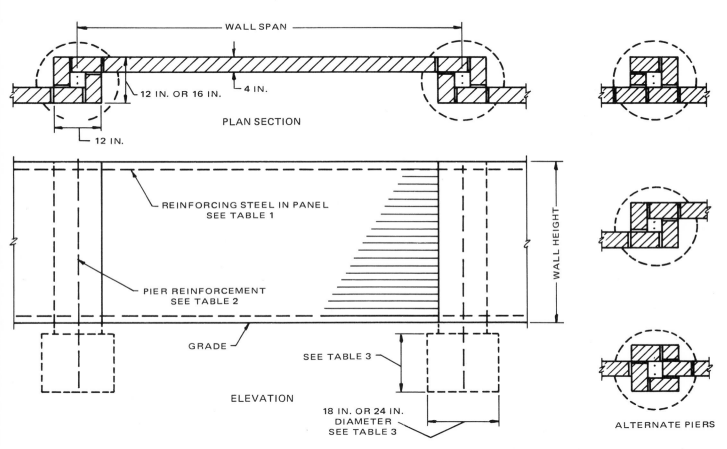

Fig. 9-76. Details of a pier and panel garden wall. See Fig. 9-77, 9-78 and 9-79 for specifications.

TABLE 1
PANEL WALL REINFORCING STEEL

WALL SPAN, FT.	VERTICAL SPACING, IN.								
	WIND LOAD, 10 PSF			WIND LOAD, 15 PSF			WIND LOAD, 20 PSF		
	A	B	C	A	B	C	A	B	C
8	45	30	19	30	20	12	23	15	9.5
10	29	19	12	19	13	8.0	14	10	6.0
12	20	13	8.5	13	9.0	5.5	10	7.0	4.0
14	15	10	6.5	10	6.5	4.0	7.5	5.0	3.0
16	11	7.5	5.0	7.5	5.0	3.0	6.0	4.0	2.5

Note: A = 2 — No. 2 bars
B = 2 — 3/16-in. diam. wires
C = 2 — 9 gage wires

Fig. 9-77. Specifications for panel wall reinforcing steel. (Brick Institute of America)

TABLE 2
PIER REINFORCING STEEL (1)

WALL SPAN FT.	WIND LOAD, 10 PSF WALL HEIGHT, FT.			WIND LOAD, 15 PSF WALL HEIGHT, FT.			WIND LOAD, 20 PSF WALL HEIGHT, FT.		
	4	6	8	4	6	8	4	6	8
8	2#3	2#4	2#5	2#3	2#5	2#6	2#4	2#5	2#5
10	2#3	2#4	2#5	2#4	2#5	2#7	2#4	2#6	2#6
12	2#3	2#5	2#6	2#4	2#6	2#6	2#4	2#6	2#7
14	2#3	2#5	2#6	2#4	2#6	2#6	2#5	2#5	2#7
16	2#4	2#5	2#7	2#4	2#6	2#7	2#5	2#6	2#7

(1)Within heavy lines 12 by 16-in. pier required. All other values obtained with 12 by 12-in. pier.

Fig. 9-78. Specifications for pier reinforcing steel.

TABLE 3
REQUIRED EMBEDMENT FOR PIER FOUNDATION (1)

WALL SPAN, FT.	WIND LOAD, 10 PSF WALL HEIGHT, FT.			WIND LOAD, 15 PSF WALL HEIGHT, FT.			WIND LOAD, 20 PSF WALL HEIGHT, FT.		
	4	6	8	4	6	8	4	6	8
8	2'–0"	2'–3"	2'–9"	2'–3"	2'–6"	3'–0"	2'–3"	2'–9"	3'–0"
10	2'–0"	2'–6"	2'–9"	2'–3"	2'–9"	3'–3"	2'–6"	3'–0"	3'–3"
12	2'–3"	2'–6"	3'–0"	2'–3"	3'–0"	3'–3"	2'–6"	3'–3"	3'–6"
14	2'–3"	2'–9"	3'–0"	2'–6"	3'–0"	3'–3"	2'–9"	3'–3"	3'–9"
16	2'–3"	2'–9"	3'–0"	2'–6"	3'–3"	3'–6"	2'–9"	3'–3"	4'–0"

(1)Within heavy lines 24-in. diam. foundation required. All other values obtained with 18-in. diam. foundation.

Fig. 9-79. Specifications for embedment for pier foundations.

economical because of the reduced panel thickness. It is also ideal for uneven terrain. Foundations are only required for the piers. These should extend below the frost depth.

The panels may be constructed with 2 x 4 forms under the first course. The forms are removed when the masonry has cured a minimum of seven days. Tables in Figs. 9-77, 9-78 and 9-79 give the specifications for building pier and panel walls for varying heights and lengths.

SERPENTINE WALLS

Serpentine walls have been used successfully for hundreds of years. The serpentine shape provides lateral strength so that it can normally be built 4 in. thick. Since the shape of the wall provides the strength, it is important that the degree of curvature be sufficient. The radius of curvature of a 4 in. wall should be no more than twice the height of the wall above the grade. The depth of curvature should be no less than 1/2 of the height. Fig. 9-80 shows the details of a typical serpentine wall.

REVIEW QUESTIONS – CHAPTER 9

1. The footing is usually wider than the foundation wall to provide _____ and _____.
2. Any foundation not placed on bed rock will settle. True or False?
3. The two general types of foundations are _____ foundations and _____ foundations.

4. _____ _____ carry light loads and are not reinforced with steel.
5. Footings which are independent and receive the loads of freestanding columns or piers are called _____ footings.
6. The usual width of a footing for an 8 in. concrete block foundation wall is _____ inches.
7. The primary reason for parging a masonry basement wall is to _____ the wall.
8. Columns, piers and pilasters are classified as similar. The _____ is built into the foundation wall.
9. The minimum length of an anchor bolt used in a concrete block masonry foundation wall is _____ inches.
10. Solid masonry walls may be built of solid or hollow masonry units or a combination of the two. True or False?
11. The "SCR brick" is designed to build a solid masonry wall _____ inches thick.
12. Two types of hollow masonry walls are _____ and _____.
13. A cavity wall is usually bonded together with _____.
14. A _____ wall is one which does not use the facing material as a load-bearing agent.
15. Name two types of retaining walls.
16. Masonry above a wall opening may be supported by a _____ or _____.
17. Identify three types of lintels.
18. Arches which do not exceed 6 ft. are classified as _____ arches.

Construction Details

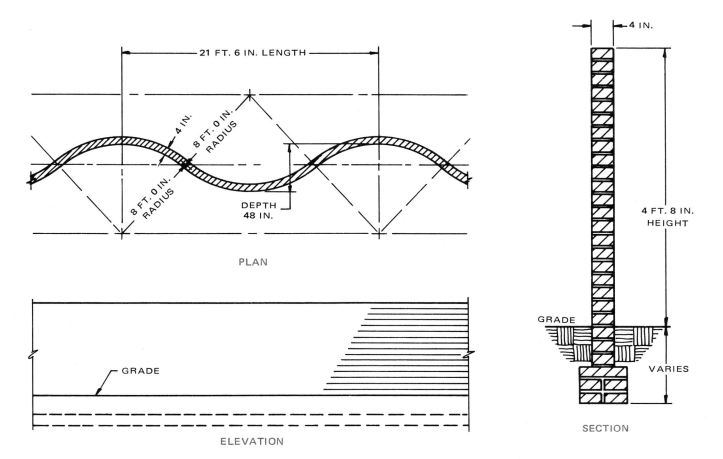

Fig. 9-80. Details of a serpentine garden wall showing the plan layout and wall sections.

19. Arches are constructed with the aid of temporary shoring or _____.
20. Name two types of arches commonly used in building construction.
21. The _____ is the horizontal support placed over the top of a window or door.
22. _____ sills are generally the same length as the window opening.
23. The proper slope for walks and patios is _____ to _____ in. per foot.
24. Treads for all outside steps are generally _____ inches wide.
25. Name four basic types of fireplaces.
26. What is the purpose of the smoke shelf?
27. The minimum wall thickness of a residential chimney is _____ inches.
28. Large squared stones used at corners and around openings of buildings for ornamental purposes are called _____.
29. Which type of garden wall provides lateral support through its shape?

Form placement for cast-in-place high rise structures calls for carefully designed materials and methods.   (Symons Corp.)

Chapter 10
FORM CONSTRUCTION

Concrete structures require forms to provide the desired shape and surface texture. Forms may be made of wood, steel, fiber glass, hardboard or other materials. The forms must be strong enough to resist the forces developed by the plastic (liquid) concrete. (Regular concrete weighs about 150 lb. per cu. ft. Poured in a form 8 ft. high, it would create a pressure of about 1200 lb. per sq. ft. along the bottom side of the form.) Also, the forms should impart the desired surface texture to the concrete structure.

FORM MATERIALS

Wood is the most popular form material. Both lumber and plywood are generally used in form construction. Construction lumber is used for frames, bracing and shoring while plywood is used for the form surface. Boards may be used instead of plywood if certain surface patterns are desired. However, plywood is manufactured in more than 40 surface textures, ranging from glass-smooth to board-and-batten panels. Its large, smooth surface; resistance to change in shape when wet and its ability to resist splitting has made it the most popular material for form construction.

Virtually any exterior-type plywood can be used for concrete formwork because it is made with waterproof glue. However, the plywood industry markets a product called *Plyform* which is specially designed for concrete forming.

Plyform is an exterior-type plywood made from special wood species and veneer grades to assure high performance. This product is available in two classes:
1. Plyform Class I
2. Plyform Class II

Class I is stronger and stiffer than Class II. Either may be purchased with a high density overlaid surface on each side.

This provides a very smooth, grainless surface that resists abrasion and moisture penetration.

Plywood is also valued as a form material because of its ability to bend. Curved surfaces can be formed easily if the thickness of the sheet and the direction of face grain is considered. Fig. 10-1 shows the minimum bending radii for plywood panels.

The rate of bending may be increased for a given thickness sheet by sawing kerfs across the inner face at right angles to the curve, Fig. 10-2. Curves may also be formed by using two or more thinner sheets bent one at a time and then fastened together.

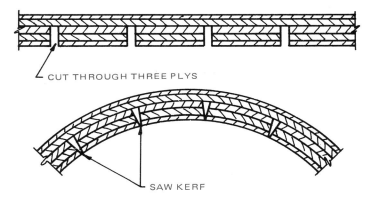

Fig. 10-2. Saw kerfs through three piles help bend short radius curves.

Hardboard which is tempered, specially treated, and 1/4 in. thick may be used as a form facing material. This material is coated with plastic to prevent water penetration. It produces a very smooth surface, but is essentially a form liner. It must be supported by a backing of lumber. Fig. 10-3 shows the recommended backing for hardboard liners.

Steel is frequently used for form construction, both frames and form facing, Figs. 10-4 and 10-5. Steel angles and other structural shapes are used for the frames of forms, Fig. 10-6. Generally, they provide greater strength and support heavier loads than wood frames. Steel frame forms may be faced with either plywood or sheet metal. Metal forms, as you might expect, have a longer life than wood forms. This offsets somewhat their greater cost.

Fiberglass form facing is increasing in use. This material is usually prefabricated in the size and shape desired. It is being

PLYWOOD THICKNESS	MEDIUM BEND RADIUS	
	PERPENDICULAR TO FACE GRAIN	PARALLEL TO FACE GRAIN
1/4 IN.	16 IN.	48 IN.
5/16 IN.	24 IN.	60 IN.
3/8 IN.	36 IN.	72 IN.
1/2 IN.	72 IN.	96 IN.
5/8 IN.	96 IN.	120 IN.
3/4 IN.	120 IN.	144 IN.

Fig. 10-1. Minimum bending radii for plywood panels used in form construction.

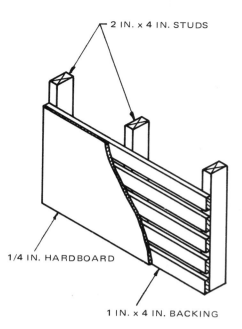

Fig. 10-3. Recommended backing for hardboard lined forms.

Fig. 10-4. Plywood facing on steel frame form module. (Symons Corp.)

Fig. 10-5. Rigid inside corner for use in corner conditions where stripping release is not required. Generally available in 2 and 4 ft. lenghts. (Symons Corp.)

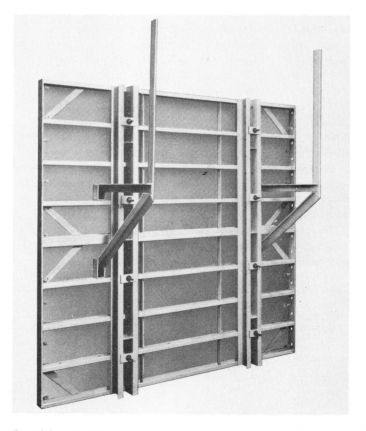

Fig. 10-6. Steel-framed, plywood-faced panels with standard walkway. (Symons Corp.)

used extensively in precast concrete plants and pan forms for ribbed concrete floors are frequently made of fiberglass.

Insulating board and rigid foam may be used as form liners but have little or no strength. Most often they are left attached to the concrete when the form is removed. Clips may be used for this purpose.

FORM DESIGNS

Forms should be designed so that they are practical and economical. They must be strong enough to resist the pressure of the plastic concrete. They must be able to retain their shape during the pouring and curing phases. Wet concrete should not leak from joints and cause fins and ridges.

Forms should be simple as possible to build and use in sizes that can be easily handled and stored. They must be designed so that they can be removed without damaging the concrete. Forms should also be safe for those who work around them.

The primary consideration in form design is usually its strength. It must support its own weight, the weight of the liquid concrete and any other loads which may be placed upon it such as wind, workers and equipment. For most general form requirements, concrete which is made with natural sand and gravel aggregates weighs about 150 pounds per cu. ft. Designs presented in this chapter are based on that weight of concrete and average conditions.

FORM ELEMENTS

Forms generally have five elements, Fig. 10-7:
1. Sheathing.
2. Studs.
3. Wales.
4. Braces or supports.
5. Ties and/or spreaders.

Sheathing gives the surface its shape and texture. It keeps the concrete in place until it hardens.

Studs support the sheathing and prevent bowing. Some types of forms do not require studs, but most wooden forms do.

Wales are used to align the forms, secure the ties and

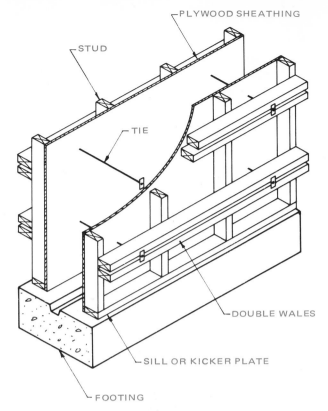

Fig. 10-7. Parts of the concrete form.

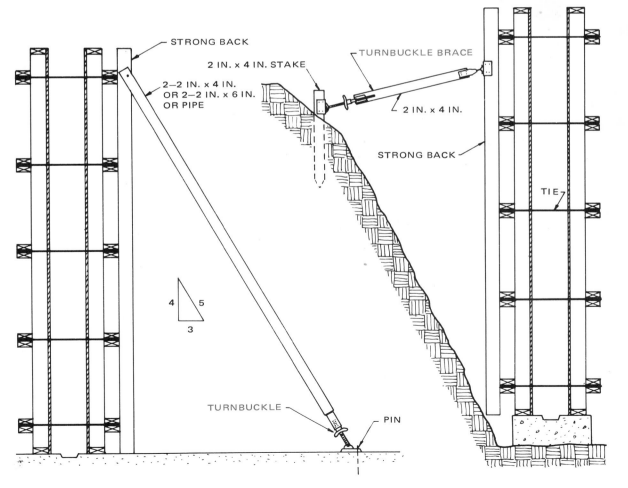

Fig. 10-8. The turnbuckle brace can be used to align the form as well as support it.

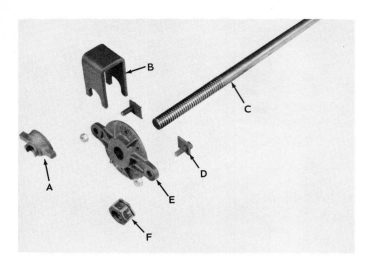

TAPERED TIE SHOWING PARTS. A—Cast knuckle nut. B—Spreader clip. C—Tapered tie rod. D—Friction clamp. E—Cast bearing washer. F—Hex knuckle nut.

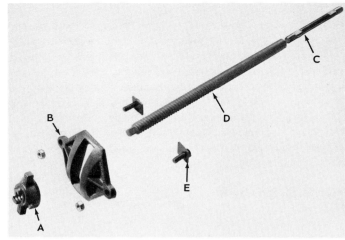

SHE-BOLT TIE WITH PARTS. A—Cast knuckle nut. B—Cast bearing washer. C—Inner unit. D—15 or 20 in. she-bolt knuckle thread. E—Friction clamp.

TIE BRACKET FOR SECURING THE TIE TO THE WALLS OR VERTICAL STUD

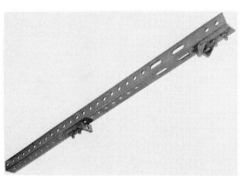

TOP TIE ANGLE

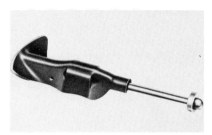

SNAP TIE EXTENSION FOR USE WITH A STRONGBACK WHICH PROVIDES EXTRA STRENGTH

Fig. 10-9. Above. Panel accessories. Below. Drawings show cross-sectional view of tie assemblies attached to the form. (Superior Concrete Accessories, Inc., Symons Corp. and Dayton Sure-Grip and Shore Co.)

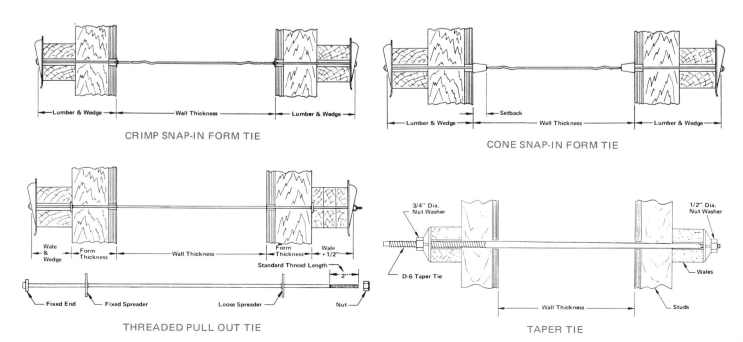

CRIMP SNAP-IN FORM TIE

CONE SNAP-IN FORM TIE

THREADED PULL OUT TIE

TAPER TIE

support the studs. Braces and supports provide lateral support against wind and other forces.

One type of brace which is easy to use is the turnbuckle, Fig. 10-8. It can be used to align the form and is only required on one side of the form. These braces are usually spaced 8 to 10 ft. apart. Ties and spreaders prevent the sides of the form from spreading or moving together, Fig. 10-9.

Wooden spreaders are removed from the inside of the form as the concrete is poured to their level. Ties remain in the concrete and become a permanent part of the structure.

When the concrete has cured and the forms are removed, the ties will project out from the surface of the concrete. They may be broken off or disconnected internally depending on the type. Fig. 10-10 shows two types of ties which may be taken apart to remove most parts from the wall.

When specifying ties, it is necessary to know the type required, the wall thickness, the actual size of sheathing, studs and wales and the required break back. *The break back is the depth the tie is broken off in the concrete.* Some ties have spreader washers and these take the place of wood spreaders.

PREFABRICATED FORMS

Prefabricated forms are pre-built usually in modular sizes. They can be fastened together on the job to produce the desired wall or column size. Prefab forms may be purchased, rented or constructed by a contractor.

The most common type of prefab form has a metal frame with a plywood facing, Figs. 10-11 and 10-12. Special connectors are used to attach the modular units together. Some have hinged corners which aid in removing the forms. The most common size of module is 2 ft. by 8 ft. The modules may be combined to make gang forms.

Gang forming is used on large jobs where repetitive forming is required. These are generally lifted with a crane since they are too heavy to be lifted by hand.

SLIP FORMS

Slip forms are used for casting very large structures of great height. They are raised slowly as the concrete is poured using

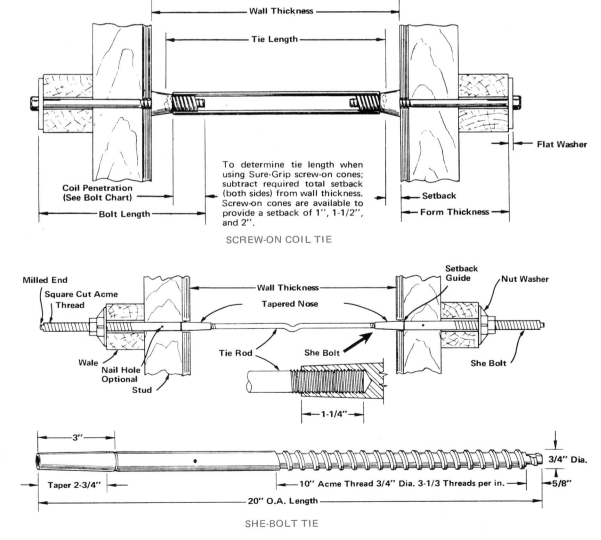

Fig. 10-10. Threaded ties such as these can be disconnected. Protruding portions can be removed from the wall. (Dayton Sure-Grip and Shore Co.)

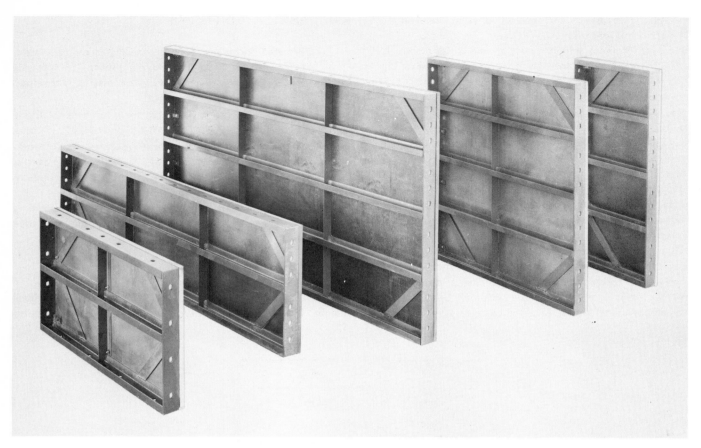

Fig. 10-11. Modular form components are produced in several sizes: 2' x 2', 2' x 4', 4' x 2', 4' x 4', 8' x 2' and 8' x 4'. Plywood facing (3/4 in.) is attached to a steel frame. (Symons Corp.)

Fig. 10-12. Strippable inside corner assemblies which provide a means of releasing forms from inside captive conditions such as counterfort walls, box conduits and vertical, square and rectangular shafts. (Symons Corp.)

jacks. An average speed is about 15 inches per hour.

Fig. 10-13 shows a section through a typical slip form. The form facing is 3/4 in. high density overlaid plywood. The wales are lumber and the yokes are steel. The form is pulled upward by the jacks which move up jackrods embedded in the

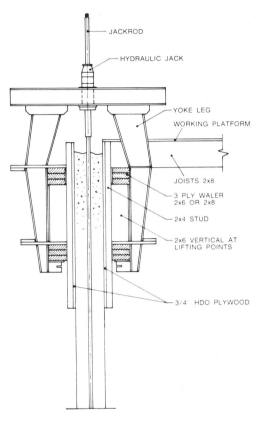

Fig. 10-13. Section through a typical slip form.
(American Plywood Assoc.)

concrete. Jackrods may be recovered after the pouring is finished by pulling them from a recovery pipe.

Slip forms are also used for casting long, low walls such as median walls between lanes of traffic.

FORM APPLICATIONS

Forms used in building construction may be grouped into those used for:
1. Footings.
2. Walls.
3. Slabs.
4. Steps.
5. Beams or columns.
6. Masonry support.

Our major attention will be directed toward forms that can be built on site.

FOOTING FORMS

Footings are not usually visible and therefore appearance is of little concern. However, they must be located accurately and built to the specified dimensions. If the soil is firm and not too porous, it may serve as the form for footings. However, inferior results are possible if the soil absorbs too much water from the concrete or soil falls into the concrete. Fig. 10-14 shows two types of earth footing forms.

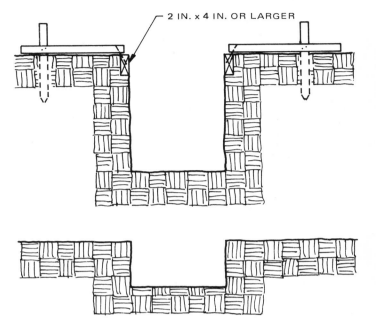

Fig. 10-14. Earth footing forms.

Concrete footing forms are usually constructed from 2 in. construction lumber the same width as the footing thickness. The boards are placed on edge, Fig. 10-15, and held in place by stakes and cross-spreaders. Soil may be used to fill cracks around the forms and provide some support. *Remember that footings must be poured on undisturbed or well-compacted earth. Do not fill in low spots under the footing with loose earth.*

Occasionally, a site will require stepped footings. Fig. 10-16 illustrates one method of forming stepped footings. The height and length of the steps should coincide with the length and

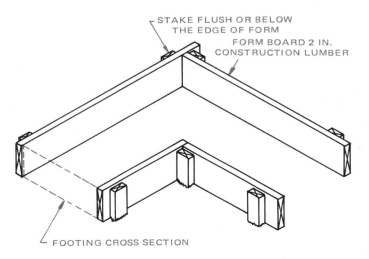

Fig. 10-15. Footing forms made of 2 in. construction lumber.

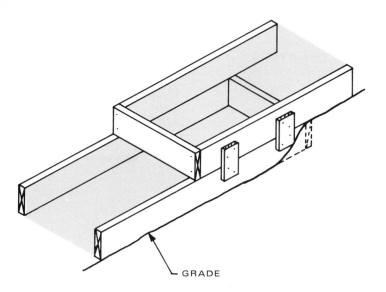

Fig. 10-16. Form for a stepped footing.

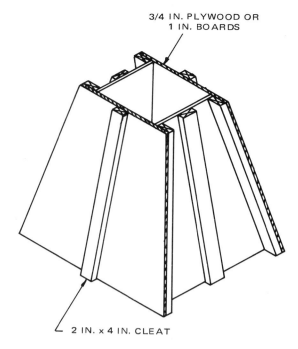

3/4 IN. PLYWOOD OR
1 IN. BOARDS

2 IN. x 4 IN. CLEAT

Fig. 10-18. Tapered footing form of 3/4 in. plywood or 1 in. boards. This form will be held in place by stakes driven into the ground.

course height of one or more masonry units being used. Always study the plans carefully.

Isolated footings for piers and columns are formed with bottomless boxes usually made with construction lumber, Fig. 10-17. The width of the boards should be the same as the thickness of the footing.

Tapered footing forms are built like a hopper. Two sides must have the exact dimensions of the footing while the other two are slightly larger so that cleats may be attached along their edges, Fig. 10-18. Side and end sections are held together with these 2 x 4 cleats. The form will try to rise as the concrete is poured so it must be securely anchored.

WALL FORMS

Wall forms may be built-in-place or prefabricated. The type used depends on the complexity of the walls, need to reuse or preference.

To build in-place forms, first attach a sole plate to the footing. Use concrete nails, power driven nails or stakes, Fig. 10-19. The sole plate provides a place to attach the studs. It should be set back from the wall line the thickness of the sheathing. Be sure that it is properly located and straight because this will determine the position of the wall.

Then cut studs to desired length and toenail them to the top of the plate as in Fig. 10-20. In a form for a typical basement wall, spacing of the studs is generally 24 in. with 3/4 in. plywood or 1 in. boards used for sheathing. Stud spacing should be decreased as the height of the wall increases.

Sheathing is nailed to the studs when they are in place. The first sheet should be leveled so that remaining sheets will be

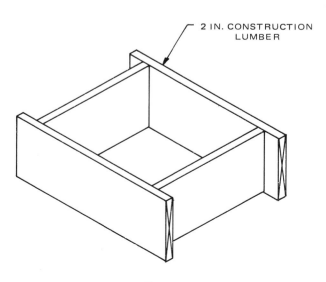

2 IN. CONSTRUCTION
LUMBER

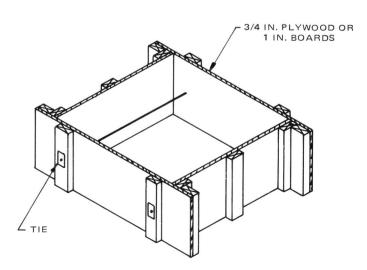

3/4 IN. PLYWOOD OR
1 IN. BOARDS

TIE

Fig. 10-17. Isolated footing forms made of 2 in. construction lumber and 3/4 in. plywood.

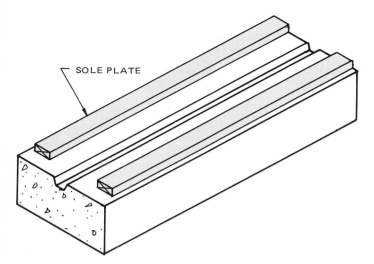

Fig. 10-19. First step in form building is to attach sole plate to footing with power-driven nails or concrete nails.

level. Holes for ties may be drilled as the sheathing is being attached. Placement of the ties determines the location of the wales, Fig. 10-21.

Wales are attached to the outside of the studs using nails, clips or wires. Either single or double wales may be used.

One end of the braces is attached to the wales. The other end is anchored in the ground or some other solid support.

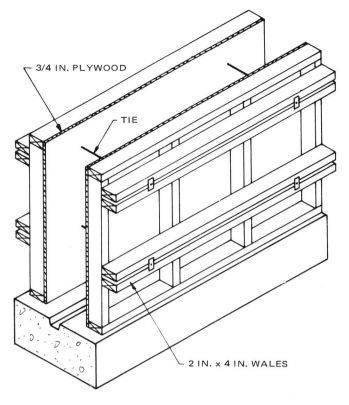

Fig. 10-21. Completed section of a wall form with sheathing, ties and wales in place.

If the wall is too narrow or deep for a person to work inside, the inside wall may be built in a horizontal position and tilted into place. Also, any reinforcing steel should be placed in position before the inside form is placed.

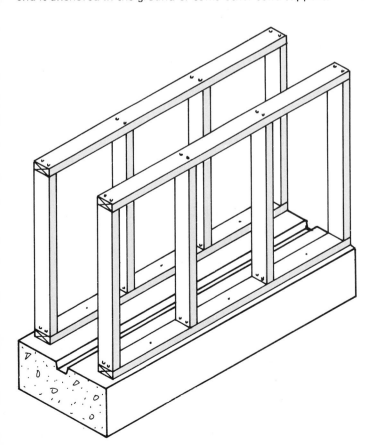

Fig. 10-20. Second step in building wall forms is to toe-nail wall studs to sole plate.

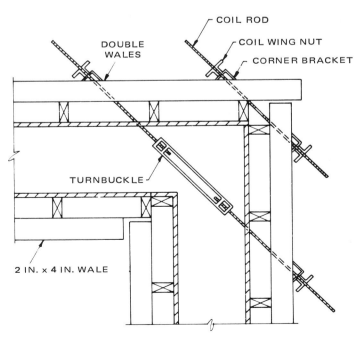

Fig. 10-22. Form units held together at corner with corner brackets and tie rods.

More and more prefab forms are being used. They may be assembled on the ground and then positioned. Bracing is required for them just as for the built-in-place forms.

Corners of forms are weak and must be given special attention. The forms must be tied together firmly to prevent concrete leakage. Corner brackets or tie rods, Fig. 10-22, may be used to give the needed strength. Some forms have a series of metal eyes fastened to the sheathing panels which allow a metal rod to secure them together. Another method of tightening the corners is to lap the wales at the corner and secure them. Wooden wedges can be driven behind the wales to close any openings.

Any openings in the wall, such as for windows, can be formed using BUCKS. Bucks are wood or steel frames set in the form between the inner and outer form to make an opening in the wall, Fig. 10-23.

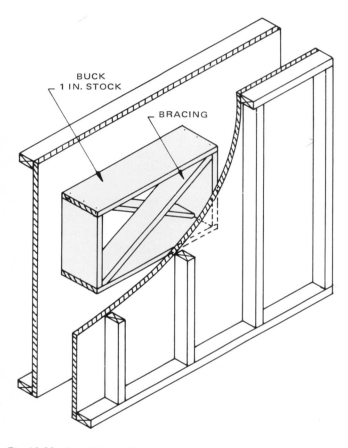

Fig. 10-23. A well braced buck in place in the form will cast a void for a window or other opening through the wall. They are braced to prevent distortion.

SLAB FORMS

Forms for drives, walks, patios or other flatwork may be wood or metal. Two-inch lumber is commonly used with wood or steel stakes for bracing.

For slabs 4 in. thick, 2 x 4 lumber is recommended, Fig. 10-24. One-inch lumber may be used, but stakes must be much closer together to keep the form straight. A 5 in. slab can be formed with 2 x 4's, but 2 x 6 materials are preferred, Fig.

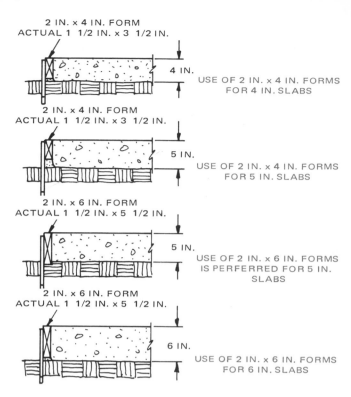

Fig. 10-24. Recommended form lumber size for typical slab thicknesses.

10-24. Slabs which are 6 in. thick require 2 x 6 lumber.

Form material for slabs should be smooth, straight and free of knot holes and other surfaces imperfections. *Remember that "dressed" (machined smooth) 1 in. lumber is actually 3/4 in. thick and 2 in. lumber is 1 1/2 in. thick. Widths are usually 1/2 in. less than the nominal size. It is important to know the actual size of materials when building forms.*

On large projects, metal forms are generally used to form slabs. They save time, stand rough handling and may be used many more times than wood.

Wood stakes are made from 1 x 2, 1 x 4, 2 x 2, or 2 x 4 material, Fig. 10-25. They are spaced about 4 ft. apart for 2 in.

Fig. 10-25. Slab form being attached to stake using form nails. Note the nail has a double head for easy removal.

thick formwork. A spacing of about 2 ft. is recommended for 1 in. thick formwork. Steel stakes are available for use with wood or metal forms. They are easier to drive and much stronger. Even though they cost more than wood stakes, they may be cheaper over a period of time due to their long life.

Stakes should be driven slightly below the top of the forms to aid in screeding and finishing the concrete slab. Drive the stakes straight and plumb so the form will be true. Use double-headed (form) nails to attach the stake to the form. Drive the nail through the stake into the form.

FORMING CURVES

Horizontal curves are easy to form if plywood, 1 in. boards, hardboard or sheet metal is used. Curves with a short radius may be formed with 1/4 or 1/2 in. plywood with the face grain vertical. If heavier lumber is desired or a sharper bend is necessary, saw kerfing will help. Wet lumber will bend easier than dry. Fig. 10-26 shows some details for forming horizontal curves.

Long gentle curves may be formed using 2 in. thick material. Staking must be very secure because it must support a heavier load. It is a good idea to use extra stakes for all types of curves.

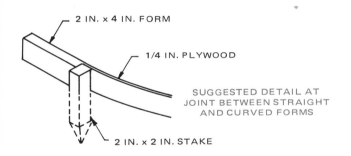

2 IN. x 4 IN. FORM

1/4 IN. PLYWOOD

SUGGESTED DETAIL AT JOINT BETWEEN STRAIGHT AND CURVED FORMS

2 IN. x 2 IN. STAKE

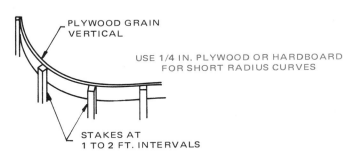

PLYWOOD GRAIN VERTICAL

USE 1/4 IN. PLYWOOD OR HARDBOARD FOR SHORT RADIUS CURVES

STAKES AT 1 TO 2 FT. INTERVALS

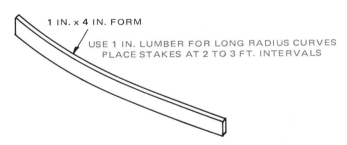

1 IN. x 4 IN. FORM

USE 1 IN. LUMBER FOR LONG RADIUS CURVES PLACE STAKES AT 2 TO 3 FT. INTERVALS

USE SAW KERFING TO BEND 2 IN. LUMBER CUT KERF 1/2 TO 2/3 THE THICKNESS. BEND SO THAT KERFS CLOSE

Fig. 10-26. Forming horizontal curves using plywood.

The proper grade can be maintained by using a mason's line or builder's level. (The builder's level is an accurate spirit level combined with a telescope on a circular base. It sits on a tripod.) When using the level, the grade is determined at a given point and a reading is taken on a rod or rule. Then the rod or rule is moved along the form line and measurements are made on the stakes indicating proper grade. When enough stakes have been sighted, the forms may be attached at the height indicated.

PERMANENT FORMS

Divider strips in a patio are an example of permanent forms. They are left in place for decorative purposes and/or to serve as control joints. These forms should be made from 1 x 4 or 2 x 4 cypress, redwood or cedar which has been treated with wood preservative.

The strips should be anchored to the concrete with 16d galvanized nails. Drive them through the board from alternate sides at 16 in. intervals. Outside forms would be nailed from only one side.

Masking tape can be used on the top edge of divider strips to protect the wood from abrasion and stain while placing the concrete. It is easily removed when the slab has set.

Another type of permanent form is a hollow tube. It may be used to cast a void in concrete slabs, beams or other applications, Fig. 10-27. The tubes are generally a fiber material and are produced in sizes from 2 1/4 in. to 36 in. in diameter. Lengths vary to meet specifications.

Fig. 10-27. Hollow tube fiber forms used to cast voids in slabs, walls, piers or other construction. (Sonoco Products. Co.)

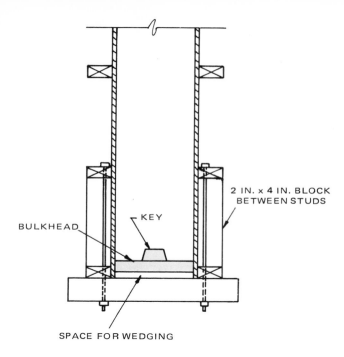

Fig. 10-28. Typical bulkhead in a concrete wall form.

CONSTRUCTION JOINTS

A large slab or wall is often poured in sections which require a construction joint. Construction joints may be formed by placing a BULKHEAD in the form. A bulkhead is a piece of material which prevents the concrete from moving past a certain point in the form, Fig. 10-28.

FORMS FOR STEPS

Two basic types of concrete steps may be constructed. One type is built directly on the slope of the ground. The other type is supported at the top and bottom with an open space under the steps.

Fig. 10-29 shows a method of forming steps built directly on sloping ground. The sides of the form are usually made from 2 x 6 or 2 x 8 lumber. The riser forms are 1 in. boards. The beveled bottom edge allows troweling of the step surface right up to the riser.

Fig. 10-30 shows one forming method for self-supporting steps. These steps must be reinforced with steel and must rest on a foundation. This form must be securely braced to hold the weight.

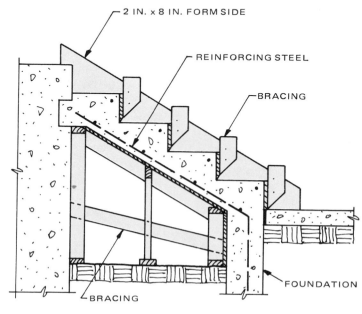

Fig. 10-30. Form for a free-standing set of concrete steps.

COLUMN FORMS

Concrete columns reinforced with steel are commonly cast in square, rectangular and round shapes. Forms for the square and rectangular columns are easily constructed from wood, Fig. 10-31. However, round column forms are usually metal, fiberboard or paper, Fig. 10-32. Round prefabricated forms are produced for various lengths and diameters. Some forms are reusable. However, forms made of fiberboard and paper are not reusable but have some advantages. They are economical, lightweight and the paper surfaces produce a very smooth concrete surface with no seams. Fiberboard forms often leave a visible seam. The form may be removed from cured concrete by sawing almost through the form material and then making the final cut with a sharp knife blade.

CENTERING FOR ARCHES

Arch centering (structure to support the masonry while it is being constructed) is usually made from wood. The ribs, Fig. 10-33, are 2 in. lumber and the lagging may be 3/4 x 2 strips of wood. The lagging should be cut 1 in. shorter than the width of the masonry wall so that it will not interfere with the

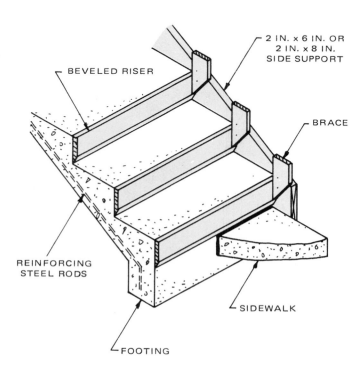

Fig. 10-29. Forming technique for steps cast on sloping terrain.

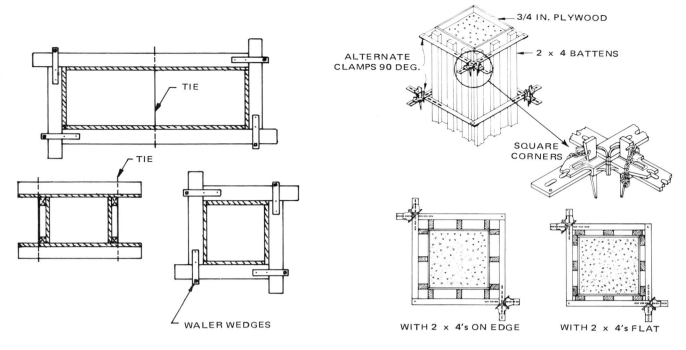

Fig. 10-31. Square and rectangular column forms.

Fig. 10-32. Column form made of fiber paper. The form is only used once because it is destroyed by the removal process. (Sonoco Products Co.)

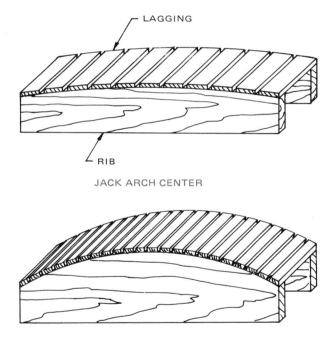

JACK ARCH CENTER

SEGMENTAL ARCH CENTER

Fig. 10-33. Centering for jack and segmental arches.

mason's line. The ribs are cut to the shape of the arch. The centering must be supported adequately for the weight it is to bear.

FORM MAINTENANCE

Form maintenance is important because it reduces the cost by extending the life of a form. Many forms are damaged during stripping (removing them from cured concrete). Do not use metal pry bars. These will damage the forms — especially

173

wood forms. Use wood wedges and tap lightly if necessary.

Forms should be inspected, cleaned, repaired and lightly oiled after they are removed. Use a stiff fiber brush on wood forms. A wire brush will raise the grain and roughen the surface. Mill oil of 100 or higher viscosity and pale in color is generally used for oiling forms. A liberal amount of oil should be applied a few days before the plywood is used. Wipe to a thin film just before using the form. Forms should be solid stacked or stacked in small packages with faces together for future use. Panels should be protected from the sun and rain.

REVIEW QUESTIONS — CHAPTER 10

1. Name three materials commonly used for form construction.
2. The weight of a cubic foot of "average" concrete is about _____ pounds.
3. Exterior plywood is made with _____ glue.
4. A specially designed plywood for form construction is called _____.
5. Steel frame forms are generally stronger than wood frame forms. True or False?
6. Forms generally have five elements. Name them.
7. What is the function of wales?
8. The most common size of prefab form module is _____.
9. _____ forms are used for casting very large structures of great height.
10. When may earth be used as a form material?
11. Stepped footings are used mainly for entrance concrete steps. True or False?
12. Wall forms rest on the _____.
13. Stud spacing for a typical basement wall form is about _____ inches.
14. _____ are wood or steel frames set in the form betweem the inner and outer form to make a void in the wall.
15. What size lumber is ordinarily used for a 4 in. thick slab?
16. The actual thickness of 1 in. lumber is _____ inch.
17. _____ curves may be formed with 2 in. thick lumber.
18. A _____ is a piece of material which prevents the concrete from moving past a certain point in the form.
19. "Centering" is a type of form used to support _____ during construction.
20. _____ may be used to coat forms.

Chapter 11
MATHEMATICS FOR MASONRY TRADES

A skilled mason or apprentice must be able to take measurements and perform various calculations related to the trade. These calculations require an understanding of certain basic mathematical concepts. The purpose of this chapter is to review some of those basic concepts and give examples of their application using trade terms and figures.

WHOLE NUMBERS

A whole number is any one of the natural numbers such as 1, 2, 5, etc. Numbers represent quantities of anything. They can be added, subtracted, multiplied or divided.

ADDITION

Addition is the process of combining two or more quantities (numbers) to find a total. The total is called the sum. Addition is indicated by the plus (+) sign and may be written as 2 + 2. The sum may be indicated by using the equal (=) sign. Example: 2 + 2 = 4. Another way of writing the same thing showing the sum of 4 is:

$$\begin{array}{r} 2 \\ + 2 \\ \hline 4 \end{array}$$

The following problems are included to refresh your memory of basic addition in trade terms.

ADDITION PROBLEMS
1. Three bricklayers working together on a job each laid the following number of brick in one day. First bricklayer laid 887, second bricklayer laid 1123, and the third bricklayer laid 1053 brick. How many brick did all three lay that day?
   Answer: 887 + 1123 + 1053 = 3063 brick.
2. A mason was paid the following sum of money for five days of work: $96, $105, $87, $97 and $85. How much money was he paid for the five days work?
   Answer: $96 + $105 + $87 + $97 + $85 = $470.
3. Weight slips on four loads of gravel delivered to one job were 6272 lb., 6098 lb., 6138 lb., and 6193 lb. What was the total weight of the gravel delivered to the job?
   Answer: 6272 + 6098 + 6138 + 6193 = 24,701 lb.

SUBTRACTION

Subtraction is the process of taking something away from the total. The portion which is left after taking some away is called the difference. The sign which indicates that one quantity (number) is to be subtracted from another is the minus (−) sign. Example: 6 − 4. In this example, 4 is being subtracted from 6. The difference is 2 or 6 − 4 = 2. Another way of writing the same thing is:

$$\begin{array}{r} 6 \\ - 4 \\ \hline 2 \end{array}$$

SUBTRACTION PROBLEMS
1. A mason ordered 75 bags of cement and used 68 bags on the job. How many bags of cement are left?
   Answer: 75 − 68 = 7 bags.
2. If a 36 in. wide chimney had two flues which were 8 in. and 16 in. wide and the thickness of the chimney walls were 4 in., how much space would be between the flues?
   Answer: 36 − (8 + 16 + 4 + 4) = 36 − 32 = 4 in.
3. A mason contracts a job for $700. Labor cost was $316 and material cost $203. Other expenses totaled $78. How much profit was made on the job?
   Answer: 700 − (316 + 203 + 78) = 700 − 597 = $103.

MULTIPLICATION

Multiplication is the process of repeated addition using the same numbers. For example, if 2 + 2 + 2 + 2 + 2 were to be summed, the shortest method would be to multiply 5 times 2 to get the total of 10. The sign used to indicate multiplication is the times (x) sign. In the previous example, 5 times 2 equals 10, would be written 5 x 2 = 10. This may also be written as:

$$\begin{array}{r} 2 \\ \times 5 \\ \hline 10 \end{array}$$

MULTIPLICATION PROBLEMS
1. If a bricklayer can lay 170 brick an hour, how many brick would he lay in four hours?
   Answer: 170 x 4 = 680 brick.

2. One type of brick cost $9 per hundred. If 14,000 brick were ordered, how much would they cost?

Answer: $9 x 140 = $1260. Note: The brick were 9¢ each, $9 per hundred or $90 per thousand. Therefore, the answer could have been determined by multiplying 9¢ x 14,000, $90 x 14 or $9 x 140.

3. If five brick covered 1 sq. ft. of wall space, how many brick would be required to cover a wall 8 ft. high by 10 ft. long?

Answer: First find the wall area by multiplying 8 x 10. The wall area is 80 sq. ft. The number of brick required is 5 x 80 = 400 brick.

DIVISION

Division is the process of finding how many times one number is contained within another number. The division symbol is ($\div$). For example, when we wish to find how many times 3 is contained in 9, we say 9 divided by 3 equals 3 or 9 $\div$ 3 = 3. The answer is called the quotient. If a number is not contained in another number an equal number of times, the amount left over is called the remainder. The following problem illustrates such a situation: 9 $\div$ 4 = 2 with 1 left over. For purposes of calculation, the problem is generally written this way:

$$4\overline{)9} \quad \text{or } 2\ 1/4$$
$$\frac{8}{1} \quad \text{remainder}$$

DIVISION PROBLEMS

1. If a set of steps had 8 risers and the total height of all the steps (total rise) was 56 in., what would the height of each step be?

Answer:
$$8\overline{)56}^{\,7} \quad \text{or 7 in.}$$

2. Four bricklayers laid 4160 brick on a job one day. What was the average number of brick laid by each bricklayer that day?

Answer: 1040 brick per bricklayer.

$$\begin{array}{r} 1040 \\ 4\overline{)4160} \\ \underline{4} \\ 16 \\ \underline{16} \\ 00 \end{array}$$

3. If a brick veneer wall requires five brick to lay up 1 sq. ft., how many square feet would 587 cover?

Answer: 117 2/5 sq. ft. of wall.

$$\begin{array}{r} 117 \\ 5\overline{)587} \\ \underline{5} \\ 8 \\ \underline{5} \\ 37 \\ \underline{35} \\ 2 \end{array}$$

FRACTIONS

A fraction is one or more parts of a whole. Fractions are written with one number over the other (1/2 or 1/4 or 3/4).

The top number is called the NUMERATOR and the bottom number is called the DENOMINATOR.

The denominator identifies the number of parts into which the whole is divided. The numerator indicates the number of parts of the whole which is of concern. In reading a fraction, the top number is always read first. For example, 1/2 would be read "one half"; 3/4 would be read "three fourths" and 3/8 would be read "three eights."

A fraction should always be reduced to its lowest denominator. For instance, 3/2 is not in correct form. It should be 1 1/2 because 2/2 = 1 and 1 + 1/2 = 1 1/2. The 1 1/2 is called a MIXED NUMBER. Always when the numerator and denominator are the same number as 1/1, 2/2, 3/3, etc. they are equal to 1.

ADDING FRACTIONS

The easiest fractions to add are those whose denominators (bottom numbers) are the same, as 1/8 + 3/8. Simply add the numerators (top numbers) together and keep the same denominator. For example, 1/8 + 3/8 = 4/8 or 1/2. (Reducing the fraction to its lowest denominator is preferred.) Another example of reducing to the lowest denominator is 8/24 = 1/3, because 24 may be divided by 8 three times.

When fractions to be added have different denominators (bottom numbers), multiply both numerator and denominator of each fraction by a number that will make the denominators equal. For example: 1/3 + 3/5 = 5/15 + 9/15. Observation indicated that 15 was the smallest number that could be divided evenly by both denominators. To complete the example, 5/15 + 9/15 = 14/15. Therefore, the sum of 1/3 and 3/5 is 14/15.

PROBLEMS IN ADDING FRACTIONS

1. What is the height of one stretcher course of brick if the brick are 2 1/4 in. high and the mortar joint is 3/8 in?

Answer: 2 1/4 + 3/8 = 2 2/8 + 3/8 = 2 5/8 in. height for one course.

2. A mason estimated the following amounts of mortar required for a job: 5 1/2 cu. yd., 11 1/3 cu. yd. and 6 1/4 cu. yd. What is total amount of mortar required for job?

Answer: 5 1/2 + 11 1/3 + 6 1/4
= 5 6/12 + 11 4/12 + 6 3/12
= 22 13/12 = 23 1/12 cu. yd. of mortar.

3. A brick wall on a residence was 8 1/16 in. thick, the furring was 3/4 in., plaster board was 3/8 in. and the plaster was 1/2 in. What was the total thickness of the wall?

Answer: 8 1/16 + 3/4 + 3/8 + 1/2 = 8 1/16 + 12/16 + 6/16 + 8/16 = 8 27/16 = 9 11/16 in.

SUBTRACTING FRACTIONS

Change all fractions to the same common denominator as was done for adding fractions. When the denominators are the same, subtract the numerators.

PROBLEMS IN SUBTRACTING FRACTIONS

1. What is the dimension of "A" in Fig. 11-1?

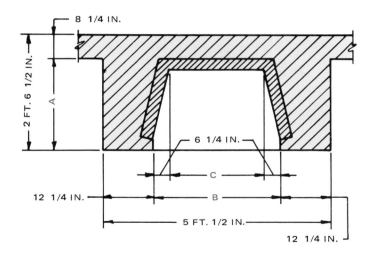

Fig. 11-1. Plan view of a fireplace with dimensions.

Answer: 2' 6 1/2" = 30 1/2". Therefore 30 1/2" − 8 1/4"
= 30 2/4" − 8 1/4" = 22 1/4". Dimension "A"
is 22 1/4" or 1' 10 1/4".
2. What is the dimension of "B" in Fig. 11-1?
Answer: 5' 1/2" = 60 1/2". Therefore 60 1/2" − (12 1/4"
+ 12 1/4") = 60 1/2" − 24 1/2" = 36" or 3'0".
Dimension "B" is 3'0".
3. What is the dimension of "C" in Fig. 11-1?
Answer: 36" (from No. 2 above) − (6 1/4 + 6 1/4) = 36" −
12 1/2" = 23 1/2". Dimension "C" is 23 1/2"
or 1' 11 1/2".

MULTIPLYING FRACTIONS

The procedure for multiplying fractions is to multiply the
numerators together to find the numerator for the answer.
Then, multiply the denominators together to find the denominator for the answer. The answer is called a PRODUCT and
the fraction is reduced to its lowest form. Example: 1/2 x 3/4
= 3/8. Whole numbers must be changed to fractions by placing
a "1" under the number. Example: 4 times 5/8 = 4/1 x 5/8 =
20/8 = 2 4/8 = 2 1/2.

PROBLEMS IN MULTIPLYING FRACTIONS
1. If standard brick are used which are 2 1/4 in. thick to lay a
wall with 3/8 in. mortar joints, what will the height of the
wall be after nine courses?
Answer: First, add the thickness of one mortar joint to the
thickness of one brick (2 1/4" + 3/8" = 2 5/8"). Then
multiply 2 5/8" times 9 to find the height. 2 5/8" x 9 =
21/8 x 9/1 = 189/8 = 23 5/8 in.
2. If a set of steps are five risers high and each riser is 7 1/4
in., what is the total rise of the steps?
Answer: 7 1/4 x 5/1 = 29/4 x 5/1 = 145/4 = 36 1/4 in.
3. What is the length of a 28 stretcher wall if each stretcher is
7 1/2 in. and the mortar joint is 1/2 in.?
Answer: 7 1/2" + 1/2" = 8". 8" x 28" = 224".
224" ÷ 12 = 18 2/3'. (2/3 x 12/1 = 24/3 = 8 in.)
Therefore the length is 18' 8".

DIVIDING FRACTIONS

The process of dividing fractions is accomplished by
inverting (turning up side down) the divisor and then
multiplying. For example, 3/8 ÷ 3/4 is solved by changing the
3/4 to 4/3. Therefore, 3/8 ÷ 3/4 = 3/8 x 4/3 = 12/24 = 1/2.

PROBLEMS IN DIVIDING FRACTIONS
1. How many risers 7 1/2 in. high would be required to
construct a flight of concrete steps 3' 1 1/2" high?
Answer: Change 3' 1 1/2" to 37 1/2".
Divide 37 1/2" by 7 1/2. 75/2 ÷ 15/2 = 75/2 x
2/15 = 150/30 = 5 risers.
2. If a brick mantel is corbeled out 4 1/2 in. in six courses,
how much does each course project past the previous
course?
Answer: 4 1/2 ÷ 6/1 = 9/2 x 1/6 = 9/12 = 3/4 in.
3. If a story pole was 8' 11 1/4" long and divided into 39
equal spaces, what is the length of each space?
Answer: 8' 11 1/4" ÷ 39 = 107 1/4" ÷ 39/1 = 429/4 x
1/39 = 429/156 = 2 117/156 = 2 3/4 in.

DECIMALS

Decimals are a type of fraction which only have numerators. The denominators are implied by the place of the last
digit to the right of the decimal point (.). Therefore: 0.1 =
1/10; 0.01 = 1/100 and 0.001 = 1/1000.

The zeros which may follow the significant digits in a
decimal number do not increase the value of the fraction.
Thus, 0.25 = 0.250 = 0.2500 and so on. The added zeros
indicate an increased accuracy, but not an increase in size.
However, zeros to the left of the number and before the
decimal are very significant. Note: 0.25 = 25/100; 0.025 =
25/1000; 0.0025 = 25/10,000 and so on.

When there is one number to the right of the decimal, it is
read as so many tenths (0.5 is five tenths). Two numbers to
the right of the decimal are read as hundredths (0.05 is five
hundredths). Three numbers to the right of the decimal
indicate thousandths and four indicate ten thousandths.

When numbers appear on both sides of the decimal, this is
an example of a DECIMAL MIXED NUMBER. The word
"and" is generally used to represent the decimal when reading
a decimal mixed number. For example, 6.125 is read as "six
and one hundred twenty five thousandths." The word "point"
is also used to denote the decimal. The quantity would then be
read as "six point one two five."

PROBLEMS IN ADDING DECIMALS
1. The cost of materials for a masonry job was as follows:
$35.50 for sand, $162.38 for cement, $9.63 for one steel
lintel and $338.19 for brick. What was total cost of the
materials for the job?
Answer: 35.50 + 162.38 + 9.63 + 338.19 = $545.70
$ 35.50
162.38
9.63   Be sure to line up the decimals!
338.19
$545.70

2. A mason wished to determine if the total cross-sectional area of three pieces of reinforcing steel met the specifications for a reinforced masonry lintel. They were 0.11 sq. in., 0.3 sq. in. and 1.000 sq. in. What was the total cross-sectional area of the three pieces of reinforcing steel?

Answer:
```
        0.11
        0.3
        1.000
        1.410 sq. in.
```

PROBLEMS IN SUBTRACTING DECIMALS

1. A mason placed a bid on a job for $1247.38. He allowed $413.75 for materials and $500.00 for labor. How much did he allow for profit and miscellaneous cost?

Answer: $1247.38 − ($413.75 + $500.00) = $1247.38 − $913.75 = $333.63.

2. A cement contractor was to stake out a foundation on a site which was 125.00 ft. wide and 200.00 ft. deep. The plot plan showed that the front of the house (placed along the width of the lot) faced south and was 50.00 ft. from the lot line. The east side of the house was 28.35 ft. from the lot line and the west side of the house was 36.65 ft. from the lot line. How wide is the front of the house?

Answer: 125.00' − (28.35' + 36.65') = 125.00' − 65.00' = 60.00 ft. or 60' 0'' as it is usually written on a construction drawing.

PROBLEMS IN MULTIPLYING DECIMALS

1. What is the cost of 20 bags of cement if each cost $3.00?

Answer: Add up the total number of spaces to the right of the decimal point in both parts of the problem and count over the same number in the answer and place the decimal there.

```
    $ 3.00
    x 20
    $60.00
```

2. How much should a mason charge for laying 400 brick if each brick costs 9¢ each and the mason got 18¢ for laying each brick? Include the cost of the brick in the problem.

Answer: (400 brick x $0.09) + (400 brick x $0.18) = $36.00 x $72.00 = $108.00.

```
    400          400          $  36.00
    x .09         x .18          +72.00
    $36.00       3200          $ 108.00
                 400
                 $72.00
```

PROBLEMS IN DIVIDING DECIMALS

1. If a mason charged $2100.00 for laying 14,000 brick, how much was received per brick?

Answer: $2100.000 ÷ 14,000 = $0.15 (15¢ each).

```
                .15
    14000 / 2100.00
            14000
            70000
            70000
                0
```

2. Clay flue lining is sold in 2 ft. lengths. What is the cost of one 2 ft. length if 56 ft. of lining costs $98.00?

Answer: 56 ÷ 2 = 28 pieces of lining. $98.00 ÷ 28 = $3.50 each.

```
              3.50
    28 / 98.00
         84
         140
         140
          00
```

CONVERTING FRACTIONS TO DECIMALS

Common fractions may be changed to decimals by adding one or more zeros to the numerator and dividing by the denominator. Thus, the fraction 3/4 may be changed to a decimal by dividing 3.00 by 4. The answer is .75. Therefore, .75 is equal to 3/4.

If the division does not come out even (with no remainder), it is carried out as far as desired. For example, the fraction 1/3 is changed to a decimal by dividing 1.000 by 3. The answer is .333 . . . .

When a mixed number such as 3 1/2 must be changed to a decimal form, the mixed number is first changed to a fraction (3 1/2 = 7/2) and then the division is performed as above:

```
         3.50
    2 / 7.00          3.50 = 3 1/2
        6
        10
        10
         0
```

PROBLEMS IN CONVERTING FRACTIONS TO DECIMALS

1. What is the labor cost of laying 12 3/4 cu. yd. of rubble stone if the cost to lay it is $9.50 per cu. yd.?

Answer: 12 3/4 = 12.75; 12.75 x $9.50 = $121.13.

2. What will the cost be for 24 1/4 cu. yd. of fill if it sells for $2.25 per cu. yd.?

Answer: 24.25 x $2.25 = $54.56.

CONVERTING DECIMALS TO FRACTIONS

A decimal may be changed to a fraction by dropping the decimal point and using the number as the numerator of the fraction. The denominator is written with a "1" and as many zeros after it as there were decimal places in the original decimal number. Thus, .125 may be changed to a fraction by writing $\frac{125}{1000}$ which equals 1/8.

PROBLEMS IN CONVERTING DECIMALS TO FRACTIONS

1. A mason read on a drawing that the reinforcing steel required for a certain situation was .375 in. in diameter. What is the fractional size of this steel?

Answer: $\frac{375}{1000}$ = 3/8 in. in diameter.

2. What is the width of a piece of steel angle which is specified as 3.875 in.? Answer should be a fraction rather than a decimal.

Answer: $3\frac{875}{1000}$ = 3 7/8 in. wide.

PERCENTAGE AND INTEREST

Percentage is a term used to denote parts of a hundred. Interest is a type of percentage. It represents the cost of borrowing money.

PERCENTAGE

Percentage is indicated by the % sign. For example, 3% of a number is three hundredths of the number. It may be indicated as 3/100, .03 or 3%.

Finding percentage is done by multiplying. The number should be written in the decimal form (.03) when computing the answer. If 4% of 1452 were desired, then 1452 would be multiplied by .04. The answer or product is 58.08.

Study the following conversions before trying the problems:

.25 = 25%		.364 = 36.4%
.75 = 75%		.5748 = 57.48%
.2 = 20%		1/4 = .25 = 25%
.7 = 70%		3/8 = .375 = 37.5%

PROBLEMS IN PERCENTAGE

1. A mason calculated that 5 cu. yd. of sand was needed for a bricklaying job. Because of waste the mason added 5%. How much sand should be ordered?
   Answer: .05 x 5 = .25 cu. yd. Therefore, 5 + .25 = 5.25 cu. yd. should be ordered.
2. It was estimated that 12,000 brick would be required for a job but an additional 2% were added because of breakage, bats and salmon brick. How many brick should be ordered?
   Answer: .02 x 12,000 = 240. Therefore, 12,000 + 240 = 12,240 brick.

INTEREST

Interest is usually considered to be the cost of borrowing money. The amount borrowed is called the principal and the percentage charged per year is the rate of interest. Interest may be figured by the year, month or day.

To illustrate, if $500 were borrowed for one year (12 months) at 8% interest, the interest to be paid would be $500.00 x .08 = $40.00. The same amount of money and the same rate of interest would cost $6.66 for two months. ($500.00 x .08) ÷ 6. The total is divided by 6 because two months is 1/6 of a year.

PROBLEMS IN INTEREST

1. If a mason deposited $1000 in a savings account for 5% interest per year, how much interest would he earn at the end of 12 months?
   Answer: $1000 x .05 = $50.00.
2. A contractor purchased a new piece of equipment for $2500 and agreed to pay 8% interest per year. How much interest would be owed at the end of the year?

Answer: $2500 x .08 = $200.00 in interest for 12 months. How much interest for six months?
Answer: $200.00 ÷ 2 = $100.00 in interest for six months.

APPLIED GEOMETRY

The properties of points, lines and planes, Fig. 11-2, are a definite concern of the mason. This section is designed to clarify terms commonly used in geometry that have a direct application to the masonry trades.

POINTS

Points are basic elements of lines. They have no width, length or height, but merely indicate a location. Fig. 11-2 shows recommended symbols for a point. Points may be identified with a name, number, letter or other symbol.

LINES

A line has only length. It is the distance between two or more points and may be either straight or curved. A horizontal line is a level line. A vertical line is a plumb line and is perpendicular to a horizontal line. Perpendicular means that the line forms a 90 degree angle (right angle) with respect to another line. The symbol for a perpendicular line is shown in Fig. 11-2.

A diagonal line is one joining two opposite angles, Fig. 11-2. Parallel lines remain a constant distance apart and never cross, Fig. 11-2. Curved lines may be either arcs with a single center or irregular curves with many centers. A circle is an example of a curve with a single center. It is a 360 degree arc. An irregular or free curve is represented in Fig. 11-2.

ANGLES

An angle is formed by two intersecting lines. An angle greater than 90 degrees is called an obtuse angle. An angle of 180 degrees is a straight angle. A right angle has 90 degrees and an acute angle has less than 90 degrees.

Angles are measured in degrees, minutes and seconds. Each degree is divided into 60 minutes and each minute is divided into 60 seconds.

CIRCLES

A circle is a continuous curve being an equal distance from its center at all times. Fig. 11-3 identifies the parts of a circle.

The circumference of a circle is the distance around it. A semi-circle is equal to half the circumference. An arc is any portion of the circumference. The diameter of a circle is the distance across a circle through its center. The radius is the distance from the center to any point on the circumference or one-half the diameter. A sector is the portion of a circle between two radii (more than one radius). A segment is the portion of a circle contained by a straight line and the circumference which it cuts off. A tangent line is a line which touches the circle, but does not cut it and is at right angles to a

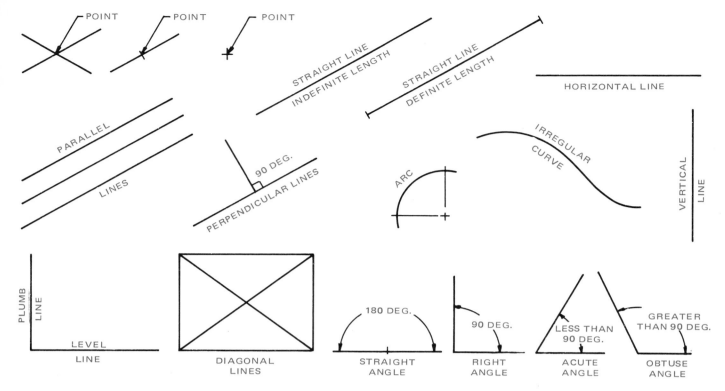

Fig. 11-2. Lines, points and planes. These elements combine to form edges, corners and surfaces of all objects.

straight line from the center.

The distance around a circle (circumference) has been found to have a direct relationship to its diameter. The circumference is equal to a constant of 3.1416 multiplied by the diameter. This constant is called "pi" and is represented by the symbol (π). Thus, the formula, C = π D.

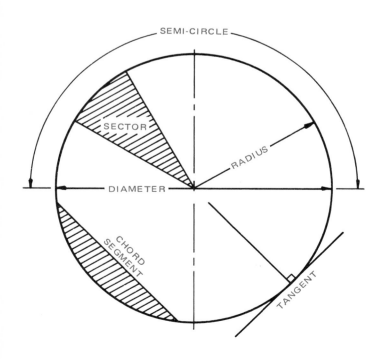

Fig. 11-3. The circle and its parts.

TRIANGULAR RELATIONSHIPS

A triangle has three angles which total 180 degrees. When two angles are known, the third may be found by subtracting the sum of the two angles from 180 degrees.

A right triangle has one angle which is 90 degrees and therefore the sum of the other two is also 90 degrees. The longest side of a right triangle is called the hypotenuse. See Fig. 11-4.

The square of the hypotenuse of a right triangle is equal to the sum of the squares of the other two sides. For example, if a right triangle had two sides equal to 3 ft. and 4 ft., then the other side (hypotenuse) would equal 5 ft. (Hypotenuse^2 = Side A^2 + Side B^2. Therefore Hyp.^{2} = $3^2 + 4^2$ = 9 + 16 = 25.) Five multiplied times itself is equal to 25. Thus, the

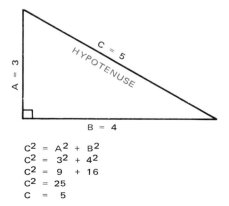

$$c^2 = A^2 + B^2$$
$$c^2 = 3^2 + 4^2$$
$$c^2 = 9 \ + 16$$
$$c^2 = 25$$
$$c \ = \ 5$$

Fig. 11-4. The length of the sides of a right triangle always have a 3:4:5 proportion.

Mathematics for Masonry Trades

AREAS OF PLANE FIGURES

PARALLELOGRAM

$A = B \times H$

TRAPEZOID

$A = \dfrac{B + C}{2} \times H$

TRIANGLE

$A = \dfrac{B \times H}{2}$

REGULAR POLYGON

$A = \dfrac{\text{SUM OF SIDES (S)}}{2} \times R$

CIRCLE

$A = \pi R^2$

$A = .7854 \times D^2$

$A = .0796 \times C^2$

ELLIPSE

$A = M \times m \times .7854$

VOLUMES OF SOLID FIGURES

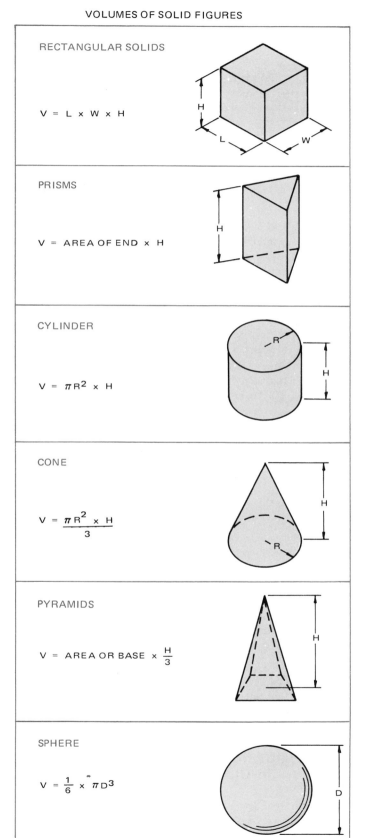

RECTANGULAR SOLIDS

$V = L \times W \times H$

PRISMS

$V = \text{AREA OF END} \times H$

CYLINDER

$V = \pi R^2 \times H$

CONE

$V = \dfrac{\pi R^2 \times H}{3}$

PYRAMIDS

$V = \text{AREA OR BASE} \times \dfrac{H}{3}$

SPHERE

$V = \dfrac{1}{6} \times \pi D^3$

Fig. 11-5. Formulas for calculating areas or volumes of typical geometric shapes.

hypotenuse is 5 ft. This is also called a 3:4:5 triangle and is very useful for laying out square corners for foundations, etc.

AREAS AND VOLUMES

Much of the math presented thus far will be used to find areas and/or volumes related to a job. Areas are two-dimensional (length x width) and are shown in square measure such as 1296 sq. in., 9 sq. ft. or 1 sq. yd. Volumes are three-dimensional quantities (length x width x height) and are shown in cubic measure (46,656 cu. in., 27 cu. ft. or 1 cu. yd.). Fig. 11-5 shows several plane and solid figures with the formulas for calculating their areas or volumes.

PROBLEMS IN CALCULATING AREAS
1. A large window opening is 72 3/8" x 60 3/8". What is the area of the opening?
   Answer: Area for a rectangle is length x height. A = L x H. Therefore, A = 72 3/8" x 60 3/8" = 72.375" x 60.375" = Area = 4369,64 sq. in. or 30.34 sq. ft. (Divide by 144.)
2. What is the area of a circular slab whose radius is 6 ft.?
   Answer: $A = \pi r^2$ (π is a constant 3.1416 or 22/7). Area = 3.1416 x 6^2. (6^2 means that the number is multiplied times itself or 6 x 6 = 36.) A = 3.1416 x 36. Area = 113.10 sq. ft. of slab.

PROBLEMS IN CALCULATING VOLUMES
1. A common brick is 8" x 2 1/4" x 3 3/4". What is its volume in cubic inches?
   Answer: Volume = L x W x H (length x width x height).
   V = 8" x 2 1/4" x 3 3/4".
   V = 8" x 2.25" x 3.75".
   V = 67.5 cu. inches.
2. How many cubic yards of concrete would be required to fill a cylindrical form 12 ft. long and 18 in. in diameter?
   Answer: Volume = πr^2 x H. (The radius is 1/2 the diameter.)
   V = 3.1416 x $(9")^2$ x 144" (must change feet to inches).
   V = 3.1416 x 81 sq. in. x 144 in.
   V = 36644.788 cu. in. or 21.21 cu. ft. or 0.79 cu. yd.
   Divide cubic inches by 1728 to determine cubic feet and divide by 27 cu. ft. to determine cubic yards. (1 cu. yd. = 27 cu. ft. = 46,656 cu. in.)

METRIC MEASUREMENT AND ENGLISH CONVERSIONS

Through the years more and more countries have begun using the metric system. The United States is changing from the English FPS (Foot-Pound-Second) system to SI metrics. It is therefore important that we become familiar with the metric units and their relationship to the familiar English units.

SI UNITS OF MEASURE

The offical name of the new metric system is "System International de Unite." Its abbreviation is "SI."

SI metric is a very convenient system which uses a base of 10 to form the various units of measure. SI metric units include the following:
1. Base units.
2. Supplemental units.
3. Derived units which have been identified by name.
4. Non-SI units.

BASE UNITS
There are seven base units to which all other units are related. These units, with the exception of the kilogram, are based on natural phenomena which may be duplicated in the laboratory and require no international prototypes which must be stored by National Standards Organizations.

The seven base units are:

Quantity	SI Unit	SI Symbol
Length	metre	m
Mass (weight)	kilogram	kg
Time	second	s
Temperature	degree Kelvin	K
Electric current	ampere	A
Luminous intensity	candela	cd
Amount of substance	mole	mol

DERIVED UNITS
Derived units are formed by multiplication or division of two or more SI base units. Some of the SI derived units are as follows:

Quantity	SI Unit	SI Symbol
Electrical charge	coulomb	C
Electrical potential (voltage)	volt	V
Electrical resistance	ohm	Ω
Electrical conductance	siemens	S
Electrical capacitance	farad	F
Electrical inductance	henry	H
Energy (heat or work)	joule	J
Force	newton	N
Frequency	hertz	Hz
Illumination	lux	lx
Luminous flux	lumen	lm
Magnetic flux	weber	Wb
Magnetic flux density	tesla	T
Power	watt	W
Pressure	pascal	Pa

SI PREFIXES
SI prefixes are formed by multiplying or dividing base units by powers of 10. This is an advantage because it eliminates insignificant digits and decimals. The accepted SI prefixes are:

Multiplication Factors		Prefix	Symbol
1 000 000 000 000	= 10^{12}	tera	T
1 000 000 000	= 10^{9}	giga	G
1 000 000	= 10^{6}	mega	M
1 000	= 10^{3}	kilo	k
100	= 10^{2}	hecto	h
10	= 10^{1}	deka	da
BASE UNITS 1	= 10^{0}		
0.1	= 10^{-1}	deci	d
0.01	= 10^{-2}	centi	c
0.001	= 10^{-3}	milli	m
0.000 001	= 10^{-6}	micro	μ
0.000 000 001	= 10^{-9}	nano	n
0.000 000 000 001	= 10^{-12}	pico	p
0.000 000 000 000 001	= 10^{-15}	femto	f
0.000 000 000 000 000 001	= 10^{-18}	atto	a

Note: Multiples of 1 000 of the base unit are to be used in preference to others.

NON-SI UNITS

Certain units of measure which are not a part of the SI system are so widely used and accepted that they have been recognized. They include the following:

Quantity	SI Name	SI Unit	Base Unit
Liquid volume	litre	L	metre
Area	square metre	m^2	metre
Solid volume	cubic metre	m^3	metre
Temperature	celsius	C	kelvin
Angle	degree	°	radian
	minute	'	radian
	second	"	radian
Time	minute	min	second
	hour	h	second
	day	d	second
Mass	tonne	t	kilogram

CONVERTING ENGLISH AND METRIC UNITS

Eventually the SI system of measurement will be adopted all over the world. However until then, the need to convert English to metric and metric to English will exist. Fig. 11-6 shows factors for converting from English to metric units. To convert a quantity from English to metric units follow the procedure listed below.

1. English measurements in fractional form should be changed to decimal form.
2. Multiply the quantity by the factor shown in Fig. 11-6.
3. Round off the result to the degree of accuracy required.

Remember: The meter is the standard of length for metric system and all divisions or multiples relate to it. For example:

1 metre	= 1000 *milli*metres	(thousandths)
1 metre	= 100 *centi*metres	(hundredths)
1 metre	= 10 *deci*metres	(tenths)
10 metres	= 1 *deka*metre	(tens)
100 metres	= 1 *hecto*metre	(hundreds)
1000 metres	= 1 *kilo*metre	(thousands)

These prefixes are used for every weight and measure in the metric system.

Fig. 11-7 shows factors for converting from metric to English units. Follow the procedure below to convert a quantity from metric to English units.

1. Multiply the quantity by the factor shown in Fig. 11-7.
2. Round off the result to the degree of accuracy required.
3. Convert the decimal part of the number to the nearest common fraction for most trade applications.

CONVERSION PROBLEMS

1. A residential structure is 40 ft. x 60 ft. What are its dimensions in metres?
   Answer: Multiply 40 ft. x 30.48 cm = 1219.2 cm.
   (1 m = 100 cm. Therefore, divide by 100.) 40 ft. = 12.192 m.
   Multiply 60 ft. x 30.48 cm = 1828.8 cm ÷ 100 = 18.288 m.
   60 ft. = 18.288 m. 40 ft. x 60 ft. = 12.192 m x 18.288 m.

LENGTHS		WEIGHTS	
1 INCH = 2.540 CENTIMETRES		1 OUNCE (AVDP) = 28.35 GRAMS	
1 FOOT = 30.48 CENTIMETRES		1 POUND = 453.6 GRAMS OR 0.4536 KILOGRAM	
1 YARD = 91.44 CENTIMETRES OR 0.9144 METRES		1 (SHORT) TON = 907.2 KILOGRAMS	
1 MILE = 1.609 KILOMETRES			

LIQUID MEASUREMENTS

AREAS		1 (FLUID) OUNCE = 0.02957 LITRE OR 28.35 GRAMS
1 SQ. IN. = 6.452 SQ. CENTIMETRES		1 PINT = 473.2 CU. CENTIMETRES
1 SQ. FT. = 929.0 SQ. CENTIMETRES OR 0.0929 SQ. METRE		1 QUART = 0.9463 LITRE
1 SQ. YD. = 0.8361 SQ. METRE		1 (US) GALLON = 3785 CU. CENTIMETRES OR 3.785 LITRES

POWER MEASUREMENTS

VOLUMES	
1 CU. IN. = 16.39 CU. CENTIMETRES	1 HORSEPOWER = 0.7457 KILOWATT
1 CU. FT. = 0.02832 CU. METRE	
1 CU. YD. = 0.7646 CU. METRE	

TEMPERATURE MEASUREMENTS

TO CONVERT DEGREES FAHRENHEIT TO DEGREES CENTIGRADE, USE THE FOLLOWING FORMULA:

$$C = 5/9 \times (F-32)$$

Fig. 11-6. English to metric conversions.

LENGTHS		WEIGHTS	
1 MILLIMETRE (mm)	= 0.03937 IN.	1 GRAM (G)	= 0.03527 OZ.
1 CENTIMETRE (cm)	= 0.3937 IN.	1 KILOGRAM (kg)	= 2.205 LB.
1 METRE (m)	= 3.281 FT. OR 1.0937 YD.	1 METRIC TON	= 2205 LB.
1 KILOMETRE (km)	= 0.6214 MILES	**LIQUID MEASUREMENTS**	
AREAS		1 CU. CENTIMETRE (cm³)	= 0.06102 CU. IN.
1 SQ. MILLIMETRE	= 0.00155 SQ. IN.	1 LITRE (1000 cm³)	1.057 QUARTS OR = 2.113 PINTS OR 61.02 CU. IN.
1 SQ. CENTIMETRE	= 0.155 SQ. IN.		
1 SQ. METRE	= 10.76 SQ. FT. OR 1.196 SQ. YD.	**POWER MEASUREMENTS**	
VOLUMES		1 KILOWATT (kw)	= 1.341 HORSEPOWER (hp)
1 CU. CENTIMETRE	= 0.06102 CU. IN.	**TEMPERATURE MEASUREMENTS**	
1 CU. METRE	= 35.31 CU. FT. OR 1.308 CU. YD.	TO CONVERT DEGREES CENTIGRADE TO DEGREES FAHRENHEIT, USE THE FOLLOWING FORMULA: $F = (9/5 \times C) + 32$	

Fig. 11-7. Metric to English conversions.

2. A metric drawing showed a recess to be 10.16 cm deep by 20.32 cm wide. What are its dimensions in inches?

Answer: Divide each dimension by 2.54 cm since 2.54 cm = 1 in. 10.16 cm ÷ 2.54 = 4 in., 20.32 cm ÷ 2.54 = 8 in. The recess is 4 in. deep by 8 in. wide.

WEIGHT AND MEASURES

Various weights and measures are used in estimating proportioning and ordering materials. The mason must be familiar with common weights and measures so that estimates will be accurate. Some of the more frequently used lengths, quantities, and their equivalents are presented to the right as a ready source for the mason. At some point in the future, these measurements will be made in metric.

144 sq. in. = 1 sq. ft.
9 sq. ft. = 1 sq. yd.
1296 sq. in. = 1 sq. yd.
1728 cu. in. = 1 cu. ft.
27 cu. ft. = 1 cu. yd.
1 cu. ft. of water weighs 62.5 lb.
1 cu. ft. of damp loose sand equals 80 lb. of dry sand.
1 cu. ft. of cement weighs approximately 100 lb.
1 cu. ft. of cement equals 1 bag; 4 bags equal 1 barrel.
1 cu. ft. of hydrated lime weighs approximately 100 lb.
1 1/4 cu. ft. of hydrated lime equals one 50 lb. sack.
1 cu. ft. of brickwork equals 120 lb.
1 cu. ft. of concrete weighs approximately 150 lb.
1 cu. ft. of packed earth weighs approximately 100 lb.
1 cu. ft. of gravel weighs approximately 110 lb.

NOMINAL SIZE OF BRICK IN. T H L	NUMBER OF BRICK PER 100 SQ. FT.	**CUBIC FEET OF MORTAR**			
		PER 100 SQ. FT.		**PER 1000 BRICK**	
		3/8 IN. JOINTS	1/2 IN. JOINTS	3/8 IN. JOINTS	1/2 IN. JOINTS
4 x 2 2/3 x 8	675	5.5	7.0	8.1	10.3
4 x 3 1/5 x 8	563	4.8	6.1	8.6	10.9
4 x 4 x 8	450	4.2	5.3	9.2	11.7
4 x 5 1/3 x 8	338	3.5	4.4	10.2	12.9
4 x 2 x 12	600	6.5	8.2	10.8	13.7
4 x 2 2/3 x 12	450	5.1	6.5	11.3	14.4
4 x 3 1/5 x 12	375	4.4	5.6	11.7	14.9
4 x 4 x 12	300	3.7	4.8	12.3	15.7
4 x 5 1/3 x 12	225	3.0	3.9	13.4	17.1
6 x 2 2/3 x 12	450	7.9	10.2	17.5	22.6
6 x 3 1/5 x 12	375	6.8	8.8	18.1	23.4
6 x 4 x 12	300	5.6	7.4	19.1	24.7
NONMODULAR BRICK					
3 3/4 x 2 1/4 x 8	655	5.8		8.8	
	616		7.2		11.7
3 3/4 x 2 3/4 x 8	551	5.0		9.1	
	522		6.4		12.2

Fig. 11-8. Modular and nonmodular brick and mortar required for single wythe walls in running or stack bond. These figures do not allow for waste and breakage.

PROBLEMS IN WEIGHTS AND MEASURES

1. What part of a square foot does the side of a clay tile represent? The side is 8″ x 12″.
   Answer: 8″ x 12″ = 96 sq. in. (1 sq. ft. = 144 sq. in.)
   96/144 = 24/36 = 2/3 sq. ft.
2. If brickwork weighs 112 lb. per cu. ft., how many tons would a 16″ x 16″ x 12′ pier weigh?
   Answer: 16″ x 16″ = 256 sq. in., 256 sq. in. x 144 in. = 36,864 cu. in. (1 cu. ft. = 1728 cu. in.) 36,864 cu. in. ÷ 1728 cu. in. = 21.33 cu. ft., 112 lb. x 21.33 cu. ft. = 2,389.33 lb.

ESTIMATING BRICK MASONRY

Except for nonmodular "standard brick" (3 3/4″ x 2 1/4″ x 8″) and some oversize brick (3 3/4″ x 2 3/4″ x 8″), almost all of the brick produced in the United States fit the 4 in. modular grid system. Modular size masonry units greatly simplify the job of estimating the number of units required for a given wall area.

The most widely used procedure for estimating brick masonry is the "wall-area" method. It consists of multiplying the quantity of materials needed per square foot by the net wall area. Net wall area is the gross area less all wall openings.

There are three standard modular joint thicknesses: 1/4 in., 3/8 in. and 1/2 in. For a given nominal size masonry unit, the number of modular masonry units per square foot of wall will be the same regardless of the mortar joint thickness if the units are laid with the joint thickness for which they were designed. However, the number of nonmodular standard brick required per square foot of wall will vary with the thickness of the mortar joint.

When estimating, determine the net quantities of all materials before adding any allowances for waste and breakage. As a general rule, 5 percent is added for brick and 10 to 25 percent for mortar. These factors may vary considerably with job conditions.

Fig. 11-8 shows the net quantities of brick and mortar necessary to construct walls one wythe in thickness with various modular and nonmodular brick sizes using either 1/2 or 3/8 in. mortar joints. Quantities shown in Fig. 11-8 are for running or stack bond which contain no headers. Bonds which

BOND	CORRECTION FACTOR
FULL HEADERS EVERY FIFTH COURSE ONLY	1/5
FULL HEADERS EVERY SIXTH COURSE ONLY	1/6
FULL HEADERS EVERY SEVENTH COURSE ONLY	1/7
ENGLISH BOND (FULL HEADERS EVERY SECOND COURSE)	1/2
FLEMISH BOND (ALTERNATE FULL HEADERS AND STRETCHERS EVERY COURSE)	1/3
FLEMISH HEADERS EVERY SIXTH COURSE	1/18
FLEMISH CROSS BOND (FLEMISH HEADERS EVERY SECOND COURSE)	1/6
DOUBLE-STRETCHER, GARDEN WALL BOND	1/5
TRIPE-STRETCHER, GARDEN WALL BOND	1/7

Fig. 11-9. Correction factors for walls in Fig. 11-8 with full headers. Add to facing and subtract from backing. Note: Correction factors are applicable only to those brick which are twice as long as they are wide.

require full headers must use the correction factor shown in Fig. 11-9. Multiwythe walls also require a correction factor for the collar joint between the wythes, Fig. 11-10.

CUBIC FEET OF MORTAR PER 100 SQ. FT. OF WALL		
1/4 IN. JOINT	3/8 IN. JOINT	1/2 IN. JOINT
2.08	3.13	4.17

Fig. 11-10. Correction factors for collar joints. Quantities are given in cu. ft. of mortar for collar joints. Note: cu. ft./1000 units = 10 x cu. ft./100 sq. ft. of wall ÷ the number of units/sq. ft. of wall.

MORTAR YIELD

Accurate mortar yield calculations are based on absolute volume. They are complex and require extensive data about the materials being used. Most masons do not have the data nor do most jobs require critical mortar yield calculations. Therefore, the following "rule-of-thumb" may generally be used: *For each 1 cu. ft. of damp loose sand, the mortar yield will be 1 cu. ft.*

PROBLEMS IN ESTIMATING BRICK MASONRY

1. How many brick (nominal size of 4″ x 2 2/3″ x 8″) will be required to veneer a wall 24′0″ long and 4′0″ high? A 3/8 in. mortar joint is to be used. The wall has no openings.
   Answer: First, find the wall area: 24 ft. x 4 ft. = 96 sq. ft. Second, find the number of brick per sq. ft. in Fig. 11-8. The table shows 675 brick/100 sq. ft. or 6.75 brick/sq. ft. Next, multiply the number of brick/sq. ft. by the wall area. 6.75 x 96 = 648 brick. Note that 5.28 cu. ft. of mortar will be required to lay these brick (96% of 5.5).
2. How many brick and how much mortar will be needed to lay a wall 40′0″ long and 8′0″ high using 6″ x 4″ x 12″ SCR brick with 1/2 in. mortar joints? The wall has three window openings 3′0″ wide by 4′0″ high.
   Answer: Gross wall area = 40 ft. x 8 ft. = 320 sq. ft. Net wall area = 320 sq. ft. − (3 ft. x 4 ft. x 3) = 320 − 36 = 284 sq. ft. Brick/sq. ft. (Fig. 11-8) = 3.0. Therefore, 284 x 3 = 852 brick. Mortar required is 7.4 per 100 sq. ft. or .074/sq. ft., .074 x 284 = 21.02 cu. ft. of mortar.

REVIEW QUESTIONS — CHAPTER 11

1. Four pieces of stone trim weighed 10.5 lb., 17.4 lb., 23.8 lb. and 5.6 lb. What is the total weight of the four pieces of trim?
2. If a mason ordered 5 1/2 cu. yd. of sand and used 3 1/4 cu. yd., how much was left?
3. If four masons each averaged laying 358 brick a day, how many brick would they average laying in five days?
4. A wall area of 23 sq. ft. required 115 brick to face it. How many brick were needed for each square foot of area?
5. What is the height of one stretcher course of concrete block if the block are 7 5/8 in. high and the mortar joint is 3/8 in.?

6. If the total thickness of a frame wall is 5 in., the studs 3 1/2 in., the wall board 3/8 in. and the weatherboard 1/2 in., how thick is the siding?

7. If one isolated footing required 1/8 yd. of concrete, how much concrete would be needed for a footing 3/4 as large?

8. If 7/8 yd. of mortar was divided among three masons, how much mortar would each get?

9. Which of the following quantities is the largest?
   a. 0.25
   b. 0.025
   c. 0.0025
   d. 2.2500

10. What is the sum of the following quantities; 0.12, 2.03, 12.004, 9.0?

11. A masonry job which lasted four days required 13.5 cu. yd. of mortar. Records show that 6.8 cu. yd. were used the first two days. How many yd. were used the last two days of the job?

12. A mason purchased 5.5 cu. yd. of washed masonry sand for $7.50 per yd. How much was paid for the sand?

13. If a mason charged $0.15 per brick and the total bill was $2100.00, how many bricks were laid?

14. What is the decimal equivalent of 3 11/16 in.?

15. What is the fractional size of a piece of steel which is 2.375 in?

16. If 3 percent were allowed for breakage on 10,000 brick, how many brick could be used?

17. How much money would be earned on a $1000 savings certificate in one year at 6 3/4 percent interest per year?

18. The length of the hypotenuse of a right triangle is desired when the other two legs are 6 ft. and 8 ft. What is the length of the hypotenuse?

19. The circumference of a circle with a diameter of 6.54 in. is:
   a. 134.4 in.
   b. 20.55 in.
   c. 18.7 in.
   d. 2.05 in.

20. The area of a 20 ft. x 20 ft. garage floor is _____ sq. ft.

21. The volume of a cylindrical pipe 6 ft. long with an inside diameter of 4 in. is:
   a. 904.8 cu. ft.
   b. 75.4 cu. ft.
   c. 75.4 cu. in.
   d. 904.8 cu. in.

22. One meter is equal to:
   a. 1000 millimetres.
   b. 100 centimetres.
   c. 10 decimetres.
   d. All of the above.

23. One inch is equal to:
   a. 2.54 cm.
   b. 36 m.
   c. 1/38 m.
   d. 100 mm.

24. How many cubic inches does 1 cu. ft. contain?

25. One cubic yard is equal to _____ cubic feet.

26. How many 4'' x 2 2/3'' x 8'' brick will be required to lay a 200 sq. ft. wall area if 675 brick are needed for each 100 sq. ft.?

Chapter 12
BLUEPRINT READING

Construction drawings describe the size, shape, location and specifications of the elements of a structure. They are frequently called "working drawings," "blueprints" or "plans." A skilled worker must be able to read these drawings and understand the information contained in them. Otherwise, the worker could not build the structure as the designer intended. The drawings use standard symbols and notes which are recognized by tradespeople in the construction field.

If you are to understand the plans, you must visualize what the architect has drawn. You must know the meaning of the lines, symbols, abbreviations and notes shown. Together, these items make up the language of the construction industry. Learning to read plans is essentially learning a new language.

LINE SYMBOLS

Lines make up a large part of the symbols used on construction drawings. They are called the "Alphabet of Lines," Fig. 12-1. A line symbol is used to communicate more precisely. It is important to learn the use of each line.

BORDER LINES

Border lines are very heavy. They are used to form a boundary for the drawing. They assure the reader that no part of the drawing has been removed. What is more, they give a "finished" appearance to the drawing.

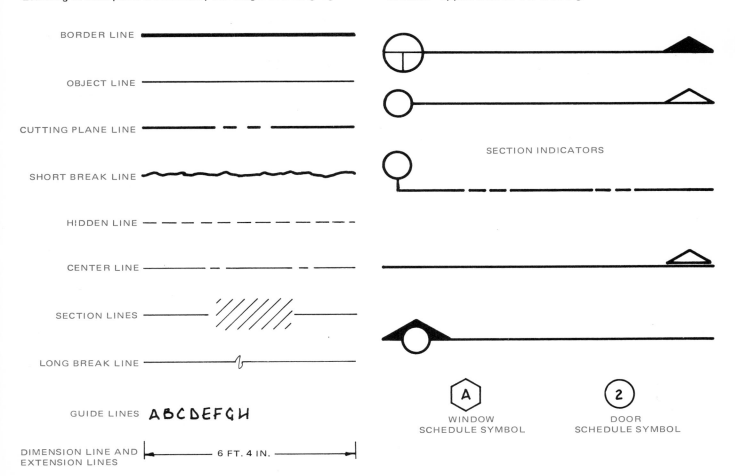

Fig. 12-1. Alphabet of Lines used on construction drawings.

OBJECT LINES

Object lines are heavy lines which show the outline of the visible features of an object. They should be easily seen since they represent important elements. Such things as walls, windows, patios and roof lines are represented by object lines.

HIDDEN LINES

Hidden lines represent an edge or intersection of two surfaces which are not visible in a given view. For example, the foundation wall and footings are represented by hidden lines on an elevation because they are below grade and therefore not visible. Also, hidden lines may be used to indicate features above the cutting plane, such as wall cabinets in a kitchen or an archway. Hidden lines are usually not as thick as object lines and are considered to be medium weight lines.

CENTER LINES

Center lines are thin lines which indicate the center of symmetrical objects. For example, a window or door may have a center line through it on the floor plan. Center lines simplify dimensioning and are used for location of features.

EXTENSION LINES

Extension lines are thin lines used to show where a dimension line ends. They extend from a portion of the object past the dimension line about 1/16 in. They "extend" the object for dimensioning purposes.

DIMENSION LINES

Dimension lines are thin lines used to show size or location of a feature of the structure. They may be placed outside or, if there is sufficient space, inside the object. All dimension lines have a dimension figure (number or letter) about halfway

between the ends. Each end has some type of termination (ending) symbol, Fig. 12-2.

LONG BREAK LINES

Long break lines are thin lines used to show that all of the part is not shown. They extend past the object about 1/16 in. on either side and have an "s" shape symbol in the center.

SHORT BREAK LINES

Short break lines are heavy lines used when part of the object is shown broken away to reveal a hidden feature. They are drawn freehand.

CUTTING PLANE LINES

Cutting plane lines are heavy lines used to indicate where the object has been sectioned to show internal features. They are generally labeled so that the proper section drawing will be identified for a specific cutting plane.

SECTION LINES

Section lines or crosshatch lines are very thin lines used to show that the feature has been sectioned. Section lines may represent a specific material or may be a general symbol. General section lines are usually drawn at 45 degrees and about 1/16 in. to 1/8 in. apart.

GUIDELINES

Guidelines are very light, thin lines used in lettering. They are for the drafter's use and help produce a neat clear drawing which is easy to read.

CONSTRUCTION LINES

Construction lines are also very light, thin lines which are drawn by the drafter in the process of making the drawing. They usually are not visible on a blueprint.

The various line symbols are identified on a simple floor plan in Fig. 12-3. Study these lines. They are the basic symbols used on all construction drawings.

SYMBOLS AND ABBREVIATIONS

The purpose of symbols and abbreviations is to conserve space. Some symbols look very much like the feature in real life. Many others do not. Fig. 12-4 shows:
1. Building material symbols.
2. Topographical symbols.
3. Plumbing symbols.
4. Climate control symbols.
5. Electrical symbols.
Many of these symbols are found on each set of construction drawings.
Symbols may represent an elevation (front, side or rear) or

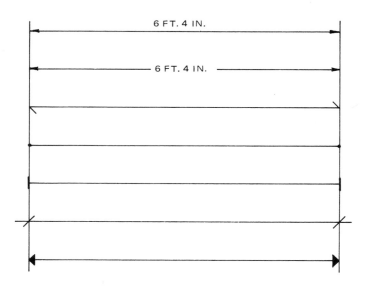

Fig. 12-2. Various methods of terminating a dimension line.

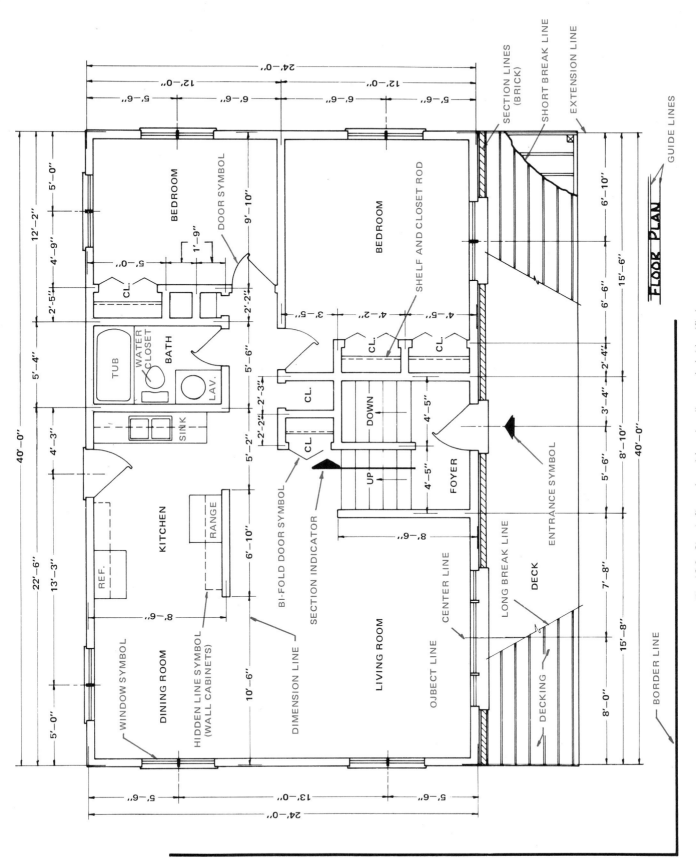

Fig. 12-3. Simple floor plan with various line symbols identified.

a section view of the material or feature. Elevation symbols are easiest to recognize because they are picture-like. Section symbols are not pictorial representations. They must be memorized or otherwise identified. *A note of caution: some symbols are used to represent several different materials or* *conditions. Care must be exercised in reading them.* For example, the face brick section symbol in Fig. 12-4 is the same as the general material symbol. The sand symbol (a series of dots) is the same in elevation and section. It is exactly like the section of cut stone, plaster and the elevation of cast concrete.

BUILDING MATERIAL SYMBOLS

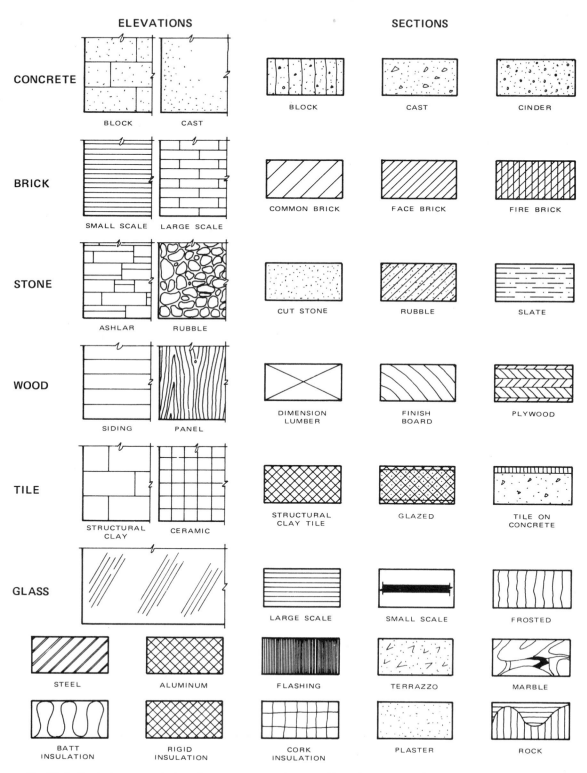

Fig. 12-4. Symbols commonly found on construction drawings. Note that some are shown in elevation and section.

TOPOGRAPHICAL SYMBOLS

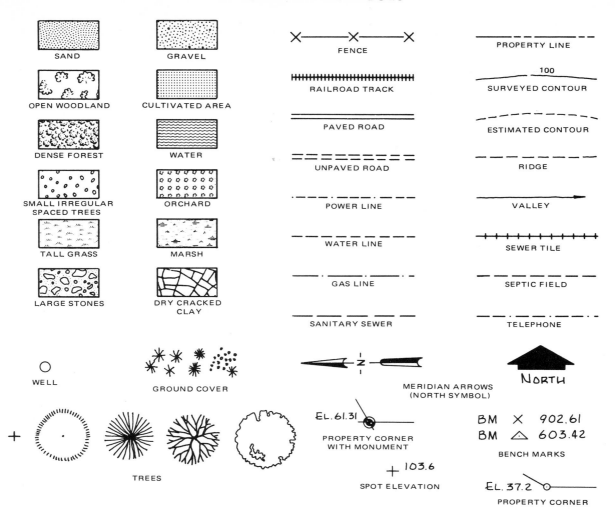

PLUMBING SYMBOLS

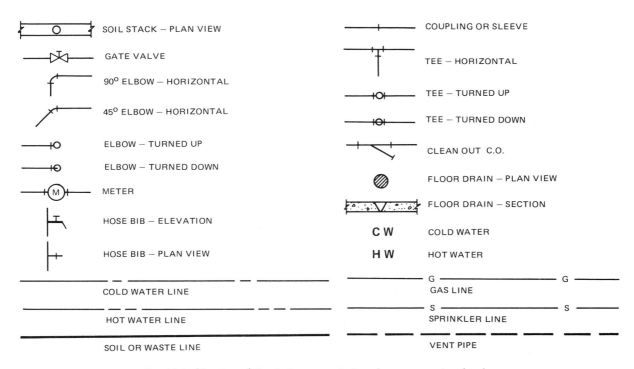

Fig. 12-4. (Continued) Symbols commonly found on construction drawings.

CLIMATE CONTROL SYMBOLS

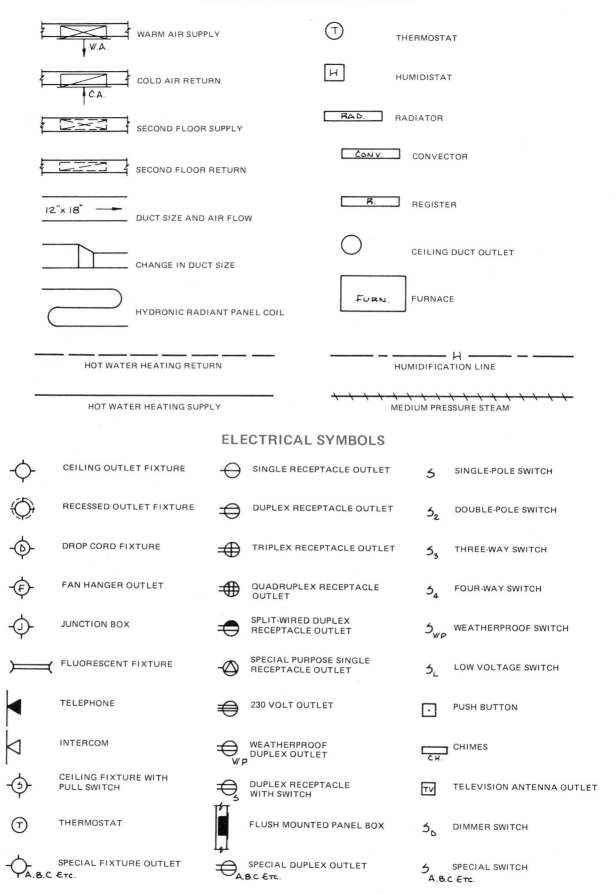

WARM AIR SUPPLY	THERMOSTAT
COLD AIR RETURN	HUMIDISTAT
SECOND FLOOR SUPPLY	RADIATOR
SECOND FLOOR RETURN	CONVECTOR
DUCT SIZE AND AIR FLOW	REGISTER
CHANGE IN DUCT SIZE	CEILING DUCT OUTLET
HYDRONIC RADIANT PANEL COIL	FURNACE
HOT WATER HEATING RETURN	HUMIDIFICATION LINE
HOT WATER HEATING SUPPLY	MEDIUM PRESSURE STEAM

ELECTRICAL SYMBOLS

CEILING OUTLET FIXTURE	SINGLE RECEPTACLE OUTLET	SINGLE-POLE SWITCH
RECESSED OUTLET FIXTURE	DUPLEX RECEPTACLE OUTLET	DOUBLE-POLE SWITCH
DROP CORD FIXTURE	TRIPLEX RECEPTACLE OUTLET	THREE-WAY SWITCH
FAN HANGER OUTLET	QUADRUPLEX RECEPTACLE OUTLET	FOUR-WAY SWITCH
JUNCTION BOX	SPLIT-WIRED DUPLEX RECEPTACLE OUTLET	WEATHERPROOF SWITCH
FLUORESCENT FIXTURE	SPECIAL PURPOSE SINGLE RECEPTACLE OUTLET	LOW VOLTAGE SWITCH
TELEPHONE	230 VOLT OUTLET	PUSH BUTTON
INTERCOM	WEATHERPROOF DUPLEX OUTLET	CHIMES
CEILING FIXTURE WITH PULL SWITCH	DUPLEX RECEPTACLE WITH SWITCH	TELEVISION ANTENNA OUTLET
THERMOSTAT	FLUSH MOUNTED PANEL BOX	DIMMER SWITCH
SPECIAL FIXTURE OUTLET A,B,C ETC.	SPECIAL DUPLEX OUTLET A,B,C ETC.	SPECIAL SWITCH A,B,C ETC.

Fig. 12-4. (Continued) Symbols commonly found on construction drawings.

Study the location of the symbol and type of drawing before deciding what the symbol represents.

Abbreviations also save space and are widely used on construction drawings. A tradesperson must be able to read and interpret all the abbreviations that have anything to do with his or her specific function. The Reference Section of this book lists many frequently-used abbreviations.

SCALE AND DIMENSIONS

Most drawings are drawn at a particular scale. Residential floor plans, elevations, foundation plans, etc. are generally drawn at 1/4" = 1' − 0" scale. This means that every 1/4 in. on the drawing is equal to 1' − 0" on the house. Details are almost always drawn at a larger scale such as 1/2" = 1' − 0" or 1" = 1' − 0." Do not be confused by the terms "size" and "scale." *Size* means inches to inches (1/4" to 1" or quarter size) and *scale* means inches to feet (1/4" to 1' − 0" or quarter scale).

For example, if the end of a building were 40 ft. long and drawn at 1/4 scale, it would be 10 in. on the drawing. However, if it were drawn at 1/4 size it would be 10 ft. long.

Drawings of large commercial buildings are many times drawn at scales other than 1/4" = 1' − 0." A scale of 1/8" = 1' − 0" is common. Be sure to look for the scale on the drawing so that you can visualize the relative size of the building elements. *As a rule, never measure a drawing to determine a length.* Look for the dimension. If the dimension is not given, then try to add or subtract other dimensions to determine the length you need. If you must measure the drawing, check it several places for accuracy.

Dimensions indicate size and location of building elements. They are very important and must be followed accurately. Dimensions are usually shown in feet and inches but they may be shown in inches if the length is less than 1 ft. or the length is some standard distance that is more easily recognized in inches.

Interior frame walls may be dimensioned several ways, Fig. 12-5. They may be dimensioned to the center of the wall, to the outside of the studs or to the outside of the finished wall. The technique used may be easily determined by remembering

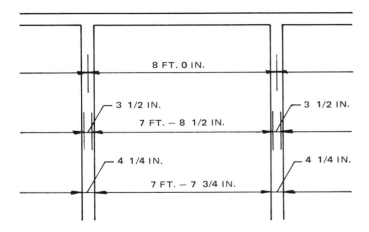

Fig. 12-5. Accepted methods of dimensioning interior frame walls.

that a 2 x 4 stud is actually 1 1/2" x 3 1/2." The thickness of the finished wall material can vary from 1/4 in. to 1 in. or more. Dimensioning to the center of the wall is the most commonly used method.

Interior masonry or concrete walls are dimensioned to the outside of the walls, Fig. 12-6. They are never dimensioned to the center.

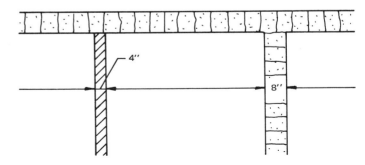

Fig. 12-6. Interior masonry and concrete walls are dimensioned to the outside of the walls and wall thickness is also shown.

Exterior walls, if they are frame, are dimensioned to the outside of the stud wall. This usually includes the sheathing but not the siding, Fig. 12-7. Again, masonry walls are dimensioned to the outside of the wall. Brick veneer is dimensioned to the outside of the stud wall.

Notes are frequently required to present information which cannot be represented by a conventional dimension or symbol. Be sure to read all notes. They contain information that will be needed to build the structure. A good procedure is to study the entire set of plans before starting work on the building. If there are any problems, these can be cleared up before work begins.

WORKING DRAWINGS

Working drawings are one group of drawings which may be required. For example, preliminary drawings are often prepared during the promotional stage of the building's design. Corrections and revisions are recorded on the preliminary drawings. They form the basis for the working drawings.

Presentation drawings are generally pictorial (picture-like) drawings. A pictorial drawing shows several faces of the structure to be built. It gives a better idea of how the structure will actually look.

There are several types of pictorial drawing. The one used most often by architects is called a perspective drawing. It is more natural in appearance — like a photograph. In a perspective you will notice that as parallel lines move away from the foreground they move closer together. If you study a photograph of a building the same thing happens.

There are many other types of working drawings needed to construct a building. Shop drawings are usually prepared by the trades participating in the project. They will use them to complete their part of the work.

Working drawings are generally prepared by the architect or designer. They are in graphic form showing the size, shape,

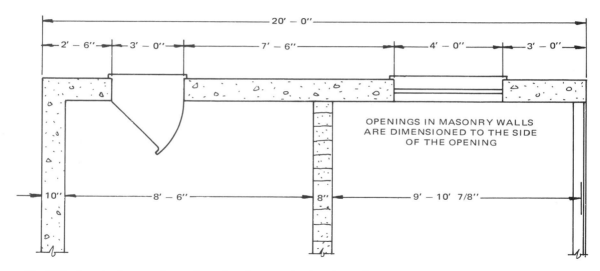

Fig. 12-7. Exterior frame walls are dimensioned to outside of stud wall. Sheathing is usually included because it is attached before lifting wall into place. Exterior masonry or concrete walls are dimensioned to the outside with thickness shown.

location, quantity and relationship of the parts of the building. They are produced in sufficient quantity so that each tradesperson has access to a set.

A set of working drawings for a typical structure might include the following:
1. Site plan.
2. Foundation plan.
3. Floor plan.
4. Elevations.
5. Electrical plan.
6. Mechanical plan.
7. Construction details.

The plans for a small bank building, Figs. 12-8 through 12-17, are included to illustrate the type of information commonly shown on the various drawings.

BRIEF DESCRIPTION OF THE DRAWINGS

The SITE PLAN shows the location of the building on the site, Fig. 12-9. It also may show utilities, topographical features, site dimensions, other buildings on the property, landscaping, walks, drives and retaining walls.

The FOUNDATION PLAN shows the foundation size and materials, Fig. 12-10. It may also give information about the excavation, waterproofing and supporting structures such as footings and/or piles. Accuracy of this drawing is very important. Upper parts of the building will depend on the layout and construction of the foundation.

The FLOOR PLAN shows all exterior and interior walls, doors, windows, patios, walks, decks, fireplaces, built-in cabinets and appliances, Fig. 12-11. A separate plan view is drawn for each floor. A floor plan is a section drawing usually cut about 4 ft. above the floor. Small buildings are usually drawn at 1/4" = 1' − 0" scale, however larger buildings may be drawn at 1/8" = 1' − 0" or some other scale.

The ELEVATIONS are drawn for each side of the structure, Fig. 12-12. The drawings show outside features such as placement and height of windows, doors, chimney and roof lines. Exterior materials are indicated as well as important vertical dimensions.

The ELECTRICAL PLAN is drawn from the floor plan, Fig. 12-13. It locates switches, convenience outlets, ceiling outlet fixtures, TV jacks, service entrance location and panel box. It also gives general information concerning circuits and special installations.

The MECHANICAL PLAN shows the plumbing, heating and/or cooling systems, Fig. 12-14. Sometimes each of these systems is shown on a separate plan. They would be designated as a Plumbing Plan, Heating Plan and Cooling Plan. Heating, cooling, humidification, dehumification and air cleaning might be drawn on a single Climate Control Plan.

CONSTRUCTION DETAILS are generally drawn of features when more information is needed on how to build them. Typical details include wall sections, sections through the entire structure, stairs, fireplaces, special masonry or other unique construction. See Fig. 12-15.

In addition to these drawings, a Roof Plan, Framing Plan, Landscaping Plan, Presentation Plan or Shop Drawings may be needed. A complex structure usually requires more detailed drawings.

A ROOF PLAN is included if the roof is complex or not clearly shown on other drawings, Fig. 12-16. It may be included in the site plan.

A FRAMING PLAN is designed to show, in detail, the framing required for a roof, floor or other framed area, Fig. 12-17.

A LANDSCAPING PLAN is sometimes combined with the site plan. It locates and identifies plants and other elements included in lanscaping.

A PRESENTATION DRAWING shows how the finished structure will appear. The two-point perspective is commonly used.

SHOP DRAWINGS provide additional information about such things as reinforcing steel, complex cabinetwork and electronic systems. This information may be furnished by suppliers or the various trades that have a part in the construction or installation.

Fig. 12-8. Three photographic views of small bank building. The following pages show working drawings used by the builders.

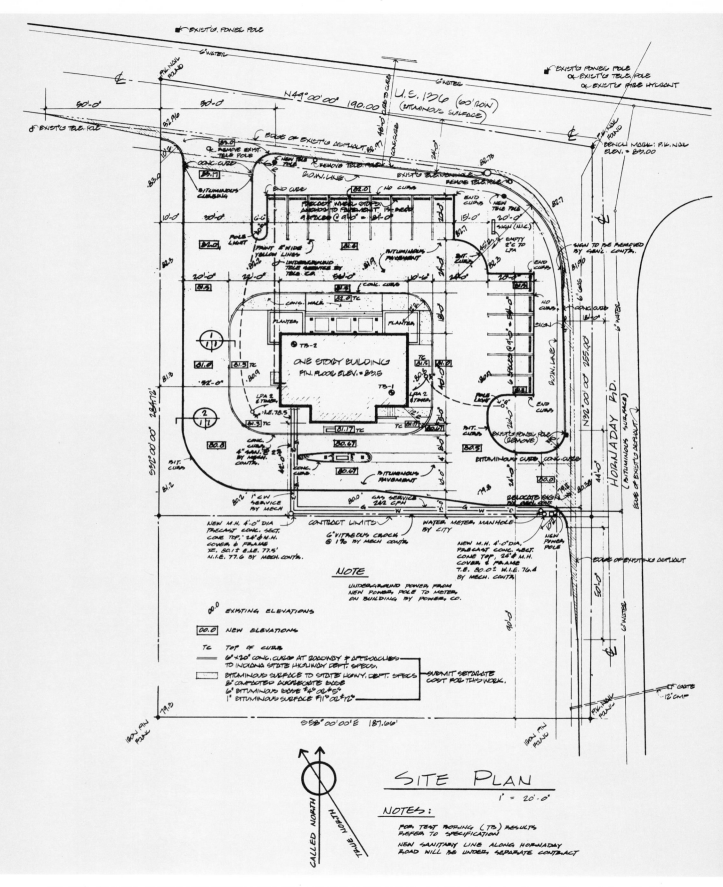

Fig. 12-9. Site plan of small bank building. This plan includes: length and bearing of lot lines; location, outline and size of building; elevation of property corners and spot elevations; meridian arrow (north symbol); streets, driveways and sidewalks; location of utilities; right-of-way lines; curbs; and scale of drawing.

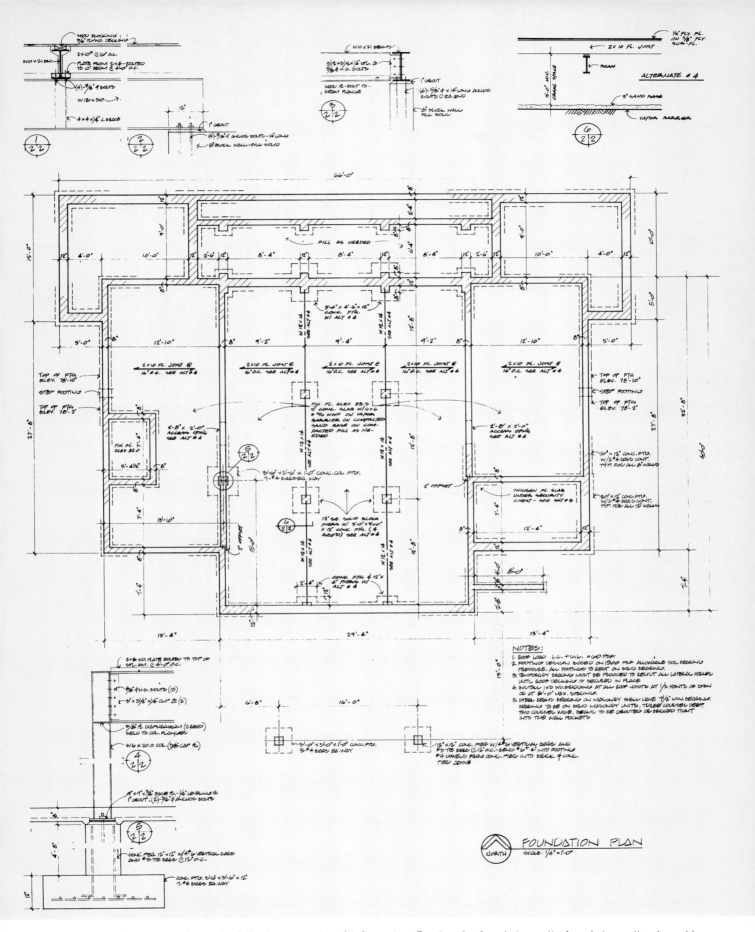

Fig. 12-10. Foundation plan shows the following categories of information: Footings for foundation walls; foundation walls; piers with footings; stepped footings; access openings in the foundation wall; beams and pilasters; direction, size and spacing of floor joists; cutting planes to show details; complete dimensions and notes; scale.

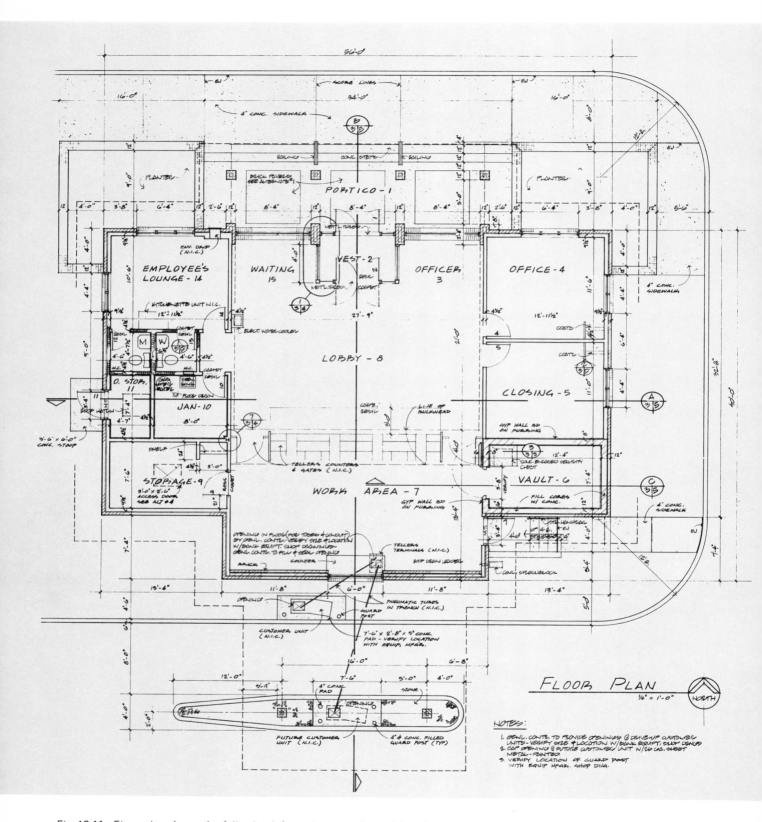

Fig. 12-11. Floor plan shows the following information: exterior and interior walls; location of windows and doors; built-in counters and fixtures; steps and sidewalks; room names; material symbols; location and size dimensions; notes; section locations; scale.

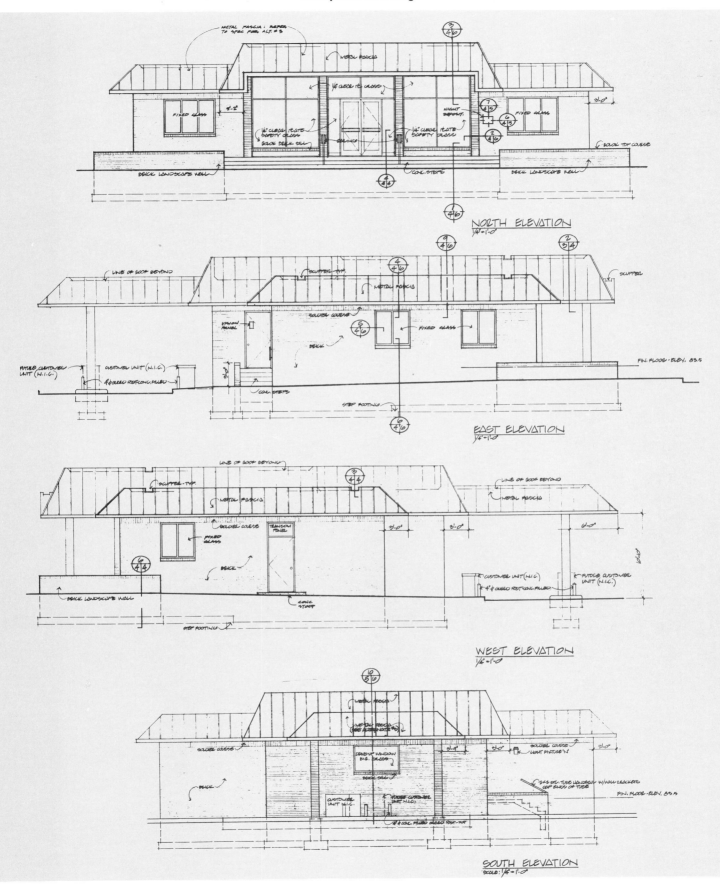

Fig. 12-12. Elevation drawings show: grade lines; finish floor level; location of exterior wall corners; windows and doors; roof features; vertical dimensions of important features; material symbols; section locations; footing and foundation depths; and scale.

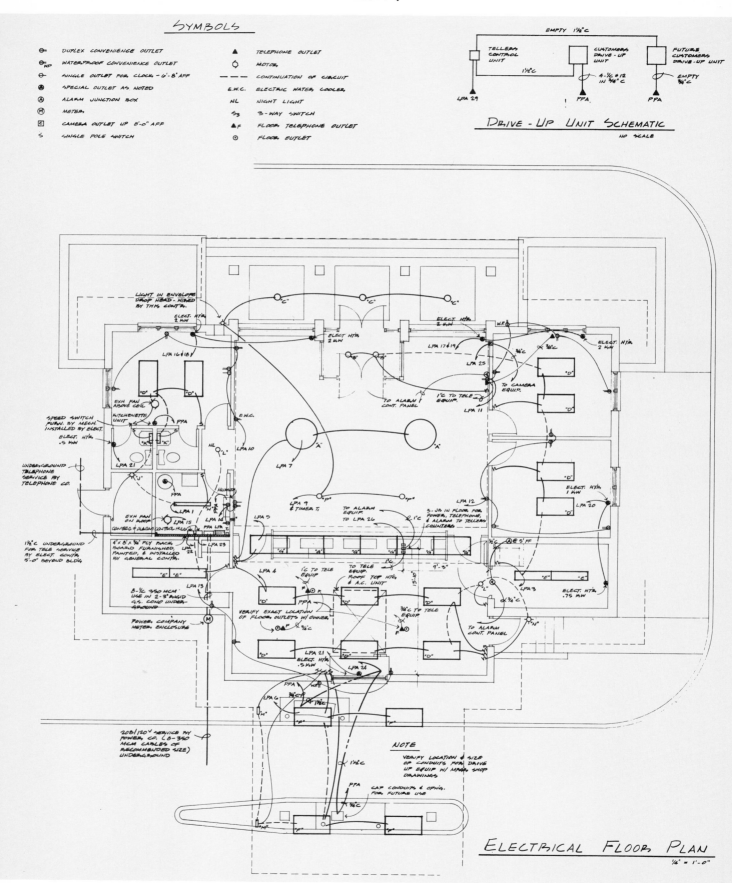

Fig. 12-13. Electrical Plan includes meter and service panels, service entrance capacity, placement and type of switches, location and type of lighting fixtures, special electrical equipment, symbols and legend, notes describing the system and the scale.

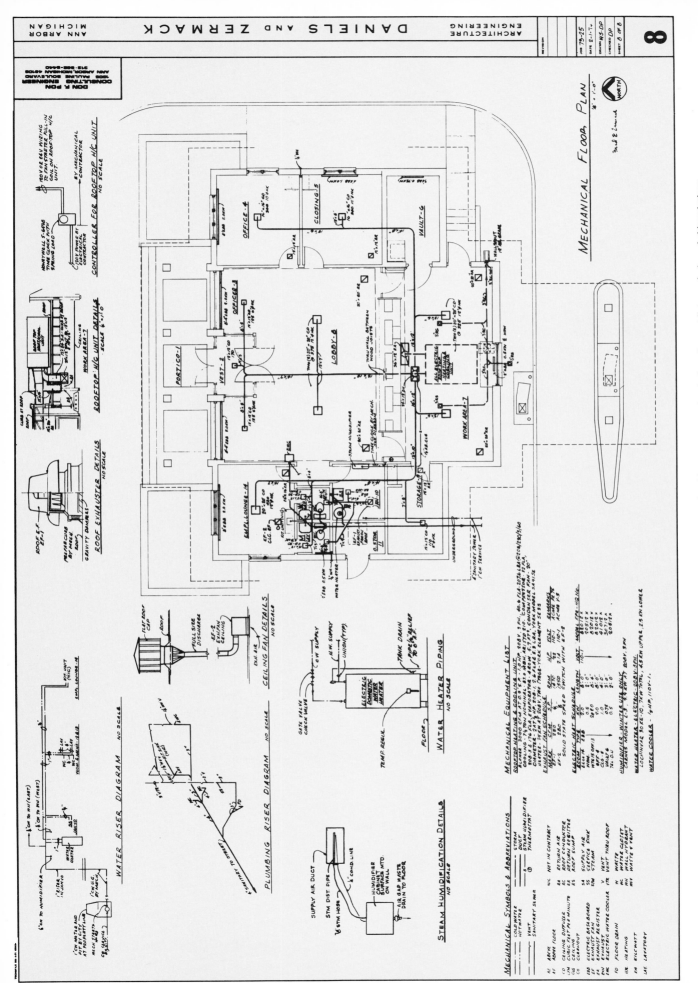

Fig. 12-14. Mechanical plan showing heating and cooling system, plumbing system, mechanical equipment list, symbols and abbreviations, various details and the scale.

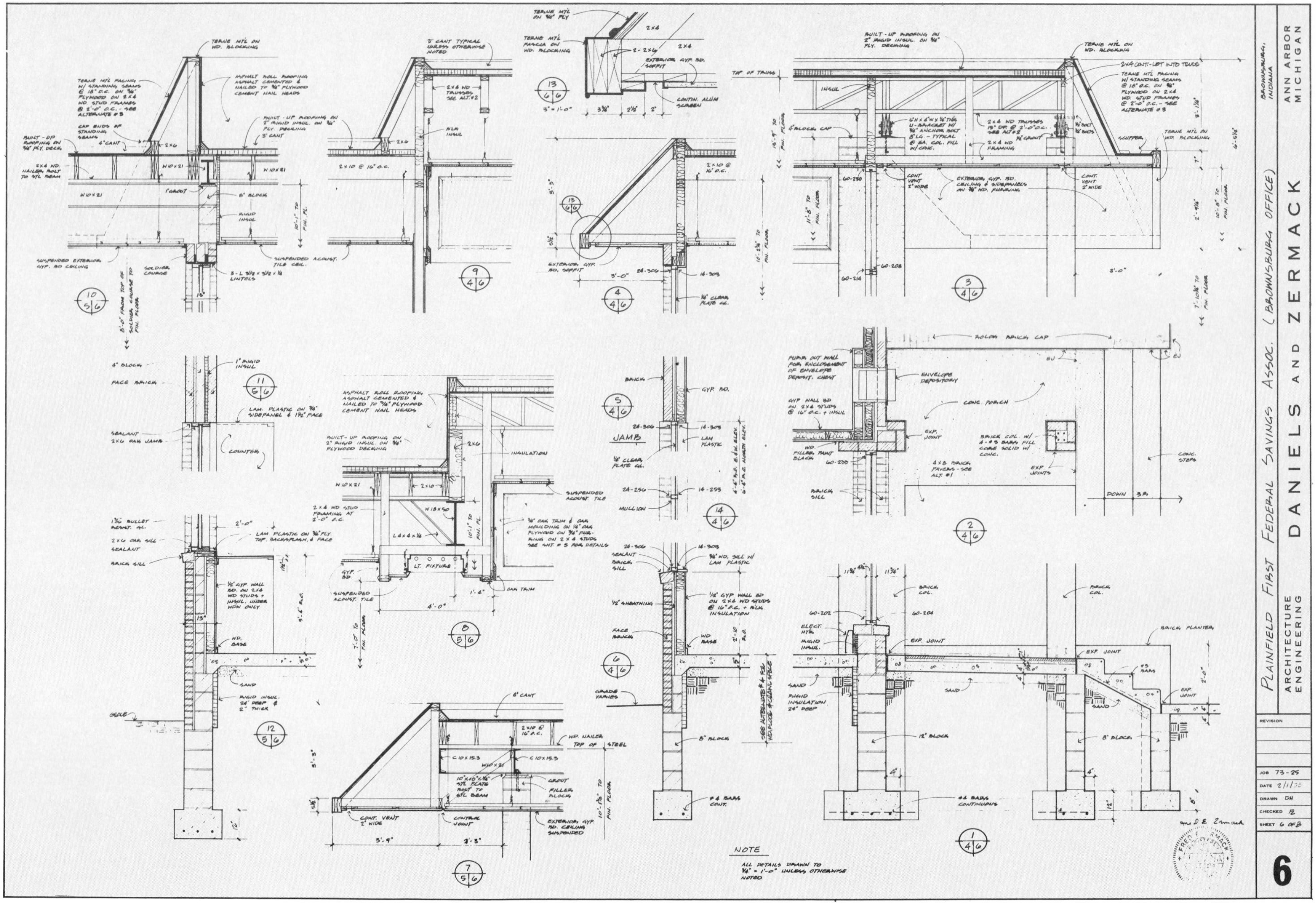

Fig. 12-15. Construction details of wall sections; roof sections; foundation and footing details; window and door heads, jambs and sills; suspended ceilings; piers and columns; depository; glazing details; scupper details and details of steps.,

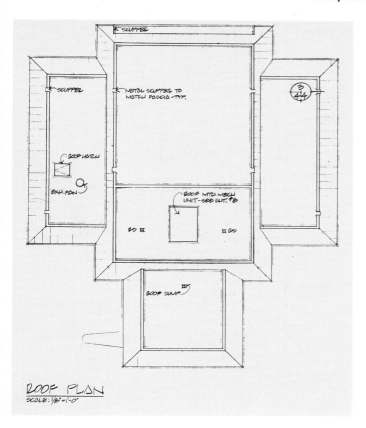

ROOF PLAN
SCALE: 1/8"=1'-0"

Fig. 12-16. Roof plan showing roof lines, scuppers, roof sump, roof mounted mechanical unit, exhaust fan location and scale.

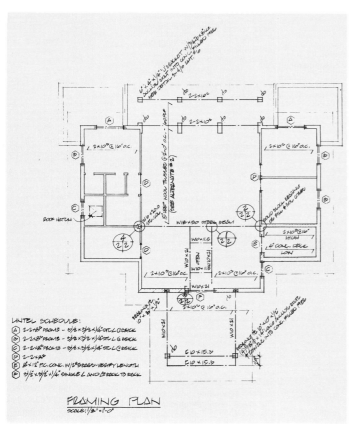

FRAMING PLAN
SCALE: 1/8"=1'-0"

Fig. 12-17. Framing plan including lintel schedule, beam sizes and locations, trusses, section locations, notes and scale.

SPECIFICATIONS

Each set of drawings is ordinarily accompanied by a set of written specifications. They are called "specs" in the trades. A set of specifications may be only a few pages long or several large volumes if the structure is large and complex.

Specifications are the written description of the construction project prepared by the architect and/or engineer. The specifications, together with the working drawings, become the Construction Documents upon which an agreement is made between all parties concerned in the project.

Many aspects of a building cannot be adequately shown or explained on the drawings. Such things must be included in the specifications. For example, quality of workmanship, responsibility for various aspects of the construction and quality of materials and fixtures are explained in the specifications.

Information in the specifications is meant to amplify (enlarge) the drawings and agree with them. Should a discrepancy occur between the specifications and the drawings, then the specs usually are accepted (take precedence) over the drawings.

Specifications are generally divided into major sections:
1. General Conditions.
2. Technical Sections.

The General Conditions deal with such things as terms of the agreement, insurance, responsibility for permits, supervision, payment for the work and temporary utilities.

The Technical Sections deal with the actual building of the structure and the tradework involved in the building. Each section is designed to include information on work to be done by a single subcontractor. Also, the arrangement of the Technical Sections is roughly similar to the sequence of work on the building. For example, the Excavation Section would be placed before the Finished Carpentry and Millwork Section. The following outline of headings is typical of smaller jobs such as a residence or small commercial building:
1. General Conditions.
2. Excavation and Grading.
3. Concrete and Masonry Work.
4. Waterproofing.
5. Miscellaneous Metals.
6. Rough Carpentry.
7. Finish Carpentry and Millwork.
8. Roofing and Sheetmetal.
9. Weatherstripping, Insulation, Caulking and Glazing.
10. Plastering or Dry-Wall Construction.
11. Ceramic Tile.
12. Flooring.
13. Painting and Decorating.
14. Plumbing.
15. Mechanical Equipment.
16. Electrical.
17. Landscaping.

A standard preprinted specification form having fill-in blanks to be completed for a specific job is shown in Fig. 12-18. This form is used by the Federal Housing Administration (FHA) and the Veterans Administration (VA). It includes most of the areas required for small buildings.

FHA Form 2005
VA Form 26-1852
Rev. 3/68

☐ Proposed Construction

☐ Under Construction

For accurate register of carbon copies, form
may be separated along above fold. Staple
completed sheets together in original order.

DESCRIPTION OF MATERIALS

No. _____
(To be inserted by FHA or VA)

Form approved.
Budget Bureau No. 63-R055.11.

Property address _____ City _____ State _____

Mortgagor or Sponsor _____
(Name)   (Address)

Contractor or Builder _____
(Name)   (Address)

INSTRUCTIONS

1. For additional information on how this form is to be submitted, number of copies, etc., see the instructions applicable to the FHA Application for Mortgage Insurance or VA Request for Determination of Reasonable Value, as the case may be.
2. Describe all materials and equipment to be used, whether or not shown on the drawings, by marking an X in each appropriate check-box and entering the information called for in each space. If space is inadequate, enter "See misc." and describe under item 27 or on an attached sheet.
3. Work not specifically described or shown will not be considered unless

required, then the minimum acceptable will be assumed. Work exceeding minimum requirements cannot be considered unless specifically described.
4. Include no alternates, "or equal" phrases, or contradictory items. (Consideration of a request for acceptance of substitute materials or equipment is not thereby precluded.)
5. Include signatures required at the end of this form.
6. The construction shall be completed in compliance with the related drawings and specifications, as amended during processing. The specifications include this Description of Materials and the applicable Minimum Construction Requirements.

1. EXCAVATION:
Bearing soil, type _____

2. FOUNDATIONS:
Footings: concrete mix _____; strength psi _____ Reinforcing _____
Foundation wall: material _____ Reinforcing _____
Interior foundation wall: material _____ Party foundation wall _____
Columns: material and sizes _____ Piers: material and reinforcing _____
Girders: material and sizes _____ Sills: material _____
Basement entrance areaway _____ Window areaways _____
Waterproofing _____ Footing drains _____
Termite protection _____
Basementless space: ground cover _____; insulation _____; foundation vents _____
Special foundations _____
Additional information: _____

3. CHIMNEYS:
Material _____ Prefabricated (make and size) _____
Flue lining: material _____ Heater flue size _____ Fireplace flue size _____
Vents (material and size): gas or oil heater _____; water heater _____
Additional information: _____

4. FIREPLACES:
Type: ☐ solid fuel; ☐ gas-burning; ☐ circulator (make and size) _____ Ash dump and clean-out _____
Fireplace: facing _____; lining _____; hearth _____; mantel _____
Additional information: _____

5. EXTERIOR WALLS:
Wood frame: wood grade, and species _____ ☐ Corner bracing. Building paper or felt _____
Sheathing _____; thickness _____; width _____; ☐ solid; ☐ spaced _____" o. c.; ☐ diagonal; _____
Siding _____; grade _____; type _____; size _____; exposure _____"; fastening _____
Shingles _____; grade _____; type _____; size _____; exposure _____"; fastening _____
Stucco _____; thickness _____"; Lath _____; weight _____ lb.
Masonry veneer _____ Sills _____ Lintels _____ Base flashing _____
Masonry: ☐ solid ☐ faced ☐ stuccoed; total wall thickness _____"; facing thickness _____"; facing material _____
Backup material _____; thickness _____"; bonding _____
Door sills _____ Window sills _____ Lintels _____ Base flashing _____
Interior surfaces: dampproofing, _____ coats of _____; furring _____
Additional information: _____
Exterior painting: material _____; number of coats _____
Gable wall construction: ☐ same as main walls; ☐ other construction _____

6. FLOOR FRAMING:
Joists: wood, grade, and species _____; other _____; bridging _____; anchors _____
Concrete slab: ☐ basement floor; ☐ first floor; ☐ ground supported; ☐ self-supporting; mix _____; thickness _____";
reinforcing _____; insulation _____; membrane _____
Fill under slab: material _____; thickness _____". Additional information: _____

7. SUBFLOORING: (Describe underflooring for special floors under item 21.)
Material: grade and species _____; size _____; type _____
Laid: ☐ first floor; ☐ second floor; ☐ attic _____ sq. ft.; ☐ diagonal; ☐ right angles. Additional information: _____

8. FINISH FLOORING: (Wood only. Describe other finish flooring under item 21.)

Location	Rooms	Grade	Species	Thickness	Width	Bldg. Paper	Finish
First floor							
Second floor							
Attic floor _____ sq. ft.							
Additional information:							

FHA Form 2005
VA Form 26-1852

1

DESCRIPTION OF MATERIALS

Fig. 12-18. Specification forms used by Federal Housing Administration and Veterans Administration.   (Continued)

DESCRIPTION OF MATERIALS

9. PARTITION FRAMING:
Studs: wood, grade, and species _____ size and spacing _____ Other _____
Additional information: _____

10. CEILING FRAMING:
Joists: wood, grade, and species _____ Other _____ Bridging _____
Additional information: _____

11. ROOF FRAMING:
Rafters: wood, grade, and species _____ Roof trusses (see detail): grade and species _____
Additional information: _____

12. ROOFING:
Sheathing: wood, grade, and species _____ ; ☐ solid; ☐ spaced _____ ” o.c.
Roofing _____ ; grade _____ ; size _____ ; type _____
Underlay _____ ; weight or thickness _____ ; size _____ ; fastening _____
Built-up roofing _____ ; number of plies _____ ; surfacing material _____
Flashing: material _____ ; gage or weight _____ ; ☐ gravel stops; ☐ snow guards
Additional information: _____

13. GUTTERS AND DOWNSPOUTS:
Gutters: material _____ ; gage or weight _____ ; size _____ ; shape _____
Downspouts: material _____ ; gage or weight _____ ; size _____ ; shape _____ ; number _____
Downspouts connected to: ☐ Storm sewer; ☐ sanitary sewer; ☐ dry-well. ☐ Splash blocks: material and size _____
Additional information: _____

14. LATH AND PLASTER
Lath ☐ walls, ☐ ceilings: material _____ ; weight or thickness _____ Plaster: coats _____ ; finish _____
Dry-wall ☐ walls, ☐ ceilings: material _____ ; thickness _____ ; finish _____ ;
Joint treatment _____

15. DECORATING: (Paint, wallpaper, etc.)

Rooms	Wall Finish Material and Application	Ceiling Finish Material and Application
Kitchen		
Bath		
Other		

Additional information: _____

16. INTERIOR DOORS AND TRIM:
Doors: type _____ ; material _____ ; thickness _____
Door trim: type _____ ; material _____ Base: type _____ ; material _____ ; size _____
Finish: doors _____ ; trim _____
Other trim (item, type and location) _____
Additional information: _____

17. WINDOWS:
Windows: type _____ ; make _____ ; material _____ ; sash thickness _____
Glass: grade _____ ; ☐ sash weights; ☐ balances, type _____ ; head flashing _____
Trim: type _____ ; material _____ Paint _____ ; number coats _____
Weatherstripping: type _____ ; material _____ Storm sash, number _____
Screens: ☐ full; ☐ half; type _____ ; number _____ ; screen cloth material _____
Basement windows: type _____ ; material _____ ; screens, number _____ ; Storm sash, number _____
Special windows _____
Additional information: _____

18. ENTRANCES AND EXTERIOR DETAIL:
Main entrance door: material _____ ; width _____ ; thickness _____ ”. Frame: material _____ ; thickness _____ ”
Other entrance doors: material _____ ; width _____ ; thickness _____ ”. Frame: material _____ ; thickness _____ ”
Head flashing _____ Weatherstripping: type _____ ; saddles _____
Screen doors: thickness _____ ”; number _____ ; screen cloth material _____ Storm doors: thickness _____ ”; number _____
Combination storm and screen doors: thickness _____ ”; number _____ ; screen cloth material _____
Shutters: ☐ hinged; ☐ fixed. Railings _____ ; Attic louvers _____
Exterior millwork: grade and species _____ Paint _____ ; number coats _____
Additional information: _____

19. CABINETS AND INTERIOR DETAIL:
Kitchen cabinets, wall units: material _____ ; lineal feet of shelves _____ ; shelf width _____
  Base units: material _____ ; counter top _____ ; edging _____
  Back and end splash _____ Finish of cabinets _____ ; number coats _____
Medicine cabinets: make _____ ; model _____
Other cabinets and built-in furniture _____
Additional information: _____

20. STAIRS:

Stair	Treads		Risers		Strings		Handrail		Balusters	
	Material	Thickness	Material	Thickness	Material	Size	Material	Size	Material	Size
Basement										
Main										
Attic										

Disappearing: make and model number _____
Additional information: _____

2

Fig. 12-18. (Continued) Standard VA and FHA specification form.

21. SPECIAL FLOORS AND WAINSCOT:

	LOCATION	MATERIAL, COLOR, BORDER, SIZES, GAGE, ETC.	THRESHOLD MATERIAL	WALL BASE MATERIAL	UNDERFLOOR MATERIAL
FLOORS	Kitchen				
	Bath				

	LOCATION	MATERIAL, COLOR, BORDER, CAP. SIZES, GAGE, ETC.	HEIGHT	HEIGHT OVER TUB	HEIGHT IN SHOWERS (FROM FLOOR)
WAINSCOT	Bath				

Bathroom accessories: ☐ Recessed; material _____ ; number _____ ; ☐ Attached; material _____ ; number _____
Additional information: _____

22. PLUMBING:

FIXTURE	NUMBER	LOCATION	MAKE	MFR'S FIXTURE IDENTIFICATION NO.	SIZE	COLOR
Sink						
Lavatory						
Water closet						
Bathtub						
Shower over tub △						
Stall shower △						
Laundry trays						

△☐ Curtain rod   △☐ Door   ☐ Shower pan: material _____
Water supply: ☐ public; ☐ community system; ☐ individual (private) system. ★
Sewage disposal: ☐ public; ☐ community system; ☐ individual (private) system. ★
★ *Show and describe individual system in complete detail in separate drawings and specifications according to requirements.*
House drain (inside): ☐ cast iron; ☐ tile; ☐ other _____   House sewer (outside): ☐ cast iron; ☐ tile; ☐ other _____
Water piping: ☐ galvanized steel; ☐ copper tubing; ☐ other _____   Sill cocks, number _____
Domestic water heater: type _____ ; make and model _____ ; heating capacity _____
_____ gph. 100° rise. Storage tank: material _____ ; capacity _____ gallons.
Gas service: ☐ utility company; ☐ liq. pet. gas; ☐ other _____   Gas piping: ☐ cooking; ☐ house heating.
Footing drains connected to: ☐ storm sewer; ☐ sanitary sewer; ☐ dry well. Sump pump; make and model _____
_____ ; capacity _____ ; discharges into _____

23. HEATING:
☐ Hot water. ☐ Steam. ☐ Vapor. ☐ One-pipe system. ☐ Two-pipe system.
  ☐ Radiators. ☐ Convectors. ☐ Baseboard radiation. Make and model _____
  Radiant panel: ☐ floor; ☐ wall; ☐ ceiling. Panel coil: material _____
  ☐ Circulator. ☐ Return pump. Make and model _____ ; capacity _____ gpm.
  Boiler: make and model _____   Output _____ Btuh.; net rating _____ Btuh.
Additional information: _____
Warm air: ☐ Gravity. ☐ Forced. Type of system _____
  Duct material: supply _____ ; return _____   Insulation _____ , thickness _____ ☐ Outside air intake.
  Furnace: make and model _____   Input _____ Btuh.; output _____ Btuh.
  Additional information: _____
☐ Space heater; ☐ floor furnace; ☐ wall heater. Input _____ Btuh.; output _____ Btuh.; number units _____
  Make, model _____   Additional information: _____
Controls: make and types _____
Additional information: _____
Fuel: ☐ Coal; ☐ oil; ☐ gas; ☐ liq. pet. gas; ☐ electric; ☐ other _____ ; storage capacity _____
  Additional information: _____
Firing equipment furnished separately: ☐ Gas burner, conversion type. ☐ Stoker: hopper feed ☐; bin feed ☐
  Oil burner: ☐ pressure atomizing; ☐ vaporizing _____
  Make and model _____   Control _____
  Additional information: _____
Electric heating system: type _____   Input _____ watts; @ _____ volts; output _____ Btuh.
  Additional information: _____
Ventilating equipment: attic fan, make and model _____ ; capacity _____ cfm.
  kitchen exhaust fan, make and model _____
Other heating, ventilating, or cooling equipment _____

24. ELECTRIC WIRING:
Service: ☐ overhead; ☐ underground. Panel: ☐ fuse box; ☐ circuit-breaker; make _____ AMP's _____ No. circuits _____
Wiring: ☐ conduit; ☐ armored cable; ☐ nonmetallic cable; ☐ knob and tube; ☐ other _____
Special outlets: ☐ range; ☐ water heater; ☐ other _____
☐ Doorbell. ☐ Chimes. Push-button locations _____   Additional information: _____

25. LIGHTING FIXTURES:
Total number of fixtures _____   Total allowance for fixtures, typical installation, $ _____
Nontypical installation _____
Additional information: _____

3   DESCRIPTION OF MATERIALS

Fig. 12-18. (Continued) Standard VA and FHA specification form.

DESCRIPTION OF MATERIALS

26. INSULATION:

Location	Thickness	Material, Type, and Method of Installation	Vapor Barrier
Roof			
Ceiling			
Wall			
Floor			

HARDWARE: *(make, material, and finish.)* _____

SPECIAL EQUIPMENT: *(State material or make, model and quantity. Include only equipment and appliances which are acceptable by local law, custom and applicable FHA standards. Do not include items which, by established custom, are supplied by occupant and removed when he vacates premises or chattles prohibited by law from becoming realty.)* _____

27. MISCELLANEOUS: *(Describe any main dwelling materials, equipment, or construction items not shown elsewhere; or use to provide additional information where the space provided was inadequate. Always reference by item number to correspond to numbering used on this form.)* _____

PORCHES:

TERRACES:

GARAGES:

WALKS AND DRIVEWAYS:

Driveway: width _____ ; base material _____ ; thickness _____ "; surfacing material _____ ; thickness _____ "

Front walk: width _____ ; material _____ ; thickness _____ ". Service walk: width _____ ; material _____ ; thickness _____ "

Steps: material _____ ; treads _____ "; risers _____ ". Cheek walls _____

OTHER ONSITE IMPROVEMENTS:

(Specify all exterior onsite improvements not described elsewhere, including items such as unusual grading, drainage structures, retaining walls, fence, railings, and accessory structures.)

LANDSCAPING, PLANTING, AND FINISH GRADING:

Topsoil _____ " thick: ☐ front yard; ☐ side yards; ☐ rear yard to _____ feet behind main building.

Lawns *(seeded, sodded, or sprigged)*: ☐ front yard _____ ; ☐ side yards _____ ; ☐ rear yard _____

Planting: ☐ as specified and shown on drawings; ☐ as follows:

_____ Shade trees, deciduous, _____ " caliper.   _____ Evergreen trees. _____ ' to _____ ', B & B.

_____ Low flowering trees, deciduous, _____ ' to _____ '   _____ Evergreen shrubs. _____ ' to _____ ', B & B.

_____ High-growing shrubs, deciduous, _____ ' to _____ '   _____ Vines, 2-year _____

_____ Medium-growing shrubs, deciduous, _____ ' to _____ '   _____

_____ Low-growing shrubs, deciduous, _____ ' to _____ '   _____

IDENTIFICATION.—This exhibit shall be identified by the signature of the builder, or sponsor, and/or the proposed mortgagor if the latter is known at the time of application.

Date_____   Signature _____

    Signature _____

FHA Form 2005
VA Form 26–1852   4   GPO 1968 o48—16—80081-1 296-152

Fig. 12-18. (Continued) Standard VA and FHA specification form.

REVIEW QUESTIONS — CHAPTER 12

1. What are three other names for construction drawings?
2. Name five line symbols which appear in the Alphabet of Lines.
3. What kind of line is used to represent the visible features of a building?
4. If a part is hidden, it is represented with a _____ line symbol.
5. A symmetrical object usually has a _____ line drawn through the center.
6. What kind of line is used to show how long a wall is?
7. When is a cutting plane line used?
8. What is the purpose of symbols and abbreviations?
9. Symbols may be shown in elevation or section. Which type is usually easiest to recognize?
10. If a plan is drawn at $1/4'' = 1' - 0''$ scale, how long on the drawing would a $40' - 0''$ wall be?
11. How long would the same $40' - 0''$ wall be if it were drawn at 1/4 size?
12. Dimensions locate _____ and _____ of building elements.
13. What is the most common method of dimensioning interior walls?
14. Where should an exterior brick veneer wall be dimensioned?
    a. To the center of the wall.
    b. To the outside of the brick.
    c. To the outside of the stud wall.
    d. To the inside of the stud wall.
    e. None of the above.
15. Notes on a drawing are really not very important because they amplify information which is already given on the drawing. True or False?
16. Who usually prepares the working drawings for a building?
17. Identify the seven drawings which usually compose a set of working drawings for a building.
18. Which plan shows the location of the structure on the site?
19. The plan which shows exterior walls, interior walls, doors and windows and patios is the _____ Plan.
20. A _____ Plan shows the footings, foundation size and materials.
21. The front view of a structure would be called an _____.
22. Which drawing shows the location of switches, convenience outlets and TV jacks?
23. Why are construction details necessary?
24. Many aspects of a building cannot be adequately shown or explained on the drawings and must be included in the _____.
25. The Construction Documents upon which an agreement is made between all parties concerned in a project is composed of the _____ and the _____.
26. Specifications are generally divided into _____ main sections.
27. Place the following specification outline headings in proper order.
    a. Miscellaneous Metals.
    b. Rough Carpentry.
    c. Concrete and Masonry Work.
    d. Excavating and Grading.
    e. Finish Carpentry and Millwork.
    f. General Conditions.
    g. Waterproofing.
    h. Roofing and Sheetmetal.

Chapter 13
A CAREER IN MASONRY

Masonry offers a rewarding career for persons who have an interest in working with tools and materials. As masons, or bricklayers, they will be skilled workers. They will understand the basic principles and practices related to the construction industry. They will earn good pay and have an opportunity for advancement. What is more, they will play an important role in building homes, schools and commercial structures. They can be proud of their skill and the fact that they can produce something that people need.

WHAT DO MASONS DO?

The term MASONRY has always referred to the craft of building with brick. Today, it has a broader meaning. It includes two types of work. The first type consists of any construction bonded together with mortar. The second type deals with cement masonry. All who lay brick, block, tile or stone are called masons. Cement masons specialize in concrete work such as slabs, footings and foundations.

Masons use masonry units such as brick, block or stone. With their tools and skills they form them into buildings or other useful structures, Fig. 13-1. This skill and knowledge must be learned and practiced if they are to be successful in their trade.

Masons must learn to proportion the ingredients of mortar, figure trade related problems, mix and spread mortar, Fig. 13-2, read blueprints, handle different kinds of masonry units and be willing to learn new techniques and procedures. They must learn to cooperate with other trades.

They will build exterior and interior walls, floors, patios, walks, columns, window and door openings, fireplaces, arches and many other building elements. Modern masonry design will also require them to use masonry saws and other power-driven equipment. And, of course they must use the age-old tools (trowel, brick hammer, level, chisels and mason's line) as well. Masons are creative persons who can continue to

Fig. 13-1. A journeyman mason who knows the trade and takes pride in workmanship.

Fig. 13-2. Mason must know how to spread mortar properly.

learn more about the trade as they progress through the years.

Cement masons will learn how to work with concrete. More specifically, they will learn the proper methods of reinforcing, placing and finishing concrete. As in any other type of masonry work, a high degree of skill is involved in quality cement mason's work.

Poured concrete construction industries employ many skilled workers as cement masons. Their job is to pour concrete into forms where they will smooth and finish it. They will work on a wide range of construction projects from floors and slabs to roofs, sidewalks, highways, dams and runways for airports. Fig. 13-3 shows workers pouring a concrete roof.

ADVANCEMENT OPPORTUNITIES

The masonry trades recognize various levels of the skill. They are:
1. Apprentice.
2. Journeyman.
3. Foreman.
4. Superintendent.

Fig. 13-3. Concrete workers pour reinforced membrane roof. Worker at left is directing the crane operator who hoists the half-yard buckets of concrete to the roof. At right finishers rough screed and rough finish the slab. Worker near top of form uses rodding hose to vibrate concrete under and around the reinforcement. (Wire Reinforcement Institute)

U.S. DEPARTMENT OF LABOR ★ Employment and Training Administration
Bureau of Apprenticeship and Training

APPRENTICESHIP AGREEMENT BETWEEN APPRENTICE AND JOINT APPRENTICESHIP COMMITTEE

CHECK APPROPRIATE BOX

☐ Vietnam-era Veteran ☐ Other Veteran ☐ Nonveteran

PRIVACY ACT STATEMENT

The information requested herein is used for apprenticeship program statistical purposes and may not be otherwise disclosed without the express permission of the undersigned apprentice.

Privacy Act of 197⸮ - P.L. 93-579

THIS AGREEMENT, entered into this *(date)* .... day of .... 19 ....

between the parties to *(Name of local apprenticeship standards)* ....

....

represented by the Joint Apprenticeship Committee, hereinafter referred to as the COMMITTEE, and

(Name of Apprentice) ...., born *(Month, Day, Year)* ....

.... hereinafter referred to as the APPRENTICE, and (if a minor) *(Name of parent*

or guardian) .... hereinafter referred to

as the GUARDIAN.

WITNESSETH THAT:

The Committee agrees to be responsible for the selection, placement, and training of said apprentice in the trade of ....

....

as work is available, and in consideration said apprentice agrees diligently and faithfully to perform the work of said trade during the period of apprenticeship, in accordance with the regulations of the Committee. The apprenticeship standards referred to herein are hereby incorporated in and made a part of this agreement.

This AGREEMENT may be terminated by mutual consent of the signatory parties, upon proper notification to the registration agency.

.SIGNATURE OF APPRENTICE

ADDRESS *(Number, Street, City, State, ZIP Code)*

SIGNATURE OF PARENT OR GUARDIAN

SIGNATURE OF JOINT APPRENTICESHIP COMMITTEE, CHAIRPERSON

SIGNATURE OF JOINT APPRENTICESHIP COMMITTEE, SECRETARY

NAME OF REGISTRATION AGENCY

SIGNATURE AND TITLE OF AUTHORIZED OFFICIAL

TRAINING DATA	
APPRENTICESHIP TERM	PROBATIONARY PERIOD
CREDIT *(By previous trade experience)*	TERM REMAINING

TO BE COMPLETED BY THE APPRENTICE

SEX ➤
(Check one)
☐ Male
☐ Female

RACE/ ETHNIC GROUP ➤
(Check one)
☐ Caucasian/White
☐ Negro/Black
☐ Oriental
☐ American Indian
☐ Spanish American
☐ Information Not Available
☐ Not Elsewhere Classified

HIGHEST EDUCATION LEVEL ➤
(Check one)
☐ 8th grade or less
☐ 9th grade or more
☐ 12th grade or more

DATE *(Month, Day, Year)*

GPO 900-664

ETA 6-111

Fig. 13-4. Apprenticeship agreement between the apprentice and the local apprenticeship and training committee.

U.S. DEPARTMENT OF LABOR ★ Employment and Training Administration Bureau of Apprenticeship and Training	PRIVACY ACT STATEMENT

U.S. DEPARTMENT OF LABOR ★ Employment and Training Administration
Bureau of Apprenticeship and Training

APPRENTICESHIP AGREEMENT
BETWEEN
APPRENTICE AND EMPLOYER

CHECK APPROPRIATE BOX

☐ Vietnam- era Veteran   ☐ Other Veteran   ☐ Nonveteran

PRIVACY ACT STATEMENT

The information requested herein is used for apprenticeship program statistical purposes and may not be otherwise disclosed without the express permission of the undersigned apprentice.

Privacy Act of 1974 - P.L. 93-579

The employer and apprentice whose signatures appear below agree to these terms of apprenticeship.

The employer agrees to the nondiscriminatory selection and training of apprentices in accordance with the Equal Opportunity Standards stated in Section 30.3 of Title 29 Code of Federal Regulations, Part 30; and in accordance with the terms and conditions of the

(Name of Apprenticeship Standards) ....
which are made a part of this agreement.

The apprentice agrees to be diligent and faithful in learning the trade in accordance with this agreement.

This AGREEMENT may be terminated by mutual consent of the parties, citing cause(s), with notification to the Registration Agency.

NAME OF APPRENTICE *(Type or Print)*

SIGNATURE OF APPRENTICE

ADDRESS *(Number, Street, City, State, ZIP Code)*

SIGNATURE OF PARENT OR GUARDIAN

NAME OF EMPLOYER AND ADDRESS *(Company)*

SIGNATURE OF AUTHORIZED COMPANY OFFICIAL

APPROVED BY JOINT APPRENTICESHIP COMMITTEE

SIGNATURE OF CHAIRPERSON OR SECRETARY | DATE

REGISTERED BY *(Name of Registration Agency)*

SIGNATURE OF AUTHORIZED OFFICIAL

TRAINING DATA

Trade	Apprenticeship Term
Probationary Period	Credit for previous experience
Term remaining	Date apprenticeship begins

TO BE COMPLETED BY THE APPRENTICE

DATE OF BIRTH
(Month, Day, Year) ►

SEX ►   ☐ Male
(Check one)   ☐ Female

RACE/
ETHNIC
GROUP ►   ☐ Caucasian/White
(Check one)   ☐ Negro/Black
☐ Oriental
☐ American Indian
☐ Spanish American
☐ Information Not Available
☐ Not Elsewhere Classified

HIGHEST
EDUCATION ►   ☐ 8th grade or less
LEVEL   ☐ 9th grade or more
(Check one)   ☐ 12th grade or more

DATE *(Mo., Day, Yr.)*

GPO 900-663

ETA 6-71B

Fig. 13-5. Apprenticeship agreement between the apprentice and the employer.

APPRENTICESHIP

An *apprentice* is a person at least 17 years old and, preferably, not over 24 years old. Apprentices are under written agreement to work at and learn the trade. The apprenticeship agreement, Fig. 13-4, is made with a local apprenticeship and training committee acting as agent of the contractor. Or the agreement may be with a contractor whose agreement is approved by the local joint committee, Fig. 13-5. An apprenticeship applicant should have completed at least two years of high school, but preferably more.

The Manpower Administration, Bureau of Apprenticeship and Training sets down the responsibilities of apprentices. They are:

1. To perform diligently and faithfully the work of the trade and other pertinent duties as assigned by the contractor in accordance with the provisions of the standards.
2. To respect the property of the contractor and abide by the working rules and regulations of the contractor and the local joint committee.
3. To attend regularly and satisfactorily complete the required hours of instruction in subjects related to the trade, as provided under the local standards.
4. To maintain such records of work experience and training received on the job and in related instruction, as may be required by the local joint committee.
5. To develop safe working habits and condust himself or herself in such manner as to assure personal safety and that of the safety of fellow workers.
6. To work for the contractor to whom assigned until the apprenticeship is completed, unless reassigned to another contractor or agreement is terminated by the local joint committee.
7. To conduct himself or herself at all times in a creditable, ethical, and moral manner, realizing that much time, money, and effort are spent to afford him or her an opportunity to become a skilled worker.

The first six months of employment after signing the apprenticeship agreement is a "trial" period. Before the end of the period, the committee will review the apprentice's ability and development. The normal term of apprenticeship is 4,500 hours of employment or about three years. Apprenticeship is divided into six periods of advancement of six months each. Pay rate is a percentage of the journeyman rate. It is based on the period of advancement.

Each apprentice is required to take related instruction away from the job for no less than 144 hours per year, each year of the apprenticeship. Subjects include blueprint reading, math, estimating, shop practice and safety. (See Reference Section for complete list.)

After successfully completing apprenticeship a Certificate of Completion of Apprenticeship is awarded, Fig. 13-6. The apprentice is now a JOURNEYMAN mason.

JOURNEYMAN

The JOURNEYMAN is an experienced craftsperson who has successfully completed an apprenticeship in the trade. At this point, the worker is a free agent and can work for any contractor. The certificate will be recognized throughout the country.

Fig. 13-6. The Certificate of Completion of Apprenticeship is awarded an apprentice upon completion of training.

FOREMAN

A FOREMAN is a jouneyman who has the responsibility of supervising a group of workers. This job requires not only a high degree of knowledge about the craft, but also the ability to supervise people.

SUPERINTENDENT

The SUPERINTENDENT is generally a foreman who has been promoted because of outstanding performance as a foreman. He or she is in charge of all the work in the field for the contractor. This includes supervising the work of the foreman and makes major decisions about the job under construction.

CONTRACTOR

CONTRACTORS are responsible for the whole job. They are the top persons in charge. They organize people and work, prepare bids for jobs, inspect work and run the business. The contractor must be knowledgeable about all phases of the business.

REVIEW QUESTIONS — CHAPTER 13

1. There are two broad categories of masons. One type of mason works with masonry units which are bonded together with mortar. Describe what the other type of mason does.
2. In addition to using the tools, name five things that a mason must learn.
3. Name four basic tools that a mason uses.
4. What kind of mason finishes concrete?
5. If you wish to become a mason, what is the beginning position?
6. About how long does apprenticeship training last?
7. How many hours of related instruction must an apprentice successfully complete each year of apprenticeship training?
8. When apprentices complete training, they become _____ masons.
9. After masons have gained experience on the job they may be promoted to _____ if they show that they can supervise people.
10. What is the title of the person who supervises foreman on a construction job?
11. Who is responsible for the whole job — the top person?

Cement finishers apply their skill to all types of structures. This worker prepares an expansion joint in a walkway so that walk will not develop unsightly cracks. (The Associated General Contractors of America)

REFERENCE SECTION

MORTAR PROPORTIONS BY VOLUME

MORTAR TYPE	PARTS BY VOLUME OF PORTLAND CEMENT* OR PORTLAND BLAST FURNACE SLAG CEMENT**	PARTS BY VOLUME OF MASONRY CEMENT	PARTS BY VOLUME OF HYDRATED LIME OR LIME PUTTY	AGGREGATE, MEASURED IN A DAMP, LOOSE CONDITION
M	1 1	1 (TYPE II) —	— 1/4	NOT LESS THAN 2 1/2 AND NOT MORE THAN 3 TIMES THE SUM OF THE VOLUMES OF THE CEMENTS AND LIME USED.
S	1/2 1	1 (TYPE II) —	— OVER 1/4 to 1/2	
N	— 1	1 (TYPE II) —	— OVER 1/2 to 1 1/4	
O	— 1	1 (TYPE I OR II) —	— OVER 1 1/4 to 1 1/2	
K	1	—	OVER 2 1/2 to 4	

* TYPES I, II, III, IA, IIA, IIIA
** TYPES IS, ISA
DATA FROM ASTM C-270

MORTAR TYPES FOR CLASSES OF CONSTRUCTION

ASTM MORTAR TYPE DESIGNATION	CONSTRUCTION SUITABILITY
M	MASONRY SUBJECTED TO HIGH COMPRESSIVE LOADS, SEVERE FROST ACTION, OR HIGH LATERAL LOADS FROM EARTH PRESSURES, HURRICANE WINDS, OR EARTHQUAKES. STRUCTURES BELOW GRADE, MANHOLES, AND CATCH BASINS.
S	STRUCTURES REQUIRING HIGH FLEXURAL BOND STRENGTH, BUT SUBJECT ONLY TO NORMAL COMPRESSIVE LOADS.
N	GENERAL USE IN ABOVE GRADE MASONRY. RESIDENTIAL BASEMENT CONSTRUCTION, INTERIOR WALLS AND PARTITIONS. CONCRETE MASONRY VENEERS APPLIED TO FRAME CONSTRUCTION.
O	NON-LOAD-BEARING WALLS AND PARTITIONS. SOLID LOAD BEARING MASONRY OF ALLOWABLE COMPRESSIVE STRENGTH NOT EXCEEDING 100 PSI.
K	INTERIOR NON-LOAD-BEARING PARTITIONS WHERE LOW COMPRESSIVE AND BOND STRENGTHS ARE PERMITTED BY BUILDING CODES.

Masonry

AREAS AND VOLUMES FOR PLANES AND SOLIDS

AREAS OF PLANE FIGURES

$A = B \times H$

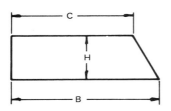

$A = \dfrac{B + C}{2} \times H$

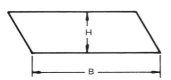

$A = \dfrac{B \times H}{2}$

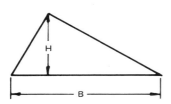

$A = \dfrac{\text{SUM OF SIDES (S)}}{2} \times R$

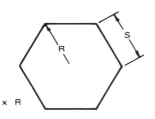

$A = \pi R^2$

$A = .7854 \times D^2$

$A = .0796 \times C^2$

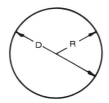

$A = M \times m \times .7854$

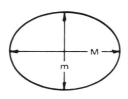

VOLUMES OF SOLID FIGURES

$V = L \times W \times H$

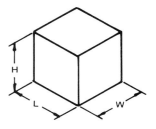

$V = \text{AREA OF END} \times H$

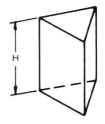

$V = \pi R^2 \times H$

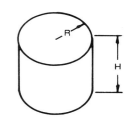

$V = \dfrac{\pi R^2 \times H}{3}$

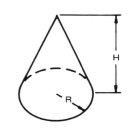

$V = \text{AREA OR BASE} \times \dfrac{H}{3}$

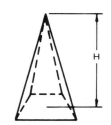

$V = \dfrac{1}{6} \times \pi D^3$

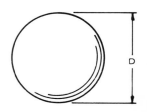

SIZES OF MODULAR BRICK

Unit Designation	*Nominal Dimensions, in.* t	h	l	*Joint Thickness in.*	*Manufactured Dimensions in.* t	h	l	*Modular Coursing in.*
Standard Modular	4	2 2/3	8	3/8	3 5/8	2 1/4	7 5/8	3C = 8
				1/2	3 1/2	2 1/4	7 1/2	
Engineer	4	3 1/5	8	3/8	3 5/8	2 13/16	7 5/8	5C = 16
				1/2	3 1/2	2 11/16	7 1/2	
Economy 8 or Jumbo Closure	4	4	8	3/8	3 5/8	3 5/8	7 5/8	1C = 4
				1/2	3 1/2	3 1/2	7 1/2	
Double	4	5 1/3	8	3/8	3 5/8	4 15/16	7 5/8	3C = 16
				1/2	3 1/2	4 13/16	7 1/2	
Roman	4	2	12	3/8	3 5/8	1 5/8	11 5/8	2C = 4
				1/2	3 1/2	1 1/2	11 1/2	
Norman	4	2 2/3	12	3/8	3 5/8	2 1/4	11 5/8	3C = 8
				1/2	3 1/2	2 1/4	11 1/2	
Norwegian	4	3 1/5	12	3/8	3 5/8	2 13/16	11 5/8	5C = 16
				1/2	3 1/2	2 11/16	11 1/2	
Economy 12 or Jumbo Utility	4	4	12	3/8	3 5/8	3 5/8	11 5/8	1C = 4
				1/2	3 1/2	3 1/2	11 1/2	
Triple	4	5 1/3	12	3/8	3 5/8	4 15/16	11 5/8	3C = 16
				1/2	3 1/2	4 13/16	11 1/2	
SCR brick	6	2 2/3	12	3/8	5 5/8	2 1/4	11 5/8	3C = 8
				1/2	5 1/2	2 1/4	11 1/2	
6-in. Norwegian	6	3 1/5	12	3/8	5 5/8	2 13/16	11 5/8	5C = 16
				1/2	5 1/2	2 11/16	11 1/2	
6-in. Jumbo	6	4	12	3/8	5 5/8	3 5/8	11 5/8	1C = 4
				1/2	5 1/2	3 1/2	11 1/2	
8-in. Jumbo	8	4	12	3/8	7 5/8	3 5/8	11 5/8	1C = 4
				1/2	7 1/2	3 1/2	11 1/2	

Available as solid units conforming to ASTM C 216- or ASTM C 62-, or, in a number of cases, as hollow brick conforming to ASTM C 652-.

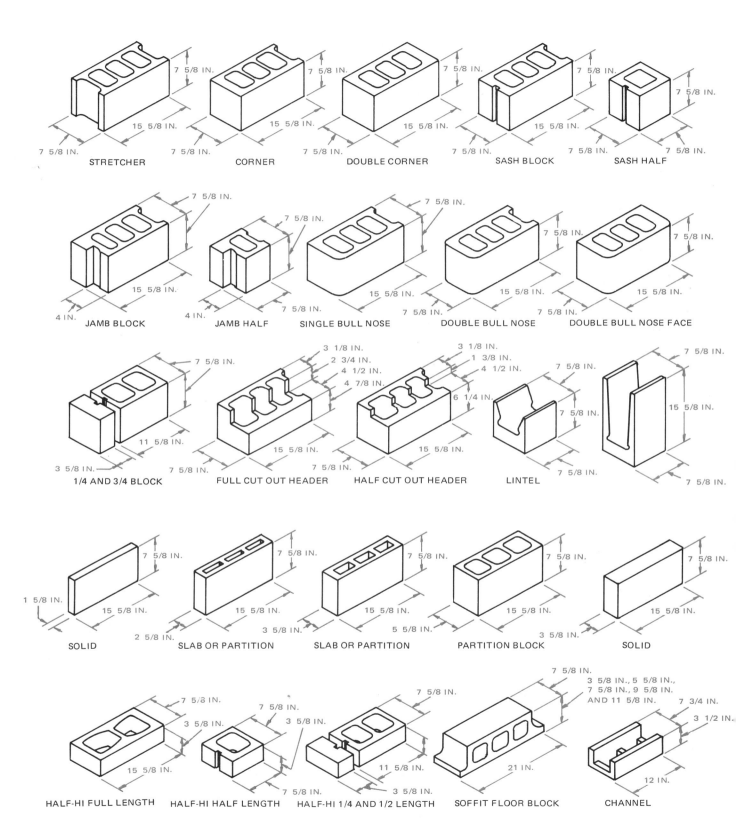

A large selection of the most popular shapes and sizes of concrete block.

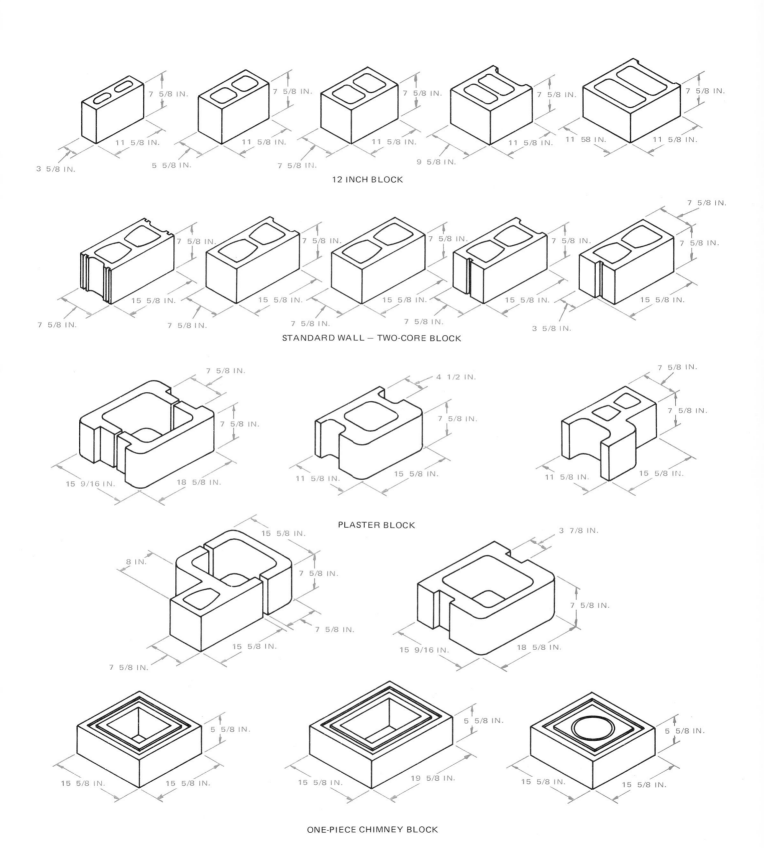

7 5/8 IN.
11 5/8 IN.
3 5/8 IN.

7 5/8 IN.
11 5/8 IN.
5 5/8 IN.

7 5/8 IN.
11 5/8 IN.
7 5/8 IN.

7 5/8 IN.
11 5/8 IN.
9 5/8 IN.

7 5/8 IN.
11 58 IN.
11 5/8 IN.
7 5/8 IN.

12 INCH BLOCK

7 5/8 IN.
15 5/8 IN.
7 5/8 IN.

7 5/8 IN.
15 5/8 IN.
7 5/8 IN.

7 5/8 IN.
15 5/8 IN.
7 5/8 IN.

7 5/8 IN.
15 5/8 IN.
7 5/8 IN.

7 5/8 IN.
7 5/8 IN.
15 5/8 IN.
3 5/8 IN.

STANDARD WALL — TWO-CORE BLOCK

7 5/8 IN.
7 5/8 IN.
15 9/16 IN.
18 5/8 IN.

4 1/2 IN.
7 5/8 IN.
11 5/8 IN.
15 5/8 IN.

7 5/8 IN.
7 5/8 IN.
11 5/8 IN.
15 5/8 IN.

PLASTER BLOCK

15 5/8 IN.
8 IN.
7 5/8 IN.
7 5/8 IN.
15 5/8 IN.
7 5/8 IN.

3 7/8 IN.
7 5/8 IN.
15 9/16 IN.
18 5/8 IN.

5 5/8 IN.
15 5/8 IN.
15 5/8 IN.

5 5/8 IN.
15 5/8 IN.
19 5/8 IN.

5 5/8 IN.
15 5/8 IN.
15 5/8 IN.

ONE-PIECE CHIMNEY BLOCK

Continued from previous page.

BASIC SHAPES OF STRUCTURAL CLAY

STRUCTURAL CLAY FACING TILE AVAILABLE SIZES		
SERIES	**NOMINAL FACE DIMENSIONS IN INCHES**	**NOMINAL THICKNESS IN INCHES**
6T	5 1/3 by 12	2, 4, 6, 8
4D	5 1/3 by 8	2, 4, 6, 8
4S	2 2/3 by 8	2, 4
8W	8 by 16	2, 4

NOMINAL MODULAR SIZES* FACE DIMENSIONS	
HEIGHT BY LENGTH IN INCHES	**HEIGHT BY LENGTH IN INCHES**
4 by 8	8 by 8
4 by 12	8 by 12
5 1/3 by 8	8 by 16
5 1/3 by 12	12 by 12

Thickness: All of the above are in nominal thicknesses of 4, 6 and 8 inches.

*Nominal sizes include the thickness of the standard mortar joint for all dimensions.

CONCRETE MIXTURES FOR VARIOUS APPLICATIONS

Concrete Fundamentals

CONSTRUCTION APPLICATION	WATER/CEMENT RATIO GAL. PER BAG	CONSISTENCY (AMOUNT OF SLUMP)	MAXIMUM SIZE OF AGGREGATE	APPROPRIATE CEMENT CONTENT BAGS PER YD.	PROBABLE 28TH DAY STRENGTH (PSI)
FOOTINGS	7	4" TO 6"	1 1/2"	5.0	2800
8 INCH BASEMENT WALL MODERATE GROUND WATER	7	4" TO 6"	1 1/2"	5.0	2800
8 INCH BASEMENT WALL SEVERE GROUND WATER	6	3" TO 5"	1 1/2"	5.8	3500
10 INCH BASEMENT WALL MODERATE GROUND WATER	7	4" TO 6"	2"	4.7	2800
10 INCH BASEMENT WALL SEVERE GROUND WATER	6	3" TO 5"	2"	5.5	3500
BASEMENT FLOOR 4 INCH THICKNESS	6	2" TO 4"	1"	6.2	3500
FLOOR SLAB ON GRADE	6	2" TO 4"	1"	6.2	3500
STAIRS AND STEPS	6	1" TO 4"	1"	6.2	3500
TOPPING OVER CONCRETE FLOOR	5	1" TO 2"	3/8"	8.0	4350
SIDEWALKS, PATIOS DRIVEWAYS, PORCHES	6	2" TO 4"	1"	6.2	3500

WELDED WIRE FABRIC STOCK SIZES

TYPE OF CONSTRUCTION	RECOMMENDED STYLE	REMARKS
BASEMENT FLOORS	6 x 6 — W1.4 x W1.4 6 x 6 — W2.1 x W2.1 6 x 6 — W2.9 x W2.9	FOR SMALL AREAS (15 FOOT MAXIMUM SIDE DIMENSION) USE 6 x 6 — W1.4 x W1.4. AS A RULE OF THUMB, THE LARGER THE AREA OR THE POORER THE SUBSOIL, THE HEAVIER THE GAGE.
DRIVEWAYS	6 x 6 — W2.9	CONTINUOUS REINFORCEMENT BETWEEN 25 TO 30 FOOT CONTRACTION JOINTS.
FOUNDATION SLABS (RESIDENTIAL ONLY)	6 x 6 — W1.4 x W1.4	USE HEAVIER GAGE OVER POORLY DRAINED SUBSOIL, OR WHEN MAXIMUM DIMENSION IS GREATER THAN 15 FEET.
GARAGE FLOORS	6 x 6 — W2.9 x W2.9	POSITION AT MIDPOINT OF 5 OR 6 INCH THICK SLAB.
PATIOS AND TERRACES	6 x 6 — W1.4 x W1.4	USE 6 x 6 — W2.1 x W2.1 IF SUBSOIL IS POORLY DRAINED.

PORCH FLOOR
  A. 6 INCH THICK SLAB
     UP TO 6 FOOT SPAN
  B. 6 INCH THICK SLAB
     UP TO 8 FOOT SPAN
	6 x 6 — W2.9 x W2.9
	
	4 x 4 — W4 x W4
		POSITION 1 INCH FROM BOTTOM FORM TO RESIST TENSILE STRESSES.

SIDEWALKS	6 x 6 — W1.4 x W1.4 6 x 6 — W2.1 x W2.1	USE HEAVIER GAGE OVER POORLY DRAINED SUBSOIL. CONSTRUCT 25 TO 30 FOOT SLABS AS FOR DRIVEWAYS.
STEPS (FREE SPAN)	6 x 6 — W2.9 x W2.9	USE HEAVIER STYLE IF MORE THAN FIVE RISERS. POSITION FABRIC 1 INCH FROM BOTTOM OF FORM.
STEPS (ON GROUND)	6 x 6 — W2.1 x W2.1	USE 6 x 6 — W2.9 x W2.9 FOR UNSTABLE SUBSOIL.

GLASS BLOCK LAYOUT TABLE

NO. OF BLOCKS	6" 5 3/4 x 5 3/4 x 3 7/8	8" 7 3/4 x 7 3/4 x 3 7/8	12" 11 3/4 x 11 3/4 x 3 7/8
1	0' – 6"	0' – 8"	1' – 0"
2	1' – 0"	1' – 4"	2' – 0"
3	1' – 6"	2' – 0"	3' – 0"
4	2' – 0"	2' – 8"	4' – 0"
5	2' – 6"	3' – 4"	5' – 0"
6	3' – 0"	4' – 0"	6' – 0"
7	3' – 6"	4' – 8"	7' – 0"
8	4' – 0"	5' – 4"	8' – 0"
9	4' – 6"	6' – 0"	9' – 0"
10	5' – 0"	6' – 8"	10' – 0"
11	5' – 6"	7' – 4"	11' – 0"
12	6' – 0"	8' – 0"	12' – 0"
13	6' – 6"	8' – 8"	13' – 0"
14	7' – 0"	9' – 4"	14' – 0"
15	7' – 6"	10' – 0"	15' – 0"
16	8' – 0"	10' – 8"	16' – 0"
17	8' – 6"	11' – 4"	17' – 0"
18	9' – 0"	12' – 0"	18' – 0"
19	9' – 6"	12' – 8"	19' – 0"
20	10' – 0"	13' – 4"	20' – 0"
21	10' – 6"	14' – 0"	21' – 0"
22	11' – 0"	14' – 8"	22' – 0"
23	11' – 6"	15' – 4"	23' – 0"
24	12' – 0"	16' – 0"	24' – 0"
25	12' – 6"	16' – 8"	25' – 0"

Masonry

STEEL REINFORCING BAR SIZES

WIRE SIZE COMPARISON

W & D SIZE NUMBER SMOOTH	DEFORMED	AREA (SQ. IN.)	NOMINAL DIAMETER (IN.)	AMERICAN STEEL & WIRE GAGE NUMBER
W31	D31	0.310	0.628	
W30	D30	.300	.618	
W28	D28	.280	.597	
W26	D26	.260	.575	
W24	D24	.240	.553	
W22	D22	.220	.529	
W20	D20	.200	.504	
		.189	.490	7/0
W18	D18	.180	.478	
		.167	.4615	6/0
W16	D16	.160	.451	
		.146	.4305	5/0
W14	D14	.140	.422	
		.122	.394	4/0
W12	D12	.120	.390	
W11	D11	.110	.374	
W10.5		.105	.366	
		.103	.3625	3/0
W10	D10	.100	.356	
W9.5		.095	.348	
W9	D9	.090	.338	
		.086	.331	2/0
W8.5		.085	.329	
W8	D8	.080	.319	
W7.5		.075	.309	
		.074	.3065	1/0
W7	D7	.070	.298	
W6.5		.065	.288	
		.063	.283	1
W6	D6	.060	.276	
W5.5		.055	.264	
		.054	.2625	2
W5	D5	.050	.252	
		.047	.244	3
W4.5		.045	.240	
W4	D4	.040	.225	4
W3.5		.035	.211	
		.034	.207	5
W3		.030	.195	
W2.9		.029	.192	6
W2.5		.025	.177	7
W2.1		.021	.162	8
W2		.020	.159	
		.017	.148	9
W1.5		.015	.138	
W1.4		.014	.135	10

SPECIFICATIONS FOR REINFORCED CONCRETE LINTELS WITH STIRRUPS

CONCRETE REINFORCED LINTELS WITH STIRRUPS FOR WALL AND FLOOR LOADS

SIZE OF LINTEL		REINFORCEMENT		WEB REINFORCEMENT NO. 6 GAGE WIRE STIRRUPS. SPACINGS FROM END OF LINTEL — BOTH ENDS THE SAME
HEIGHT IN.	WIDTH IN.	CLEAR SPAN OF LINTEL FT.	TOP	BOTTOM	
7 5/8	7 5/8	3	NONE	2 – 1/2-IN. ROUND	NO STIRRUPS REQUIRED
7 5/8	7 5/8	4	NONE	2 – 3/4-IN. ROUND	3 STIRRUPS, SP. :2,3,3 IN.
7 5/8	7 5/8	5	2 – 3/8-IN. ROUND	2 – 7/8-IN. ROUND	5 STIRRUPS, SP. :2,3,3,3,3 IN.
7 5/8	7 5/8	6	2 – 1/2-IN. ROUND	2 – 7/8-IN. ROUND	6 STIRRUPS, SP. :2,3,3,3,3,3 IN.
7 5/8	7 5/8	7	2 – 1-IN. ROUND	2 – 1-IN. ROUND	9 STIRRUPS, SP. :2,3,3,3,3,3,3,3,3 IN.

SOURCE: PORTLAND CEMENT ASSOCIATION

Reference Section

SPECIFICATIONS FOR ONE-PIECE REINFORCED CONCRETE LINTELS

CONCRETE REINFORCED ONE-PIECE LINTELS WITH WALL LOAD ONLY

| SIZE OF LINTEL | | CLEAR SPAN OF LINTEL FT. | BOTTOM REINFORCEMENT | |
HEIGHT IN.	WIDTH IN.		SIZE DESIGNATION OF BARS	SIZE OF BARS
5 3/4	7 5/8	UP TO 7	NO. 2	3/8-IN. ROUND DEFORMED
5 3/4	7 5/8	7 TO 8	NO. 2	5/8-IN. ROUND DEFORMED
7 5/8	7 5/8	UP TO 8	NO. 2	3/8-IN. ROUND DEFORMED
7 5/8	7 5/8	8 TO 9	NO. 2	1/2-IN. ROUND DEFORMED
7 5/8	7 5/8	9 TO 10	NO. 2	5/8-IN. ROUND DEFORMED

SPECIFICATIONS FOR REINFORCED CONCRETE SPLIT LINTELS

CONCRETE REINFORCED SPLIT LINTELS WITH WALL LOAD ONLY

| SIZE OF LINTEL | | CLEAR SPAN OF LINTEL FT. | BOTTOM REINFORCEMENT | |
HEIGHT IN.	WIDTH IN.		SIZE DESIGNATION OF BARS	SIZE OF BARS
5 3/4	3 5/8	UP TO 7	NO. 1	3/8-IN. ROUND DEFORMED
5 3/4	3 5/8	7 TO 8	NO. 1	5/8-IN. ROUND DEFORMED
7 5/8	3 5/8	UP TO 8	NO. 1	3/8-IN. ROUND DEFORMED
7 5/8	3 5/8	8 TO 9	NO. 1	1/2-IN. ROUND DEFORMED
7 5/8	3 5/8	9 TO 10	NO. 1	5/8-IN. ROUND DEFORMED

GRADE-USE GUIDE FOR CONCRETE FORMS*

| Use these terms when you specify plywood | DESCRIPTION | Typical Grade-trademarks | VENEER GRADE | |
			Faces	Inner Plys
B-B PLYFORM Class I & II**	Specifically manufactured for concrete forms. Many reuses. Smooth, solid surfaces. Edge-sealed. Mill-oiled unless otherwise specified.	B-B PLYFORM CLASS I EXTERIOR PS 1-66 DFPA TESTED QUALITY 000	B	C
High Density Overlaid PLYFORM Class I & II**	Hard, semi-opaque resin-fiber overlay, heat-fused to panel faces. Smooth surface resists abrasion. Up to 200 reuses. Edge-sealed. Light oiling recommended between pours.	HDO · PLYFORM·I · EXT·DFPA · PS 1-66	B	C Plugged
STRUCTURAL I PLYFORM**	Especially designed for engineered applications. All Group 1 species. Stronger and stiffer than PLYFORM Class I and II. Recommended for high pressures where face grain is parallel to supports. Also available with High Density Overlay faces.	STRUCTURAL I B-B PLYFORM CLASS I EXTERIOR PS 1-66 DFPA TESTED QUALITY 000	B	C or C Plugged
Special Overlays, proprietary panels and Medium Density Overlaid plywood specifically designed for concrete forming**	Produce a smooth uniform concrete surface. Generally mill treated with form release agent. Check with manufacturer for design specifications, proper use, and surface treatment recommendations for greatest number of reuses.			

*Commonly available in 5/8" and 3/4" panel thicknesses (4'x8' size).
**Check dealer for availability in your area.

SPECIFICATIONS FOR REINFORCED MASONRY RETAINING WALLS

8" WALLS

HEIGHT OF WALL	WIDTH OF FOOTING	THICKNESS OF FOOTING	DIST. TO FACE OF WALL	SIZE & SPACING OF VERTICAL RODS IN WALL	SIZE & SPACING OF HORIZONTAL RODS IN FOOTING
3'−4''	2'−4''	9''	8''	3/8'' @ 32''	3/8'' @ 27''
4'−0''	2'−9''	9''	10''	1/2'' @ 32''	3/8'' @ 27''
4'−8''	3'−3''	10''	12''	5/8'' @ 32''	3/8'' @ 27''
5'−4''	3'−8''	10'	14''	1/2'' @ 16''	1/2'' @ 30''
6'−0''	4'−2''	12''	15''	3/4'' @ 24''	1/2'' @ 25''

12" WALLS

HEIGHT OF WALL	WIDTH OF FOOTING	THICKNESS OF FOOTING	DIST. TO FACE OF WALL	SIZE & SPACING OF VERTICAL RODS IN WALL	SIZE & SPACING OF HORIZONTAL RODS IN FOOTING
6'−8''	4'−6''	12''	16''	3/4'' @ 24''	1/2'' @ 22''
7'−4''	4'−10''	12''	18''	7/8'' @ 32''	5/8'' @ 26''
8'−0''	5'−4''	12''	20''	7/8'' @ 24''	5/8'' @ 21''
8'−8''	5'−10''	14''	22''	7/8'' @ 16''	3/4'' @ 26''
9'−4''	6'−4''	14''	24''	1'' @ 8''	3/4'' @ 21''

ENGLISH TO METRIC CONVERSIONS

LENGTHS

1 INCH = 2.540 CENTIMETRES
1 FOOT = 30.48 CENTIMETRES
1 YARD = 91.44 CENTIMETRES OR
0.9144 METRES
1 MILE = 1.609 KILOMETRES

AREAS

1 SQ. IN. = 6.452 SQ. CENTIMETRES
1 SQ. FT. = 929.0 SQ. CENTIMETRES OR
0.0929 SQ. METRE
1 SQ. YD. = 0.8361 SQ. METRE

VOLUMES

1 CU. IN. = 16.39 CU. CENTIMETRES
1 CU. FT. = 0.02832 CU. METRE
1 CU. YD. = 0.7646 CU. METRE

WEIGHTS

1 OUNCE (AVDP) = 28.35 GRAMS
1 POUND = 453.6 GRAMS OR
0.4536 KILOGRAM
1 (SHORT) TON = 907.2 KILOGRAMS

LIQUID MEASUREMENTS

1 (FLUID) OUNCE = 0.02957 LITRE OR
28.35 GRAMS
1 PINT = 473.2 CU. CENTIMETRES
1 QUART = 0.9463 LITRE
1 (US) GALLON = 3785 CU. CENTIMETRES
OR 3.785 LITRES

POWER MEASUREMENTS

1 HORSEPOWER = 0.7457 KILOWATT

TEMPERATURE MEASUREMENTS

TO CONVERT DEGREES FAHRENHEIT TO DEGREES CENTIGRADE, USE THE FOLLOWING FORMULA:

$$C = 5/9 \times (F\ 32)$$

METRIC TO ENGLISH CONVERSIONS

LENGTHS

1 MILLIMETRE (mm) = 0.03937 IN.
1 CENTIMETRE (cm) = 0.3937 IN.
1 METRE (m) = 3.281 FT. OR
1.0937 YD.
1 KILOMETRE (km) = 0.6214 MILES

AREAS

1 SQ. MILLIMETRE = 0.00155 SQ. IN.
1 SQ. CENTIMETRE = 0.155 SQ. IN.
1 SQ. METRE = 10.76 SQ. FT. OR
1.196 SQ. YD.

VOLUMES

1 CU. CENTIMETRE = 0.06102 CU. IN.
1 CU. METRE = 35.31 CU. FT. OR
1.308 CU. YD.

WEIGHTS

1 GRAM = 0.03527 OZ.
1 KILOGRAM (kg) = 2.205 LB.
1 METRIC TON = 2205 LB.

LIQUID MEASUREMENTS

1 CU. CENTIMETRE (cm^3) = 0.06102 CU. IN.

1 LITRE (1000 cm^3) = 1.057 QUARTS OR
2.113 PINTS OR
61.02 CU. IN.

POWER MEASUREMENTS

1 KILOWATT (kw) = 1.341 HORSEPOWER (hp)

TEMPERATURE MEASUREMENTS

TO CONVERT DEGREES CENTIGRADE TO DEGREES FAHRENHEIT, USE THE FOLLOWING FORMULA:

$$F = (9/5 \times C) + 32$$

QUANTITIES OF BRICK AND MORTAR NECESSARY TO CONSTRUCT WALL ONE WYTHE IN THICKNESS

MODULAR BRICK

NOMINAL SIZE OF BRICK IN. T H L	NUMBER OF BRICK PER 100 SQ. FT.	CUBIC FEET OF MORTAR			
		PER 100 SQ. FT.		PER 1000 BRICK	
		3/8 IN. JOINTS	1/2 IN. JOINTS	3/8 IN. JOINTS	1/2 IN. JOINTS
4 x 2 2/3 x 8	675	5.5	7.0	8.1	10.3
4 x 3 1/5 x 8	563	4.8	6.1	8.6	10.9
4 x 4 x 8	450	4.2	5.3	9.2	11.7
4 x 5 1/3 x 8	338	3.5	4.4	10.2	12.9
4 x 2 x 12	600	6.5	8.2	10.8	13.7
4 x 2 2/3 x 12	450	5.1	6.5	11.3	14.4
4 x 3 1/5 x 12	375	4.4	5.6	11.7	14.9
4 x 4 x 12	300	3.7	4.8	12.3	15.7
4 x 5 1/3 x 12	225	3.0	3.9	13.4	17.1
6 x 2 2/3 x 12	450	7.9	10.2	17.5	22.6
6 x 3 1/5 x 12	375	6.8	8.8	18.1	23.4
6 x 4 x 12	300	5.6	7.4	19.1	24.7

NONMODULAR BRICK

NOMINAL SIZE OF BRICK IN. T H L	NUMBER OF BRICK PER 100 SQ. FT.	3/8 IN. JOINTS	1/2 IN. JOINTS	3/8 IN. JOINTS	1/2 IN. JOINTS
3 3/4 x 2 1/4 x 8	655	5.8		8.8	
	616		7.2		11.7
3 3/4 x 2 3/4 x 8	551	5.0		9.1	
	522		6.4		12.2

CORRECTION FACTORS FOR BONDS WITH FULL HEADERS

BOND	CORRECTION FACTOR
FULL HEADERS EVERY FIFTH COURSE ONLY	1/5
FULL HEADERS EVERY SIXTH COURSE ONLY	1/6
FULL HEADERS EVERY SEVENTH COURSE ONLY	1/7
ENGLISH BOND (FULL HEADERS EVERY SECOND COURSE)	1/2
FLEMISH BOND (ALTERNATE FULL HEADERS AND STRETCHERS EVERY COURSE)	1/3
FLEMISH HEADERS EVERY SIXTH COURSE	1/18
FLEMISH CROSS BOND (FLEMISH HEADERS EVERY SECOND COURSE)	1/6
DOUBLE-STRETCHER, GARDEN WALL BOND	1/5
TRIPE-STRETCHER, GARDEN WALL BOND	1/7

MINIMUM THICKNESS AND MAXIMUM SPAN FOR CONCRETE MASONRY SCREEN WALLS

Construction	Minimum Nominal Thickness, (t)	Maximum Distance Between Lateral Supports (Height or Length, Not Both)			
		Nominal Thickness of Wall, inches			Other Nominal Thicknesses (t)
		4	6	8	
Nonload-Bearing Reinforced:*					
Exterior	4"	10'-0"	15'-0"	20'-0"	30 t
Interior	4"	16'-0"	24'-0"	32'-0"	48 t
Nonreinforced:					
Exterior	4"	6'-8"	10'-0"	13'-4"	20 t
Interior	4"	12'-0"	18'-0"	24'-0"	36 t
Loadbearing:					
Reinforced*	6"	Not recommended	12'-6"	16'-8"	25 t
Nonreinforced	6"	Not recommended	9'-0"	12'-0"	18 t

*Total steel area, including joint reinforcement, not less than 0.002 times the gross cross-sectional area of the wall, not more than two-thirds of which may be used in either vertical or horizontal direction.

VERTICAL COURSING TABLE FOR NON-MODULAR BRICK[1]

No. of Courses	*Nominal Height (h) of Unit*[2]				
	2"	2 2/3"	3 1/5"	4"	5 1/3"
1	0'- 2"	0'- 2 11/16"	0'- 3 3/16"	0'-4"	0'- 5 5/16"
2	0'- 4"	0'- 5 5/16"	0'- 6 3/8"	0'-8"	0'-10 11/16"
3	0'- 6"	0'- 8"	0'- 9 5/8"	1'-0"	1'- 4"
4	0'- 8"	0'-10 11/16"	1'- 0 13/16"	1'-4"	1'- 9 5/16"
5	0'-10"	1'- 1 5/16"	1'- 4"	1'-8"	2'- 2 11/16"
6	1'- 0"	1'- 4"	1'- 7 3/16"	2'-0"	2'- 8"
7	1'- 2"	1'- 6 11/16"	1'-10 3/8"	2'-4"	3'- 1 5/16'
8	1'- 4"	1'- 9 5/16"	2'- 1 5/8"	2'-8"	3'- 6 11/16'
9	1'- 6"	2'- 0"	2'- 4 13/16"	3'-0"	4'- 0"
10	1'- 8"	2'- 2 11/16"	2'- 8"	3'-4"	4'- 5 5/16"
11	1'-10"	2'- 5 5/16"	2'-11 3/16"	3'-8"	4'-10 11/16"
12	2'- 0"	2'- 8"	3'- 2 3/8"	4'-0"	5'- 4"
13	2'- 2"	2'-10 11/16"	3'- 5 5/8"	4'-4"	5'- 9 5/16"
14	2'- 4"	3'- 1 5/16"	3'- 8 13/16"	4'-8"	6'- 2 11/16"
15	2'- 6"	3'- 4"	4'- 0"	5'-0"	6'- 8"
16	2'- 8"	3'- 6 11/16"	4'- 3 3/16"	5'-4"	7'- 1 5/16"
17	2'-10"	3'- 9 5/16"	4'- 6 3/8"	5'-8"	7'- 6 11/16"
18	3'- 0"	4'- 0"	4'- 9 5/8"	6'-0"	8'- 0"
19	3'- 2"	4'- 2 11/16"	5'- 0 13/16"	6'-4"	8'- 5 5/16"
20	3'- 4"	4'- 5 5/16"	5'- 4"	6'-8"	8'-10 11/16"
21	3'- 6"	4'- 8"	5'- 7 3/16"	7'-0"	9'- 4"
22	3'- 8"	4'-10 11/16"	5'-10 3/8"	7'-4"	9'- 9 5/16"
23	3'-10"	5'- 1 5/16"	6'- 1 5/8"	7'-8"	10'- 2 11/16"
24	4'- 0"	5'- 4"	6' 4 13/16"	8'-0"	10'- 8"
25	4'- 2"	5'- 6 11/16"	6'- 8"	8'-4"	11'- 1 5/16"
26	4'- 4"	5'- 9 5/16"	6'-11 3/16"	8'-8"	11'- 6 11/16"
27	4'- 6"	6'- 0"	7'- 2 3/8"	9'-0"	12'- 0"
28	4'- 8"	6'- 2 11/16"	7'- 5 5/8"	9'-4"	12'- 5 5/16"
29	4'-10"	6'- 5 5/16"	7'- 8 13/16"	9'-8"	12'-10 11/16"
30	5'- 0"	6'- 8"	8'- 0"	10'-0"	13'- 4"
31	5'- 2"	6'-10 11/16"	8'- 3 3/16"	10'-4"	13'- 9 5/16"
32	5'- 4"	7'- 1 5/16"	8'- 6 3/8"	10'-8"	14' 2 11/16"
33	5'- 6"	7'- 4"	8'- 9 5/8"	11'-0"	14'- 8"
34	5'- 8"	7'- 6 11/16"	9'- 0 13/16"	11'-4"	15'- 1 5/16"
35	5'-10"	7'- 9 5/16"	9'- 4"	11'-8"	15'- 6 11/16"
36	6'- 0"	8'- 0"	9'- 7 3/16"	12'-0"	16'- 0"
37	6'- 2"	8'- 2 11/16"	9'-10 3/8"	12'-4"	16'- 5 5/16"
38	6'- 4"	8'- 5 5/16"	10'- 1 5/8"	12'-8"	16'-10 11/16"
39	6'- 6"	8'- 8"	10'- 4 13/16"	13'-0"	17'- 4"
40	6'- 8"	8'- 10 11/16"	10'- 8"	13'-4"	17'- 9 5/16"
41	6'-10"	9'- 1 5/16"	10'-11 3/16"	13'-8"	18'- 2 11/16"
42	7'- 0"	9'- 4"	11'- 2 3/8"	14'-0"	18'- 8"
43	7'- 2"	9'- 6 11/16"	11'- 5 5/8"	14'-4"	19'- 1 5/16"
44	7'- 4"	9'- 9 5/16"	11'- 8 13/16"	14'-8"	19'- 6 11/16"
45	7'- 6"	10'- 0"	12'- 0"	15'-0"	20'- 0"
46	7'- 8"	10'- 2 11/16"	12'- 3 3/16"	15'-4"	20'- 5 5/16"
47	7'-10"	10'- 5 5/16"	12'- 6 3/8"	15'-8"	20'-10 11/16"
48	8'- 0"	10'- 8"	12'- 9 5/8"	16'-0"	21'- 4"
49	8'- 2"	10'-10 11/16"	13'- 0 13/16"	16'-4"	21'- 9 5/16"
50	8'- 4"	11'- 1 5/16"	13'- 4"	16'-8"	22'- 2 11/16"
100	16'- 8"	22'- 2 11/16"	26'- 8"	33'-4"	44'- 5 5/16"

[1] Brick positioned in wall as stretchers.

[2] For convenience in using table, nominal 1/3", 2/3" and 1/5" heights of units have been changed to nearest 1/16". Vertical dimensions are from bottom of mortar joint to bottom of mortar joint.

No. of Courses	2¼-in. High Units		2⅝-in. High Units		2¾-in. High Units	
	⅜″ Joint	½″ Joint	⅜″ Joint	½″ Joint	⅜″ Joint	½″ Joint
1	0′- 2⅝″	0′- 2¾″	0′-3″	0′- 3⅛″	0′- 3⅛″	0′- 3¼″
2	0′- 5¼″	0′- 5½″	0′-6″	0′- 6¼″	0′- 6¼″	0′- 6½″
3	0′- 7⅞″	0′- 8¼″	0′-9″	0′- 9⅜″	0′- 9⅜″	0′- 9¾″
4	0′-10½″	0′-11″	1′-0″	1′- 0½″	1′- 0½″	1′- 1″
5	1′- 1⅛″	1′- 1¾″	1′-3″	1′- 3⅝″	1′- 3⅝″	1′- 4¼″
6	1′- 3¾″	1′- 4½″	1′-6″	1′- 6¾″	1′- 6¾″	1′- 7½″
7	1′- 6⅜″	1′- 7¼″	1′-9″	1′- 9⅞″	1′- 9⅞″	1′-10¾″
8	1′- 9″	1′-10″	2′-0″	2′- 1″	2′- 1″	2′- 2″
9	1′-11⅝″	2′- 0¾″	2′-3″	2′- 4⅛″	2′- 4⅛″	2′- 5¼″
10	2′- 2¼″	2′- 3½″	2′-6″	2′- 7¼″	2′- 7¼″	2′- 8½″
11	2′- 4⅞″	2′- 6¼″	2′-9″	2′-10⅜″	2′-10⅜″	2′-11¾″
12	2′- 7½″	2′- 9″	3′-0″	3′- 1½″	3′- 1½″	3′- 3″
13	2′-10⅛″	2′-11¾″	3′-3″	3′- 4⅝″	3′- 4⅝″	3′- 6¼″
14	3′- 0¾″	3′- 2½″	3′-6″	3′- 7¾″	3′- 7¾″	3′- 9½″
15	3′- 3⅜″	3′- 5¼″	3′-9″	3′-10⅞″	3′-10⅞″	4′- 0¾″
16	3′- 6″	3′- 8″	4′-0″	4′- 2″	4′- 2″	4′- 4″
17	3′- 8⅝″	3′-10¾″	4′-3″	4′- 5⅛″	4′- 5⅛″	4′- 7¼″
18	3′-11¼″	4′- 1½″	4′-6″	4′- 8¼″	4′- 8¼″	4′-10½″
19	4′- 1⅞″	4′- 4¼″	4′-9″	4′-11⅜″	4′-11⅜″	5′- 1¾″
20	4′- 4½″	4′- 7″	5′-0″	5′- 2½″	5′- 2½″	5- 5″
21	4′- 7⅛″	4′- 9¾″	5′-3″	5′- 5⅝″	5′- 5⅝″	5′- 8¼″
22	4′- 9¾″	5′- 0½″	5′-6″	5′- 8¾″	5′- 8¾″	5′-11½″
23	5′- 0⅜″	5′- 3¼″	5′-9″	5′-11⅞″	5′-11⅞″	6′- 2¾″
24	5′- 3″	5′- 6″	6′-0″	6′- 3″	6′- 3″	6′- 6″
25	5′- 5⅝″	5′- 8¾″	6′-3″	6′- 6⅛″	6′- 6⅛″	6′- 9¼″
26	5′- 8¼″	5′-11½″	6′-6″	6′- 9¼″	6′- 9¼″	7′- 0½″
27	5′-10⅞″	6′- 2¼″	6′-9″	7′- 0⅜″	7′- 0⅜″	7′- 3¾″
28	6′- 1½″	6′- 5″	7′-0″	7′- 3½″	7′- 3½″	7′- 7″
29	6′- 4⅛″	6′- 7¾″	7′-3″	7′- 6⅝″	7′- 6⅝″	7′-10¼″
30	6′- 6¾″	6′-10½″	7′-6″	7′- 9¾″	7′- 9¾″	8′- 1½″
31	6′- 9⅜″	7′- 1¼″	7′-9″	8′- 0⅞″	8′- 0⅞″	8′- 4¾″
32	7′- 0″	7′- 4″	8′-0″	8′- 4″	8′- 4″	8′- 8″
33	7′- 2⅝″	7′- 6¾″	8′-3″	8′- 7⅛″	8′- 7⅛″	8′-11¼″
34	7′- 5¼″	7′- 9½″	8′-6″	8′-10¼″	8′-10¼″	9′- 2½″
35	7′- 7⅞″	8′- 0¼″	8′-9″	9′- 1⅜″	9′- 1⅜″	9′- 5¾″
36	7′-10½″	8′- 3″	9′-0″	9′- 4½″	9′- 4½″	9′- 9″
37	8′- 1⅛″	8′- 5¾″	9′-3″	9′- 7⅝″	9′- 7⅝″	10′- 0¼″
38	8′- 3¾″	8′- 8½″	9′-6″	9′-10¾″	9′-10¾″	10′- 3½″
39	8′- 6⅜″	8′-11¼″	9′-9″	10′- 1⅞″	10′- 1⅞″	10′- 6¾″
40	8′- 9″	9′- 2″	10′-0″	10′- 5″	10′- 5″	10′-10″
41	8′-11⅝″	9′- 4¾″	10′-3″	10′- 8⅛″	10′- 8⅛″	11′- 1¼″
42	9′- 2¼″	9′- 7½″	10′-6″	10′-11¼″	10′-11¼″	11′- 4½″
43	9′- 4⅞″	9′-10¼″	10′-9″	11′- 2⅜″	11′- 2⅜″	11′- 7¾″
44	9′- 7½″	10′- 1″	11′-0″	11′- 5½″	11′- 5½″	11′-11″
45	9′-10⅛″	10′- 3¾″	11′-3″	11′- 8⅝″	11′- 8⅝″	12′- 2¼″
46	10′- 0¾″	10′- 6½″	11′-6″	11′-11¾″	11′-11¾″	12′- 5½″
47	10′- 3⅜″	10′- 9¼″	11′-9″	12′- 2⅞″	12′- 2⅞″	12′- 8¾″
48	10′- 6″	11′- 0″	12′-0″	12′- 6″	12′- 6″	13′- 0″
49	10′- 8⅝″	11′- 2¾″	12′-3″	12′- 9⅛″	12′- 9⅛″	13′- 3¼″
50	10′-11¼″	11′- 5½″	12′-6″	13′- 0¼″	13′- 0¼″	13′- 6½″
100	21′-10½″	22′-11″	25′-0″	26′- 0½″	26′- 0½″	27′- 1″

[1] Brick positioned in wall as stretchers. Vertical dimensions are from bottom of mortar joint to bottom of mortar joint.

WEIGHTS AND MEASURES

Water									Cement						Water-Cement Ratio		
Weight, lb.	U.S. gallons	Imp. gallons	Weight, lb.	U.S. gallons	Imp. gallons	Weight, lb.	U.S. gallons	Imp. gallons	U.S. bags	Can. bags	Weight, lb.	U.S. bags	Can. bags	Weight, lb.	W/C, by weight	Can. w/c gal./bag	U.S. w/c gal./bag
10	1.2	1.0	185	22.2	18.5	360	43.2	36.0	1.0	1.1	94	4.5	4.8	423	.30	2.6	3.4
15	1.8	1.5	190	22.8	19.0	365	43.8	36.5	1.1	1.2	103	4.6	4.9	432	.31	2.7	3.5
20	2.4	2.0	195	23.4	19.5	370	44.4	37.0	1.2	1.3	113	4.7	5.0	442	.32	2.8	3.6
25	3.0	2.5	200	24.0	20.0	375	45.0	37.5	1.3	1.4	122	4.8	5.2	451	.33	2.9	3.7
30	3.6	3.0	205	24.6	20.5	380	45.6	38.0	1.4	1.5	132	4.9	5.3	461	.34	3.0	3.8
35	4.2	3.5	210	25.2	21.0	385	46.2	38.5	1.5	1.6	141	5.0	5.4	470	.35	3.1	3.9
40	4.8	4.0	215	25.8	21.5	390	46.8	39.0	1.6	1.7	150	5.1	5.5	479	.36	3.2	4.1
45	5.4	4.5	220	26.4	22.0	395	47.4	39.5	1.7	1.8	160	5.2	5.6	489	.37	3.2	4.2
50	6.0	5.0	225	27.0	22.5	400	48.0	40.0	1.8	1.9	169	5.3	5.7	498	.38	3.3	4.3
55	6.6	5.5	230	27.6	23.0	405	48.6	40.5	1.9	2.0	179	5.4	5.8	508	.39	3.4	4.4
60	7.2	6.0	235	28.2	23.5	410	49.2	41.0	2.0	2.1	188	5.5	5.9	517	.40	3.5	4.5
65	7.8	6.5	240	28.8	24.0	415	49.8	41.5	2.1	2.3	197	5.6	6.0	526	.41	3.6	4.6
70	8.4	7.0	245	29.4	24.5	420	50.4	42.0	2.2	2.4	207	5.7	6.1	536	.42	3.7	4.7
75	9.0	7.5	250	30.0	25.0	425	51.0	42.5	2.3	2.5	216	5.8	6.2	545	.43	3.8	4.8
80	9.6	8.0	255	30.6	25.5	430	51.6	43.0	2.4	2.6	226	5.9	6.3	555	.44	3.9	5.0
85	10.2	8.5	260	31.2	26.0	435	52.2	43.5	2.5	2.7	235	6.0	6.4	564	.45	3.9	5.1
90	10.8	9.0	265	31.8	26.5	440	52.8	44.0	2.6	2.8	244	6.1	6.6	573	.46	4.0	5.2
95	11.4	9.5	270	32.4	27.0	445	53.4	44.5	2.7	2.9	254	6.2	6.7	583	.47	4.1	5.3
100	12.0	10.0	275	33.0	27.5	450	54.0	45.0	2.8	3.0	263	6.3	6.8	592	.48	4.2	5.4
105	12.6	10.5	280	33.6	28.0	455	54.6	45.5	2.9	3.1	273	6.4	6.9	602	.49	4.3	5.5
110	13.2	11.0	285	34.2	28.5	460	55.2	46.0	3.0	3.2	282	6.5	7.0	611	.50	4.4	5.6
115	13.8	11.5	290	34.8	29.0	465	55.8	46.5	3.1	3.3	291	6.6	7.1	620	.51	4.5	5.7
120	14.4	12.0	295	35.4	29.5	470	56.4	47.0	3.2	3.4	301	6.7	7.2	630	.52	4.6	5.9
125	15.0	12.5	300	36.0	30.0	475	57.0	47.5	3.3	3.5	310	6.8	7.3	639	.53	4.6	6.0
130	15.6	13.0	305	36.6	30.5	480	57.6	48.0	3.4	3.7	320	6.9	7.4	649	.54	4.7	6.1
135	16.2	13.5	310	37.2	31.0	485	58.2	48.5	3.5	3.8	329	7.0	7.5	658	.55	4.8	6.2
140	16.8	14.0	315	37.8	31.5	490	58.8	49.0	3.6	3.9	338	7.1	7.6	667	.56	4.9	6.3
145	17.4	14.5	320	38.4	32.0	495	59.4	49.5	3.7	4.0	348	7.2	7.7	677	.57	5.0	6.4
150	18.0	15.0	325	39.0	32.5	500	60.0	50.0	3.8	4.1	357	7.3	7.8	686	.58	5.1	6.5
155	18.6	15.5	330	39.6	33.0	505	60.6	50.5	3.9	4.2	367	7.4	7.9	696	.59	5.2	6.7
160	19.2	16.0	335	40.2	33.5	510	61.2	51.0	4.0	4.3	376	7.5	8.1	705	.60	5.3	6.8
165	19.8	16.5	340	40.8	34.0	515	61.8	51.5	4.1	4.4	385	7.6	8.2	714	.61	5.3	6.9
170	20.4	17.0	345	41.4	34.5	520	62.4	52.0	4.2	4.5	395	7.7	8.3	724	.62	5.4	7.0
175	21.0	17.5	350	42.0	35.0	525	63.0	52.5	4.3	4.6	404	7.8	8.4	733	.63	5.5	7.1
180	21.6	18.0	355	42.6	35.5	530	63.6	53.0	4.4	4.7	414	7.9	8.5	743	.64	5.6	7.2

Note: Larger quantities can be secured by addition of above figures.

Equivalent weights and measures

1 gal. of water, U.S. = 8.33 lb.
1 gal. of water, imp. = 10.00 lb.
1 cu.ft. of water = 62.40 lb.
1 bag of cement, U.S. = 94.00 lb.
1 bag of cement, Can. = 87.50 lb.

Metric conversions

To convert from:	To:	Multiply by:
U.S. gal.	cu. meters	0.003785
imp. gal.	cu. meters	0.004546
U.S. bags	kilograms	42.637
Can. bags	kilograms	39.689
pounds	kilograms	0.4536
cu.ft.	cu. meters	0.02832
cu.yd.	cu. meters	0.7646

CONSTRUCTION DETAILS-SCREEN WALLS

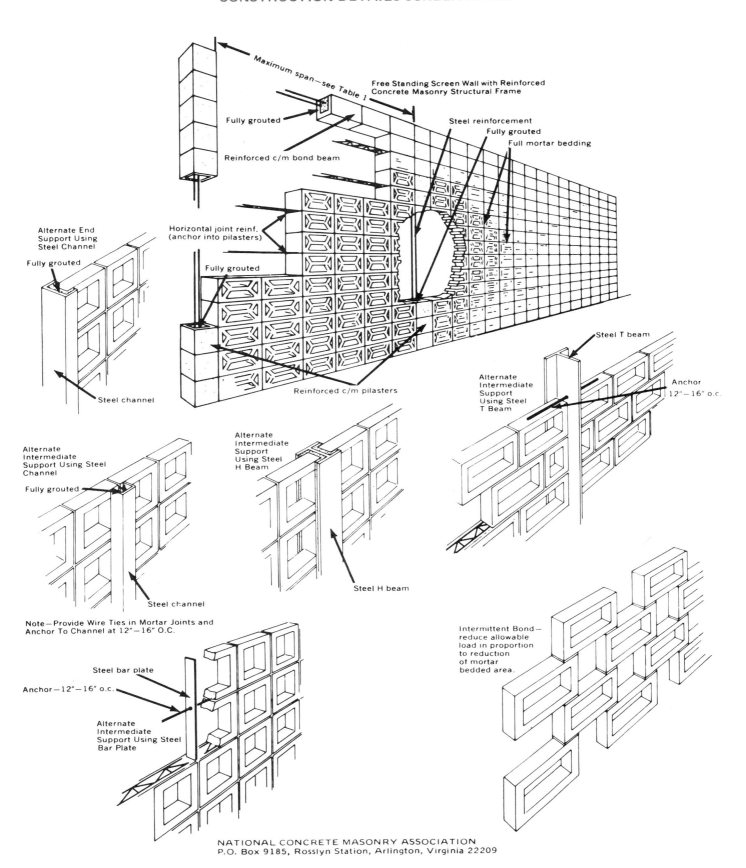

Maximum span—see Table 1

Free Standing Screen Wall with Reinforced Concrete Masonry Structural Frame

Fully grouted

Reinforced c/m bond beam

Steel reinforcement
Fully grouted
Full mortar bedding

Alternate End Support Using Steel Channel

Fully grouted

Horizontal joint reinf. (anchor into pilasters)

Fully grouted

Steel channel

Reinforced c/m pilasters

Steel T beam

Alternate Intermediate Support Using Steel T Beam

Anchor 12"—16" o.c.

Alternate Intermediate Support Using Steel Channel

Fully grouted

Alternate Intermediate Support Using Steel H Beam

Steel channel

Steel H beam

Note—Provide Wire Ties in Mortar Joints and Anchor To Channel at 12"—16" O.C.

Steel bar plate

Anchor—12"—16" o.c.

Alternate Intermediate Support Using Steel Bar Plate

Intermittent Bond—reduce allowable load in proportion to reduction of mortar bedded area.

NATIONAL CONCRETE MASONRY ASSOCIATION
P.O. Box 9185, Rosslyn Station, Arlington, Virginia 22209

SHAPES AND SIZES OF STRUCTURAL CLAY TILE

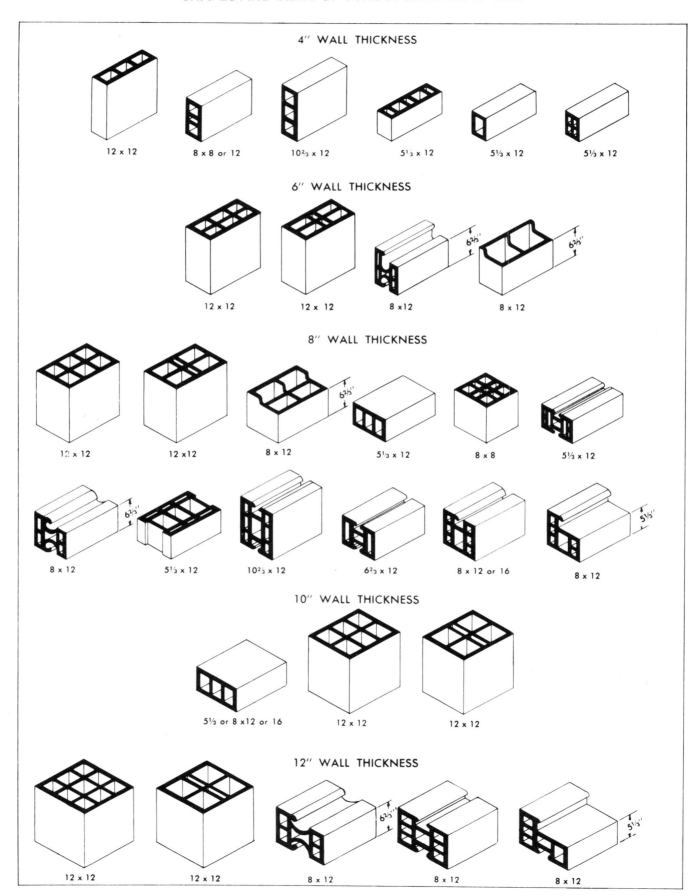

4″ WALL THICKNESS

12 x 12   8 x 8 or 12   10²⁄₃ x 12   5¹⁄₃ x 12   5¹⁄₃ x 12   5¹⁄₃ x 12

6″ WALL THICKNESS

12 x 12   12 x 12   8 x 12   8 x 12

8″ WALL THICKNESS

12 x 12   12 x 12   8 x 12   5¹⁄₃ x 12   8 x 8   5¹⁄₃ x 12

8 x 12   5¹⁄₃ x 12   10²⁄₃ x 12   6²⁄₃ x 12   8 x 12 or 16   8 x 12

10″ WALL THICKNESS

5¹⁄₃ or 8 x12 or 16   12 x 12   12 x 12

12″ WALL THICKNESS

12 x 12   12 x 12   8 x 12   8 x 12   8 x 12

MOST WIDELY AVAILABLE SIZES OF STRUCTURAL CLAY TILE

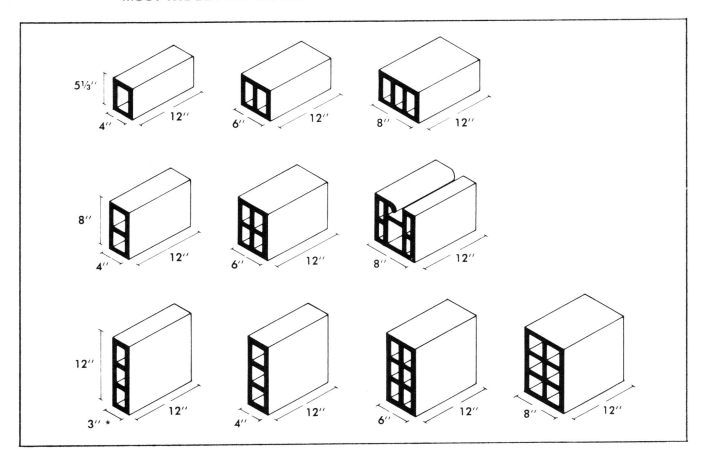

LINE SYMBOLS

BORDER LINE

OBJECT LINE

CUTTING PLANE LINE

SHORT BREAK LINE

HIDDEN LINE

CENTER LINE

SECTION LINES

LONG BREAK LINE

GUIDE LINES ABCDEFGH

DIMENSION LINE AND
EXTENSION LINES

6 FT. 4 IN.

SECTION INDICATORS

WINDOW
SCHEDULE SYMBOL A

DOOR
SCHEDULE SYMBOL 2

BUILDING MATERIAL SYMBOLS

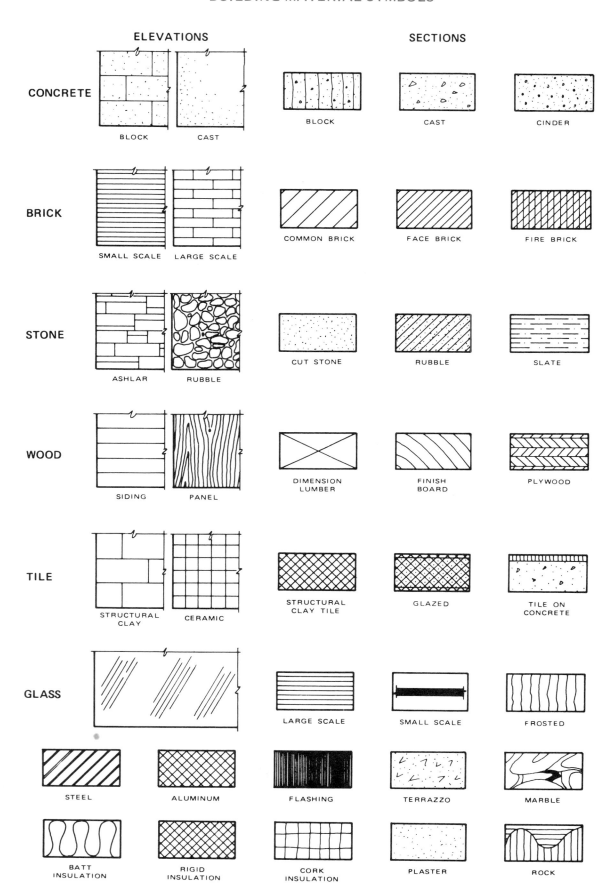

TOPOGRAPHICAL SYMBOLS

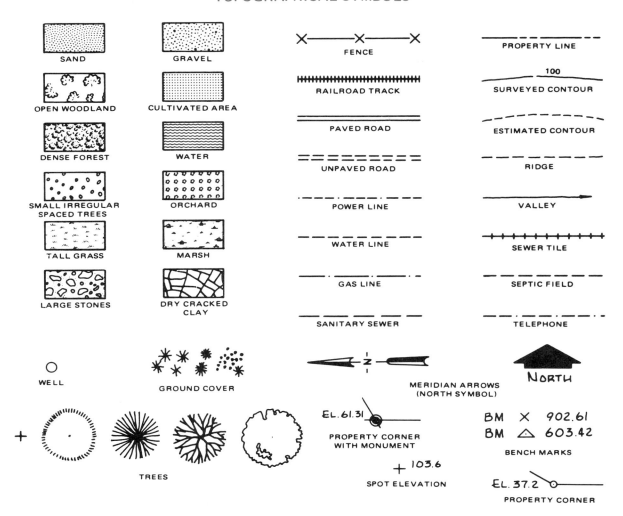

PLUMBING SYMBOLS

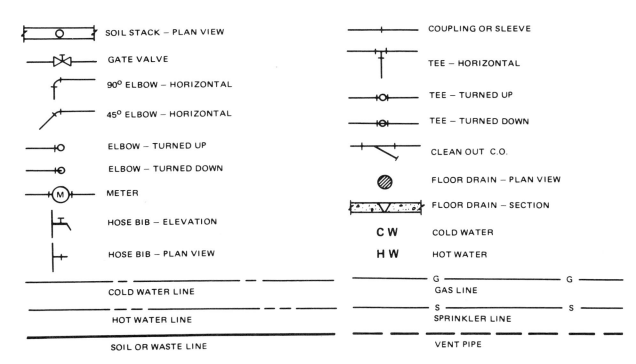

Masonry

CLIMATE CONTROL SYMBOLS

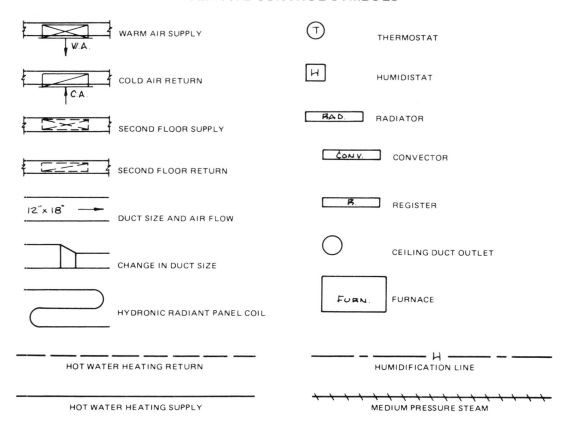

WARM AIR SUPPLY	THERMOSTAT
COLD AIR RETURN	HUMIDISTAT
SECOND FLOOR SUPPLY	RADIATOR
SECOND FLOOR RETURN	CONVECTOR
DUCT SIZE AND AIR FLOW	REGISTER
CHANGE IN DUCT SIZE	CEILING DUCT OUTLET
HYDRONIC RADIANT PANEL COIL	FURNACE
HOT WATER HEATING RETURN	HUMIDIFICATION LINE
HOT WATER HEATING SUPPLY	MEDIUM PRESSURE STEAM

ELECTRICAL SYMBOLS

CEILING OUTLET FIXTURE	SINGLE RECEPTACLE OUTLET	SINGLE-POLE SWITCH
RECESSED OUTLET FIXTURE	DUPLEX RECEPTACLE OUTLET	DOUBLE-POLE SWITCH
DROP CORD FIXTURE	TRIPLEX RECEPTACLE OUTLET	THREE-WAY SWITCH
FAN HANGER OUTLET	QUADRUPLEX RECEPTACLE OUTLET	FOUR-WAY SWITCH
JUNCTION BOX	SPLIT-WIRED DUPLEX RECEPTACLE OUTLET	WEATHERPROOF SWITCH
FLUORESCENT FIXTURE	SPECIAL PURPOSE SINGLE RECEPTACLE OUTLET	LOW VOLTAGE SWITCH
TELEPHONE	230 VOLT OUTLET	PUSH BUTTON
INTERCOM	WEATHERPROOF DUPLEX OUTLET	CHIMES
CEILING FIXTURE WITH PULL SWITCH	DUPLEX RECEPTACLE WITH SWITCH	TELEVISION ANTENNA OUTLET
THERMOSTAT	FLUSH MOUNTED PANEL BOX	DIMMER SWITCH
SPECIAL FIXTURE OUTLET A.B.C Etc.	SPECIAL DUPLEX OUTLET A.B.C Etc.	SPECIAL SWITCH A.B.C Etc.

234

PATTERNS FOR CONCRETE MASONRY

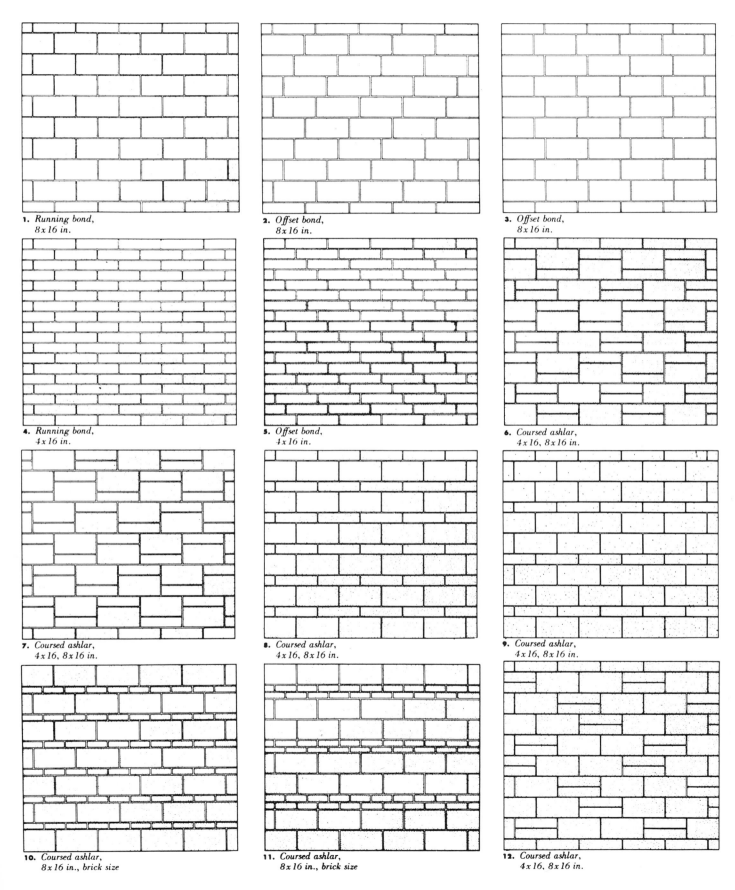

1. *Running bond,*
   8 x 16 in.

2. *Offset bond,*
   8 x 16 in.

3. *Offset bond,*
   8 x 16 in.

4. *Running bond,*
   4 x 16 in.

5. *Offset bond,*
   4 x 16 in.

6. *Coursed ashlar,*
   4 x 16, 8 x 16 in.

7. *Coursed ashlar,*
   4 x 16, 8 x 16 in.

8. *Coursed ashlar,*
   4 x 16, 8 x 16 in.

9. *Coursed ashlar,*
   4 x 16, 8 x 16 in.

10. *Coursed ashlar,*
    8 x 16 in., brick size

11. *Coursed ashlar,*
    8 x 16 in., brick size

12. *Coursed ashlar,*
    4 x 16, 8 x 16 in.

PATTERNS FOR CONCRETE MASONRY

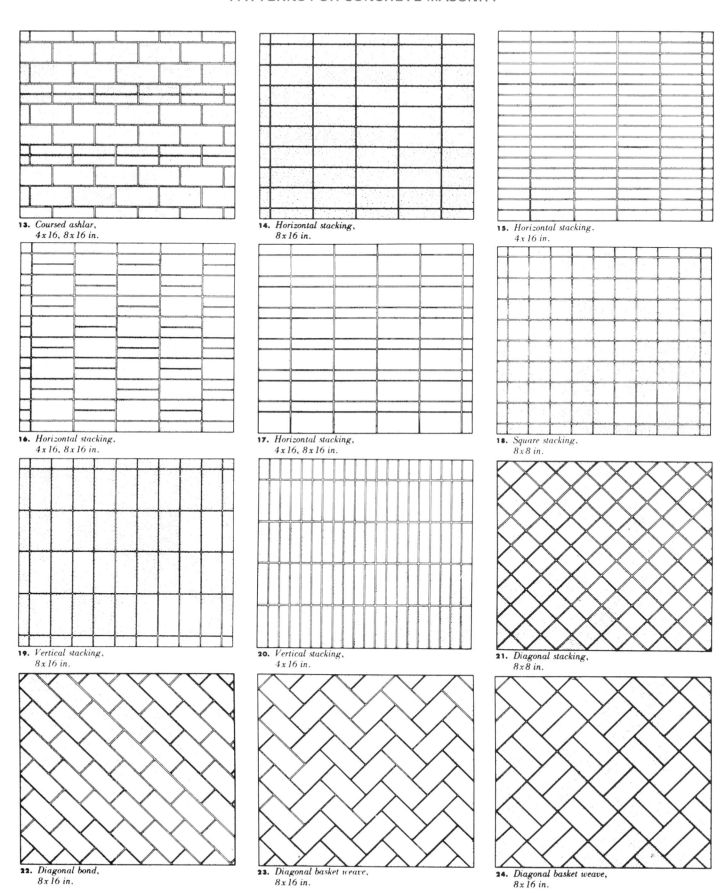

13. *Coursed ashlar,*
4 x 16, 8 x 16 in.

14. *Horizontal stacking,*
8 x 16 in.

15. *Horizontal stacking,*
4 x 16 in.

16. *Horizontal stacking,*
4 x 16, 8 x 16 in.

17. *Horizontal stacking,*
4 x 16, 8 x 16 in.

18. *Square stacking,*
8 x 8 in.

19. *Vertical stacking,*
8 x 16 in.

20. *Vertical stacking,*
4 x 16 in.

21. *Diagonal stacking,*
8 x 8 in.

22. *Diagonal bond,*
8 x 16 in.

23. *Diagonal basket weave,*
8 x 16 in.

24. *Diagonal basket weave,*
8 x 16 in.

PATTERNS FOR CONCRETE MASONRY

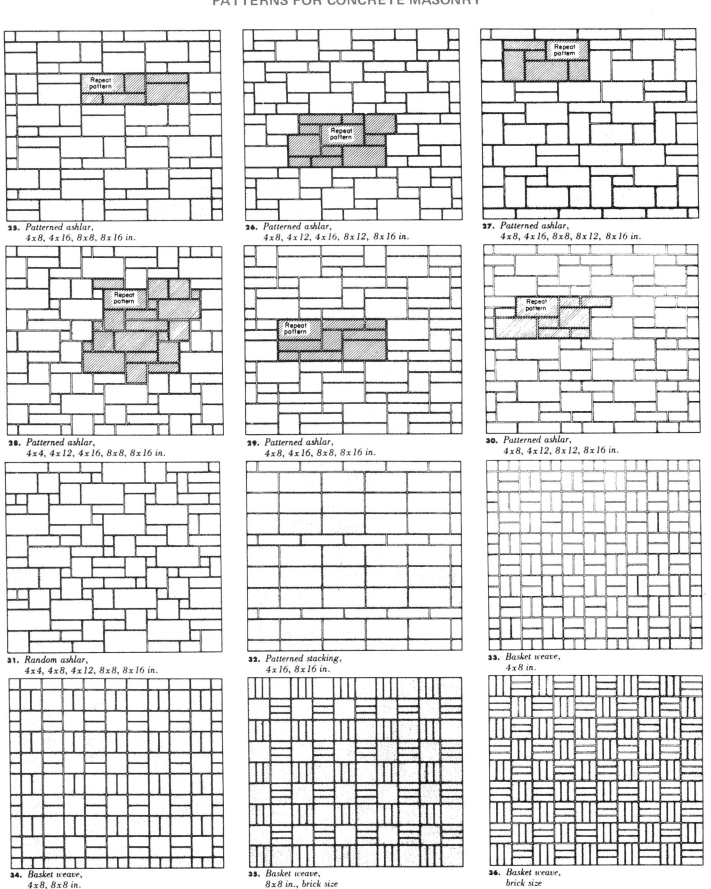

25. *Patterned ashlar,*
4 x 8, 4 x 16, 8 x 8, 8 x 16 in.

26. *Patterned ashlar,*
4 x 8, 4 x 12, 4 x 16, 8 x 12, 8 x 16 in.

27. *Patterned ashlar,*
4 x 8, 4 x 16, 8 x 8, 8 x 12, 8 x 16 in.

28. *Patterned ashlar,*
4 x 4, 4 x 12, 4 x 16, 8 x 8, 8 x 16 in.

29. *Patterned ashlar,*
4 x 8, 4 x 16, 8 x 8, 8 x 16 in.

30. *Patterned ashlar,*
4 x 8, 4 x 12, 8 x 12, 8 x 16 in.

31. *Random ashlar,*
4 x 4, 4 x 8, 4 x 12, 8 x 8, 8 x 16 in.

32. *Patterned stacking,*
4 x 16, 8 x 16 in.

33. *Basket weave,*
4 x 8 in.

34. *Basket weave,*
4 x 8, 8 x 8 in.

35. *Basket weave,*
8 x 8 in., brick size

36. *Basket weave,*
brick size

PATTERNS FOR CONCRETE MASONRY

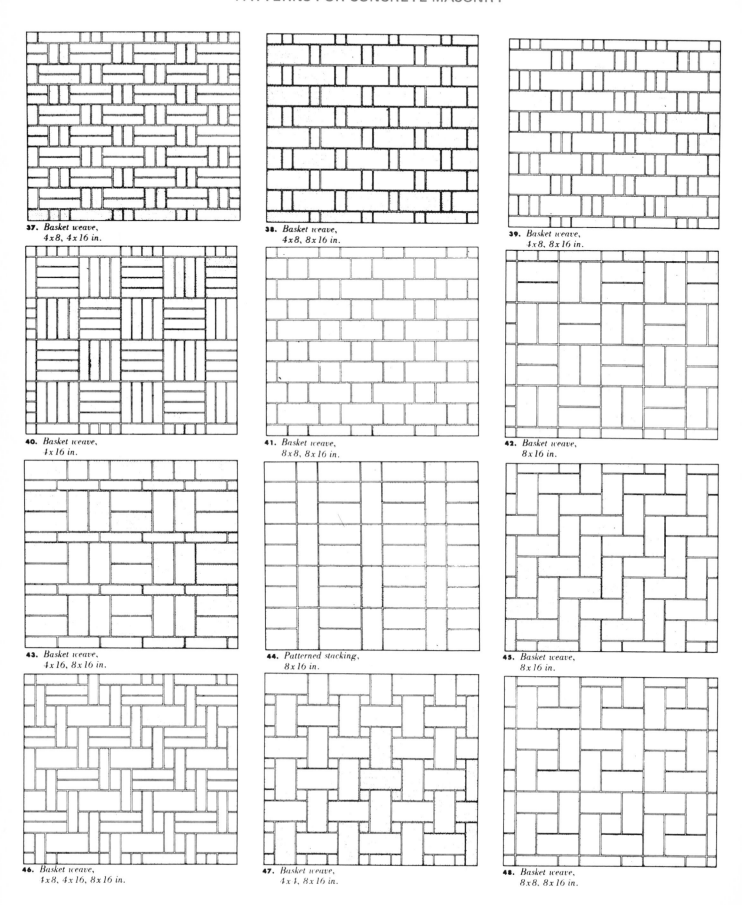

37. *Basket weave,*
4 x 8, 4 x 16 in.

38. *Basket weave,*
4 x 8, 8 x 16 in.

39. *Basket weave,*
4 x 8, 8 x 16 in.

40. *Basket weave,*
4 x 16 in.

41. *Basket weave,*
8 x 8, 8 x 16 in.

42. *Basket weave,*
8 x 16 in.

43. *Basket weave,*
4 x 16, 8 x 16 in.

44. *Patterned stacking,*
8 x 16 in.

45. *Basket weave,*
8 x 16 in.

46. *Basket weave,*
4 x 8, 4 x 16, 8 x 16 in.

47. *Basket weave,*
4 x 4, 8 x 16 in.

48. *Basket weave,*
8 x 8, 8 x 16 in.

ABBREVIATIONS

aggr	aggregate	frbk	firebrick	ptn	partition
alum.	aluminum	fdr	fire door	d	penny
L	angle	fp	fireplace	perp	perpendicular
arch.	architectural	fprf	fireproof	pc	piece
asb	asbestos	flge	flange	plas	plaster
asph	asphalt	flr	floor	pl	plate
AT.	asphalt tile	flg	flooring	Pl Gl	plate glass
avg	average	fluor	fluorescent	lb	pound, pounds
bbl	barrel, barrels	fl	flush	psi	pounds per square inch
basmt	basement	' or ft	foot, feet	P/C	poured concrete
B	bathroom	ft lb	foot pound	prect	precast
BR	bedroom	ftg	footing	prefab	prefabricated
BM	bench mark	fdn	foundation	PC	pull chain
bev	beveled	fr	frame	PB	push button
BP	blueprint	gal.	gallon, gallons	r or rad	radius
BRKT	bracket	galv	galvanized	recp	receptacle
br	brass	GA	gage	rect	rectangle
BRK	brick	gl	glass	ref	refrigerator
Btu	British thermal unit	gr	grade	reg	register
bro	bronze	g	granite	reinf	reinforce
Bldg	building	grtg	grating	ret	return
BUZ	buzzer	gyp	gypsum	RH	right hand
CAB	cabinet	hdw	hardware	R	riser
CP	candlepower	hdwd	hardwood	rm	room
CI	cast iron	hgt	height	rgh	rough
clk	caulk	hor	horizontal	rnd	round
clg	ceiling	HB	hose bibb	rub.	rubber
cem	cement	hw	hot water	sad.	saddle
c or CL	center line	C	hundred	sch	schedule
cm	centimetre, centimetres	I	I beam	smls	seamless
cer	ceramic	'' or in.	inches, inches	sec	second
chfr	chamfer	incl	include	''	second (angular measure)
chkd	checked	ID	inside diameter	shthg	sheathing
cin bl	cinder block	insul	insulation	shlp	ship lap
cir	circular	int	interior	sh	shower
CIRC	circumference	jt	joint	sdg	siding
c — o	cleanout	j	junction	sc	sill cock
clr	clear	jb	junction box	S	sink
CL	closet	kg	kilogram	sl	slate
col	column	km	kilometre	sp	soil pipe
com	common	kw	kilowatt, kilowatts	S	south
con	concrete	kwh	kilowatt-hour	SE	southeast
conc b	concrete block	k	kip (1000 lb.)	SW	southwest
conc clg	concrete ceiling	K	kitchen	sq	square
conc fl	concrete floor	ldg	landing	sq ft	square foot, feet
cnd	conduit	lth	lath	sst	stainless steel
const	construction	Lau	laundry	Stp	standpipe
cop	copper	LT	laundry tray	stl	steel
cu	cubic	lav	lavatory	stiff.	stiffener
cu ft	cubic foot, feet	LDR	leader	stn	stone
cu in.	cubic inch, inches	ld	leader drain	SP	sump pit
cu yd	cubic yard, yards	LH	left-hand	susp clg	suspending ceiling
DP	dampproofing	lgth	length	sw	switch
det	detail	lev	level	T	tee
diag	diagram	lt	light	tel	telephone
dia	diameter	lwc	lightweight concrete	TC	terra cotta
dim.	dimension	ls	limestone	ter	terrazzo
DR	dining room	lin ft	linear feet	thermo	thermostat
d — h	double hung	L CL	linen closet	thk	thick, thickness
dn	down	lino	linoleum	thd	thread
ds	downspout	Ll	live load	T & G	tongue and groove
dr	drain	LR	living room	tr	tread
dwg	drawing	l	lumen	trnbkl	turnbuckle
drwn	drawn	mip	malleable iron pipe	unfin	unfinished
dw	dry well	mfr	manufacture	V	vent, ventilator
E	east	mr	marble	vent.	ventilation
elec	electric	MO	masonry opening	vert	vertical
el	elevation	matl	material	vol	volume
elev	elevator	max	maximum	WV	wall vent
ent	entrance	mm	millimetre, millimetres	wp	waterproofing
equip.	equipment	min	minimum	w	watt, watts
est	estimate	(')	minute (angular measure)	whr	watt-hour
exc	excavate	moldg	molding	WS	weather stripping
exist.	existing	ni	nickle	wp	weatherproof
exp bt	expansion bolt	nom	nominal	whp	weep hole
exp jt	expansion joint	N	north	W	west
EXT	extension	No.	number	WF	wide flange
ext	exterior	oc	off center	wth	width
xh	extra heavy	opg	opening	wdw	window
fab	fabricate	oz	ounce, ounces	w/	with
F	fahrenheit	out.	outlet	wf	wood frame
fig.	figure	OD	out diameter	wrt	wrought
fil	fillet	oa	overall	WI	wrought iron
fin.	finish	ovfl	overflow	yd	yard, yards
FAO	finish all over	ovhd	overhead	z	zinc

Masonry

OUTLINE OF SUGGESTED RELATED INSTRUCTION COURSE FOR APPRENTICE BRICKLAYERS

The following is a suggested course outline of related instruction to supplement on-the-job training and the work processes recommended as necessary for a 4500-hour apprenticeship program.

FIRST AND SECOND YEARS

A. Masonry Material
1. Masonry Units

   History, description, manufacture, classification, types, special units, structural characteristics, physical properties, color, texture, and uses for:
   a. Clay and shale brick.
   b. Fire brick.
   c. Sandlime brick.
   d. Concrete masonry units.
   e. Tile (structural and facing).
   f. Stone (granite, limestone, sandstone, marble).
   g. Acid brick.
   h. Glass block.
   i. Terra cotta.

2. Mortar

   Properties, description, uses, workability, water retentivity, bond, durability, and admixtures for:
   a. Hydrated lime.
   b. Cement lime.
   c. Cement mortar.
   d. Prepared masonry cement mortar.
   d. Special mortar for:
      Fire brick.
      Glass block.
      Acid brick.
      Stone (granite, marble, etc.).

3. Sand

   Classification, description, selection, tests, types and uses.

B. Tools and Equipment

Use, care, operation, and safe practices for:
1. Brick trowel.
2. Brick hammer, blocking chisels, six-foot rules, levels, and jointing tools.
3. Story-pole and spacing rule.
4. Stone setting:
   a. Wood wedges.
   b. Setting tools.
   c. Caulking gun.
   d. Chain hoists.
   e. Cranes.
   f. Hangers.
5. Accessories:
   a. Wall ties.
   b. Expansion strips.
   c. Clip angles.
   d. Nailing blocks.
   e. Reinforced steel for grouted walls and lintels.
   g. Flashing materials.
   h. Anchor bolts.
   i. Steel bearing plates.
6. Welding Equipment.

C. Trade Arithmetic
1. Review of the fundamental operations of arithmetic, including:
   a. Fractions.
   b. Decimals.
   c. Conversions.
   d. Weights.
   e. Measures.
2. Reading the rule:
   a. Six-foot rule.
   b. Spacing rule.

D. Plan, Blueprint Reading and Trade Sketching
1. Fundamentals of plan and blueprint reading:
   a. Types of plans.
   b. Kinds of plans.
   c. Conventions.
   d. Symbols.
   e. Scale representation.
   f. Dimensions.
2. Trade Sketching:
   a. Tools (types).
   b. Straight line sketching.
   c. Circles and arcs.
   d. Making a working sketch.

E. Construction Details
1. Trade terms, motion study, bonds (structural and pattern), laying of units, points, etc., for:
   a. Walls.
   b. Footings.
   c. Pilasters, columns, and piers.
   d. Chases.
   e. Recesses (corbelling).
   f. Chimneys and fireplaces.
2. Cleaning, caulking, and pointing.
3. Reinforced masonry lintels.

F. Shop Practices
1. Spreading mortar.
2. Laying bricks to line — building inside and outside corners for a 4, 8, and 12-inch wall.
3. Layout and erect:
   a. Walls and corners with:
      Flemish and Dutch bond.
      Tile backing.
      Pilasters and chase.
      A cavity.
      Reinforced grouted brick.
   b. Brick piers.
   c. Chimneys — single and double flues.
4. Setting sills, copings, and quoins.

G. Safety

THIRD YEAR

A. Tools and Equipment
1. Use, care, operation, and safe practices for:
   a. Builder's level and transit.
   b. Frames, beams, lintels, and rods.
   c. Welding equipment.

B. Blueprint Reading
1. Specifications.
2. Job layout.
3. Shop drawings.
4. Modular measure.

C. Construction Details
1. Arch construction.
2. Modular masonry.
3. Firebox construction.
4. Layout of story poles and batter boards.

D. Estimating
1. Mortar.
2. Masonry units (modular and nonmodular types).
3. Concrete footings.

E. Shop Practice
1. Layout and erect:
   a. Reinforced masonry lintels.
   b. Story pole and batter boards.
   c. Fireplaces (with and without steel fireplace forms).
   d. Project with glazed tile leads and panels.
   e. Project with marble or granite setting, adhesive terra cotta, glass block.
   f. Modular wall.
   g. Circular corner.

F. Prefabricated Masonry Panels
1. Layout.
2. Assembly.
3. Welding and erection.
4. Installation.
5. Caulking, painting, and cleaning.

G. Application of Insulating Materials For Masonry Walls
H. Safety

WORK EXPERIENCES

To acquire the necessary skills of the trade in its various categories, the apprentice shall (as near as possible) be provided with employment in the following categories and amounts:

Type of Experience	Approximate Hours:
1. Laying of bricks, including:	2250
a. Mixing mortar, cement, and patent mortar; spreading mortar; bonding and tying.	
b. Building footings and foundations.	
c. Plain exterior brickwork (straight wall work, backing up brickwork).	
d. Building arches, groins, columns, piers, and corners.	
e. Planning and building chimneys, fireplaces and flues, and floors and stairs.	
f. Building masonry panels.	
g. Welding.	
2. Laying of stone, including:	450
a. Cutting and setting of rubblework or stonework.	
b. Setting of cut-stone trimmings.	
c. Butting ashlar.	
3. Pointing, cleaning, and caulking, including:	150
a. Pointing brick and stone, cutting and raking joints.	
b. Cleaning stone, brick, and tile (water, acid, sandblast).	
c. Caulking stone, brick, and glass block.	
4. Installation of building units, such as:	1275
a. Tile block cutting and setting.	
b. Cutting, setting, and pointing of cement blocks, artificial stone, glass blocks, and cork and styrofoam panels.	
c. Blockarching.	
5. Fireproofing, including:	225
a. Building party walls (partition tile, gypsum blocks, glazed tile).	
b. Standardized firebrick.	
c. Specialties.	
6. Care and use of tools and equipment, including:	150
a. Trowels.	
b. Brickhammer.	
c. Plumb rule.	
d. Scaffolds.	
e. Cutting saws.	
f. Welding equipment.	
Total	4500

All such work shall be performed under the supervision of a journeyman. Supervision should not be of such nature as to prevent the development of responsibility and initiative.

An employer who is to train apprentices shall meet the qualifying requirements as set forth in the local bargaining agreement, and shall be able to provide the necessary work experience for training.

GLOSSARY

ABSORBENT: Capable of taking in water or moisture.

ABSORPTION: The weight of water a brick or tile unit absorbs when immersed in either cold or boiling water for a stated length of time. It is expressed as a percentage of the weight of the dry unit.

ABSORPTION RATE: The weight of water absorbed when a brick is partially immersed for one minute, usually expressed in either grams or ounces per minute. Also called suction or initial rate of absorption.

ABUTMENT: A supporting wall carrying the end of a bridge or span and sustaining the pressure of the earth next to it. Also: a skewback and the masonry which supports it.

ACCELERATOR: A material which speeds hardening of concrete or mortar.

ACID-RESISTANT BRICK: Brick suitable for use in contact with chemicals, usually used along with acid-resistant mortars.

ADHESION TYPE CERAMIC VENEER: The inner sections of ceramic veneer held in place by adhesion of mortar to unit and backing. No metal anchors are required.

ADMIXTURES: Materials added to mortar as water-repellent or coloring agents or to retard or hasten setting.

ADOBE BRICK: Large roughly-molded, sun-dried clay brick of varying size.

AGGREGATE: Inert particles which are mixed with portland cement and water to form concrete, mortar and the like.

AIR-ENTRAINING AGENT: A material used to trap air in mortar to improve its workability and durability.

AIR SPACE: A cavity or space in the wall, or between building materials.

ALL STRETCHER BOND: Bond showing only stretchers on the face of the wall, each stretcher divided evenly over the stretchers below it.

ALUMINA: A mineral contained in clay used for brick-making.

AMERICAN BOND: That bond consisting of from five to seven stretcher courses between headers.

ANCHOR: A piece of material usually metal, used to attach building parts (e.g., plates, joists, trusses, etc.) to masonry. Also: Masonry materials or a metal tie used to secure stone in place.

ANCHORED TYPE CERAMIC VENEER: Thicker sections of ceramic veneer held in place by grout and wire anchors connected to backing wall.

ANGLE BRICK: Any brick shaped to an oblique angle to fit a salient (sharp) corner.

ANGLE IRON: A structural piece of steel in the form of a 90 degree angle used, in certain situations, to support brickwork.

APRON: A plain or molded piece of finish below the stool of the window covering the rough edge of the plastering.

APRON WALL: That part of a panel wall between window sill and wall support.

ARCADE: A range of arches, supported either on columns or on piers and detached or attached to the wall.

ARCH: A curved compressive structural member, spanning openings or recesses; also built flat. Also a form of construction in which a number of units span an opening by transferring vertical loads laterally to adjacent units and thus to the supports. An arch is normally classified by the curve of its intrados.

ARCH AXIS: The median line of the arch ring.

ARCH BRICK: Wedge-shaped brick for special use in an arch. Also: Extremely hard-burned brick from an arch of a scove kiln.

ARCH BUTTRESS: Sometimes called a flying buttress; an arch springing from a buttress or pier.

ARCHITECTURAL TERRA COTTA: Hard-burned, glazed or unglazed clay building units, plain or ornamental, machine-extruded or hand-molded and generally larger in size than brick or facing tile.

AREA WALL: The masonry surrounding or partly surrounding an area. Also: The retaining wall around basement windows below grade.

ARRIS: The external edge formed by two surfaces, whether plane or curved, meeting each other.

ARTIFICAL: Made to resemble a natural product—synthetic.

ASHLAR: A squared or cut block of stone, usually of rectangular dimensions. Also: A flat-faced surface generally square or rectangular having sawed or dressed beds and joints.

> **Coursed Ashlar:** Ashlar set to form continuous horizontal joints.
>
> **Random Ashlar:** Ashlar set with stones of varying length and height so that neither vertical nor horizontal joints are continuous.
>
> **Stacked Ashlar:** Ashlar set to form continuous vertical joints.

ASHLAR LINE: The main line of the surface of a wall of the superstructure.

Glossary

ASHLAR MASONRY: Masonry composed of rectangular units of burned clay or shale, or stone. Generally larger in size than brick and properly bonded, it has sawed, dressed or squared beds and joints laid in mortar. Often the unit size varies to provide a random pattern, which is called random ashlar.

ASTM: American Society for Testing and Materials.

AXIS: The spindle or center of any rotating object. In a sphere, an imaginary line through the center.

BACK ARCH: A concealed arch carrying the backing of a wall where the exterior facing is carried by a lintel.

BACK FILLING: (a) Rough masonry built behind a facing or between two faces. (b) Filling over the extrados of an arch. (c) Brickwork in spaces between structural timbers, sometimes called brick nogging.

BACKBOARD: A temporary board on the outside of a scaffold.

BACKING: That portion of a masonry wall built in the rear of the face and bonded to the face; usually a cheaper class of masonry.

BACKING UP: Laying the inside portion of the wall after the facing of wall has been built header high.

BACKUP: That part of a masonry wall behind the exterior facing.

BALANCED: Made symmetrical and in correct proportion.

BASEMENT: The lower part of a house or building, usually below the ground.

BAT: A piece of brick.

BATTER: (a) Recessing or sloping masonry back in successive courses; the opposite of corbel. (b) The slope or inclination of the face or back of a wall from a vertical plane. (c) The slope backwards of the face of the wall, opposite to "overhand."

BATTER STICK: A tapering stick used in connection with a plumb rule for building battering surfaces.

BAY: Any division or compartment of an arcade, roof, etc. Thus, each space from pillar to pillar in a cathedral, is called a bay or severy.

BAY WINDOW: Any window projecting outward from the wall of a building. It can be either square, semicircular or polygonal. If it is carried on projecting corbels, it is called an oriel window.

BEAM: A piece of timber, iron, steel or other material which is supported at either end by walls, columns or posts. It is used to support heavy flooring or the weight over an opening.

BEARING BLOCKS: Small blocks of stone built in a wall to support the ends of particular beams.

BEARING WALL: One which supports a vertical load in addition to its own weight.

BED JOINT: (a) The horizontal layer of mortar on which a masonry unit is laid. (b) A horizontal joint, or one perpendicular to the line of pressure. (c) A joint between two horizontal courses of brick.

BEDFORD LIMESTONE: A certain formation of limestone rock which is called Bedford, because it was first found at the city of Bedford, Ohio.

BELT COURSE: A narrow, continuous horizontal course of masonry, sometimes slightly projected such as window sills. Sometimes called string course or sill course. Also: A continuous horizontal course, marking a division in the wall plane.

BELTSTONES OR COURSES: Horizontal bands or zones of stone encircling a building or extending through a wall.

BENCHES: Brick in that part of the kiln next to the fire that are generally baked to vitrification.

BEVEL: The angle that one surface or line makes with another, when they are not at right angles.

BINDERS: Brick which extends only a part of the distance across the wall.

BLIND BOND: Bond used to tie the front course to the wall in pressed brickwork where it is not desirable that any headers should be seen in the face work.

BIND HEADER: A concealed brick header in the interior of a wall, not showing on the faces.

BLOCK AND CROSS BOND: A combination of the two bonds. The face of the wall is in cross bond and the backing in block bond.

BLOCKING: A method of bonding two adjoining or intersecting walls, not built at the same time, by means of offsets whose vertical dimensions are not less than 8 in.

BLOCKING COURSE: A course of stone placed on top of a cornice crowning the walls.

BODY BRICK: The best brick in the kiln. The brick that are baked hardest with the least distortion.

BOND: (a) Tying various parts of a masonry wall by lapping units one over another or by connecting with metal ties. (b) Patterns formed by exposed faces of units. (c) Adhesion between mortar or grout and masonry units of reinforcement.

BOND BEAM: A horizontal reinforced concrete masonry beam designed to strengthen a masonry wall and reduce the probability of cracking.

BOND COURSE: The course consisting of units which overlap more than one wythe of masonry.

BOND STONE: Stones projecting laterally (straight back) into the backup wall used to tie the wall together.

BONDER: A bonding unit.

BREAKING JOINT: Any arrangement of masonry units which prevents continuous vertical joints from occurring in adjoining courses.

BREAST OF A WINDOW: The masonry forming the back of the recess and the parapet under the window sill.

BREAST WALL: One built to prevent the falling of a vertical face cut into the natural soil.

BRICK: A solid masonry unit of clay or shale formed into a rectangular prism while plastic and burned or fired in a kiln.

BRICK AND BRICK: A method of laying brick so that units touch each other with only enough mortar to fill surface irregularities.

BRICK ASHLAR: Walls with ashlar facing backed with bricks.

BRICK VENEER: The outside facing of brickwork used to cover a wall built of other material; usually refers to brick walls covering a frame building.

BRICKWORK: Masses of wall built of bricks laid in mortar.

BROKEN RANGE: Masonry construction in which the continuity of the courses are broken at intervals.

BUGGED FINISH: A smooth finish produced by grinding with power sanders.

BUILDING BRICK: Brick for building purposes not especially treated for texture or color. Formerly called common brick.

BULL HEADER: A brick laid on its edge showing only the end of the face of the wall.

BULLNOSE: Convex rounding of a member, such as the front edge of a stair tread, a concrete block or window sill.

BULL STRETCHER: A brick laid on edge so as to show the broad side of the brick on the face of the wall.

BUTTERED JOINT: A very thin mortar joint made by scraping a small quantity of mortar with the trowel on all edges of the brick and laying it without the usual mortar bed.

BUTTERING: Placing mortar on a masonry unit with a trowel.

BUTTRESS: Masonry projecting from a wall and intended to strengthen the wall against the thrust of a roof or vault. Also: A vertical, projecting piece of stone or brick masonry built in front of a wall to strengthen it.

BUTTRESS, FLYING: A detached buttress or pier of masonry at some distance from a wall, and connected to it by an arch or a portion of an arch, so as to discharge the thrust of a roof or vault on some point.

CAMBER: The relatively small rise of a jack arch.

CAPITAL: Column cap.

CAVITY WALL: A wall built of masonry units arranged to provide a continuous air space 2 to 3 in. thick. Facing and backing wythes are connected with rigid metal ties.

C/B RATIO: The ratio of the weight of water absorbed by a masonry unit during immersion in cold water to weight absorbed during immersion in boiling water. An indication of the probable resistance of brick to freezing and thawing. Also called saturation coefficient.

CELL: One of the hollow spaces in building tile.

CEMENT: A burned mixture of clay and limestone pulverized (crushed) for making mortar or concrete.

CENTERING : Temporary formwork for the support of masonry arches or lintels during construction.

CERAMIC COLOR GLAZE: An opaque colored glaze of satin or gloss finished obtained by spraying the clay body with a compound of metallic oxides, chemicals and clays. It is burned at high temperatures, fusing glaze to body making them inseparable.

CERAMIC VENEER: Architectural terra cotta, characterized by large face dimensions and thin sections.

CHAIN BOND: The building into masonry of iron bars, chain or heavy timbers.

CHASE: A continuous recess built into a wall to receive pipes, ducts, etc.

CHAT SAWED: Description of a textured stone finish, obtained by using chat sand in the gang sawing process.

CHIMNEY LINING: Fire clay or terra cotta material made for use inside of a chimney.

CLAY MORTAR MIX: Finely ground clay which is used as a plasticizer for masonry mortars.

CLEANOUT: An opening in the first course of a reinforced concrete masonry wall for removal of mortar protrusions and droppings. Also: An opening under a fireplace for removing ashes.

CLEAR CERAMIC GLAZE: Same as ceramic color glaze except that it is translucent or slightly tinted with a gloss finish.

CLINKER BRICK: A hard-burned brick whose shape is distorted or bloated due to nearly complete vitrification.

CLIP: A portion of a brick cut to length.

CLIP COURSE: The course of brick resting on a clip joint.

CLIP JOINT: A joint of abnormal thickness to bring the course up to the required height. In no case should a clip joint be over 1/2 in. thick.

CLIPPED HEADER: A bat placed to look like a header for purposes of establishing a pattern. Also called false header.

CLOSER: The last masonry unit laid in a course. It may be whole or a portion of a unit.

CLOSURE: A quarter or three-quarter brick to close when required; the end of a course, as distinguished from a half brick.

CODE (BUILDING): A set of laws or regulations governing the location, materials and workmanship in the construction of buildings.

COLLAR JOINT: The vertical, longitudinal (often mortared) joint between wythes of masonry.

COMMON BRICK: See building brick.

COMMON AMERICAN BOND: The bond in which from five to seven stretcher courses are laid between headers.

COMMON BOND: Several courses of stretchers followed by one course of either Flemish or full headers.

COMMON BRICKWORK: The wall built out of the ordinary and cheaper classes of brick, where appearance is not an important consideration.

COMPOSITE WALL: Any bonded wall with wythes constructed of different masonry units.

CONCRETE: A mixture of portland cement, aggregates and water.

CONCRETE BRICK: A solid concrete masonry unit usually not larger than 4" by 4" by 12".

CONCRETE MASONRY UNIT: A hollow or solid unit made of portland cement and suitable aggregates. Units are often referred to by the type of aggregate used in their manufacture: cinder block, lightweight block, etc.

COPING: The material or masonry units forming a cap or finish on top of a wall, pier, pilaster, chimney, etc. It protects masonry below from penetration of water from above.

CORBEL: A shelf or ledge formed by projecting successive courses of masonry out from the face of the wall.

CORBEL OUT: To build out one or more courses of brick or stone from the face of a wall to form a support for timbers.

CORBEL TABLE: A projecting cornice or parapet, supported by a range of corbels a short distance apart, which carry a molding above which is a plain piece of projecting wall forming a parapet, and covered by a coping. Sometimes small arches are thrown across from corbel to corbel to carry the projection.

CORNER BLOCK: A concrete masonry unit with a flat end for construction of the end or corner of a wall.

CORNICE: The projection at the top of a wall finished by a blocking course, common in classic architecture. Also: A molded projecting stone at the top of an entablature.

COURSE: (a) One of the continuous horizontal layers of units, bonded with mortar in masonry. (b) A horizontal row of brick in a wall. (c) Each separate layer in stone, brick or other masonry.

COURSE BED: Stone, brick or other building material in position upon which other material is to be laid.

COVE: A concave molding used in the composition of a cornice.

CRAMPS: Bars of iron having their ends turned at right angles to the body or the bar in order to enter holes in the faces of adjacent stones to hold them in place.

CROSS BOND: Bond in which the joints of the second stretcher course come in the middle of the first; a course composed of headers and stretchers intervening.

CROSS JOINT: The joint between the two ends of the brick.

CROWDING THE LINE: Laying the bricks in such a way as to prevent the line or string from being clear of the face of the brickwork. In other words, building with a tendency to make the wall overhand.

CROWN: The apex of the arch ring. In symmetrical arches the crown is at mid-span.

CULLING: Sorting brick for size, color and quality.

CULLS: The brick rejected in culling.

CURTAIN WALL: A nonbearing wall. Built for the enclosure of a building, it is not supported at each story.

CUT STONE: Finished, dimensioned stone ready to set in place.

DAMP COURSE: A course of layer of impervious material which prevents capillary entrance of moisture from the ground or a lower course. Often called damp check.

DAMPPROOFING: One or more coatings of a compound that is impervious to water. Usually applied to the back of stone or face of back of wall.

DEFORMED BAR: A reinforcing bar with irregular surfacing for producing a better bond with grout that can be obtained with a smooth bar.

DENTIL: The cogged or tooth-like members which project under a cornice; they are used for decorative effect.

DEPTH: The depth of any arch is the dimension which is perpendicular to the tangent of the axis. The depth of a jack arch is taken to be its greatest vertical dimension.

DIAGONAL BOND: A form of raking bond where the bricks are laid in an oblique direction in the center of a thick wall or in paving.

DIAPER: Any continuous pattern in brickwork of which the various bonds are examples. It is usually applied, however, to diamond or other diagonal patterns.

DIMENSIONED STONE: Stone precut and shaped to specified sizes.

DOG'S TOOTH: Brick laid with their corners projecting from the wall face.

DOWELS: Straight metal bars used to connect two sections of

masonry; also a two-piece instrument for lifting stones.

DRAFT: A margin on the surface of a stone cut approximately to the width of the chisel.

DRIP: A projecting piece of material, shaped to throw off water and prevent its running down the face of wall or other surface. Also: A slot cut in the bottom of a projected stone, to interrupt the capillary attraction of rain water.

DRY-PRESS BRICK: Brick formed in molds under high pressures from relatively dry clay (5 to 7 percent moisture content).

DRY STONE WALLS: Walls which are of any of the classes of masonry with the exception that the mortar is omitted. They should be built according to the principles laid down for the class to which they belong.

DUTCH ARCH: A flat arch whose voussoirs (wedge-shaped pieces) are laid parallel to the skewback on each side of the center.

DUTCH BOND: The arrangement of bricks forming a modification of Old English Bond made by introducing a header as the second brick in every alternate stretching course, with a three-quarter brick beginning the other stretching courses.

DWARF WALL: A wall or partition which does not extend to the ceiling.

ECONOMY BRICK: Brick whose nominal dimensions are 4" by 4" by 8".

EDGESET: A brick set on its narrow side instead of on its flat side.

EFFLORESCENCE: A deposit of white powder or crust on the surface of brickwork which is due to soluble salts in the mortar or the brick drawn out to the surface by moisture.

ENCLOSURE WALL: An exterior nonbearing wall in skeleton frame construction. It is anchored to columns, piers or floors, but not necessarily built between columns or piers nor wholly supported at each story.

END-CONSTRUCTION TILE: Tile designed to be laid with axes of the cells vertical.

ENGINEERED BRICK: Brick whose nominal dimensions are 3.2" by 4" by 8".

ENGLISH BOND: Usually called Old English Bond, the bond which is made by alternate courses of stretchers and headers with a 2 in. piece or closer next to the corner header. Also: Alternate courses of headers and stretchers.

ENGLISH CROSS BOND: A variation of English bond made by putting one header next to the corner stretcher on alternate stretching courses.

EXPANSION ANCHOR: A metal expandable unit inserted into a drilled hole that grips stone by expansion.

EXPANSION JOINT: A vertical joint or space to allow for expansion due to temperature changes.

EXTERIOR WALL: Any outside wall or vertical enclosure of a building other than a party wall.

EXTRADOS: The convex curve which bounds the upper extremities of the arch.

FACE: The exposed surface of a wall or masonry unit. Also: The surface of a unit designed to be exposed in the finished masonry.

FACE BRICK: A well-burned brick especially prepared, selected and handled to provide attractive appearance in the face of a wall.

FACE SHELL: The side wall of a hollow concrete masonry unit.

FACED WALL: One in which facing and backing are bonded to exert common action under load.

FACING: Any material, forming a part of a wall, used as a finished surface. Also: The projecting courses at the base of a wall for the purpose of distributing the weight over an increased area; a footing.

FACING BRICK: Brick made especially for facing purposes, often treated to produce surface texture. They are made of selected clays, or are treated to produce desired color.

FACING TILE: Tile for exterior and interior masonry with exposed faces.

FAT MORTAR: Mortar containing a high percentage of cementitious components. It is a sticky mortar which adheres to a trowel.

FIELD: The expanse of wall between openings, corners, etc., principally composed of stretchers.

FILLING IN: The process of building in the center of the wall between the face and back.

FILTER BLOCK: A hollow, vitrified clay masonry unit, sometimes salt-glazed, designed for trickling filter floors in sewage disposal plants.

FIRE BRICK: Brick made of refractory ceramic material which will resist high temperatures.

FIRE CLAY: A grade of clay that can stand a great amount of heat without softening or burning up; therefore, used for fire bricks.

FIRE STOP: A projection of brickwork on the walls between the joists to prevent the spread of fire or vermin.

FIRE DIVISION WALL: Any wall which subdivides a building so as to resist the spread of fire. It is not necessarily continuous through all stories to and above the roof.

FIRE WALL: Any wall which subdivides a building to resist the spread of fire and which extends continuously from the foundation through the roof.

FIREPROOFING: Any material or combination of materials protecting structural members to increase their fire resistance.

FIREPROOFING TILE: Tile designed for protecting structural members against fire.

FIXED ARCH: An arch whose skewback is fixed in position and inclination. Plain masonry arches are, by nature of their construction, fixed arches.

FLARE HEADER: A header of darker color than the field of the wall.

FLASHING: A thin impervious material placed in mortar joints and through air spaces in masonry to prevent water penetration and/or provide water drainage.

FLAT ARCH: An arch whose top and bottom are flat or practically so. It involves the same principles of stress and strain as a segmental arch, but has voussoir (wedge-shaped pieces) extended up and down to reach the level lines.

FLAT STRETCHER COURSE: A course of stretchers set on edge and exposing their flat sides on the surface of the wall. Frequently done with brick finished for the purpose of the flat side, such as enameled or glazed brick.

FLEMISH BOND: The arrangement of bricks made by alternating headers and stretchers in each course. The position of each header is in the center of the stretcher above and below.

FLEMISH BOND (DOUBLE): The arrangement of the bricks which gives Flemish bond on both sides of the wall.

FLEMISH CROSS BOND: Any bond having alternate courses of Flemish headers and stretcher courses. The Flemish headers being plumb over each other and the alternate stretcher courses being crossed over each other.

FLEMISH DOUBLE CROSS BOND: A bond with odd numbered courses made up of stretchers divided evenly over each other. The even numbered courses are made up of Flemish headers in various locations with reference to the plumb of each other.

FLEMISH GARDEN BOND: Bricks laid so that each course has a header to every three or four stretchers.

FLOAT: A flat, broad-bladed wood or metal hand tool.

FLOATING CONCRETE: Leveling a fresh concrete surface with a float. Also, creating sufficient surface paste needed for troweling the concrete. Floating is performed immediately after a concrete surface has been consolidated and struck off.

FLOOR BRICK: Smooth dense brick, highly resistant to abrasion, used as finished floor surfaces.

FLOOR TILE: Structural units for floor and roof slab construction.

FLUE: A passage in a chimney especially for the exit of smoke and gases. One or more may be enclosed in the same chimney.

FLUE LINING: A smooth one-celled hollow tile for protecting flues.

FLUSH: Having the surface even with the adjoining surface.

FLUSHED: Filled up to the surface.

FOOTING: The broadened base of a foundation wall or other superstructure.

FOUNDATION WALL: That portion of a load-bearing wall below the level of the adjoining grade, or below first floor beams or joists.

FRAME HIGH: The height of the top of the window or door frames; the level at which the lintel is to be laid.

FROG: A depression in the bed surface of a brick. Sometimes called a panel.

FTI: Facing Tile Institute.

FULL HEADER: A course consisting of all headers.

FURRING: A method of finishing the interior face of a masonry wall to provide space for insulation, prevent moisture transmittance, or to provide a level surface of finishing.

FURRING TILE: Tile designed for lining the inside of exterior walls and carrying no superimposed loads.

GANG SAW: A machine with multiple blades used to saw rough quarry blocks into slabs.

GARDEN WALL BOND: A name given to any bond particularly adapted to walls two tiers thick. A bond consisting of one header to three stretchers in every course.

GAGE: To measure for a particular purpose. Some tools, for particular measurement may be termed as gage; as for example, a "gage stock" with courses of brick marked thereon.

GAGE BRICK: Brick which have been ground or otherwise produced to accurate dimensions. Also: A tapered arch brick.

GLAZED TILE: Tile which is finished with glasslike surface.

GOTHIC ARCH: A pointed arch made of two segments whose points meet at the crown.

GREEN BRICKWORK: Brickwork in which the mortar has not yet set.

GROUNDS: Nailing strips placed in masonry walls as a means of attaching trim or furring.

GROUT: A cementitious component of high water-cement ratio, permitting it to be poured into spaces within masonry walls. Grout consists of portland cement, lime and aggregate. It is often formed by adding water to mortar.

HACKING: The procedure of stacking brick in a kiln or on a kiln car. Also: Laying brick with the bottom edge set in from the plane surface of the wall.

HARD-BURNED: Nearly vitrified clay products which have been fired at high temperatures. They have relatively low absorptions and high compressive strengths.

HEAD JOINT: The vertical mortar joint between ends of masonry units. Often called cross joint.

HEADER: A masonry unit which overlaps two or more adjoining wythes of masonry to tie them together. Often called bonder. Also: A brick laid on its flat side across the thickness of the wall so as to show the end of the brick on the surface of the wall; also a stone having its greatest dimension at right angles to the face of the wall.

HEADER BLOCK: Concrete masonry units made with part of one side of the height removed to provide space for bonding with adjoining units such as brick.

HEADER BOND: Bond showing only headers on the face, each header divided evenly on the header under it.

HEADER COURSE: A course composed entirely of headers.

HEADER HIGH: The height up to the top of the course directly under a header course.

HEADER JOINT: A joint between the ends of two bricks in the same course; also vertical joint.

HEADER TILE: Tile containing recesses for brick headers in masonry faced walls.

HEADING COURSE: A continuous bonding course of header brick. Also called header course.

HEART BONDS: When two headers meet in the middle of the wall and the joint between them is covered by another header.

HEARTH: That portion of a fireplace level with the floor, upon which the fire is built. The rear portion extending into the fire opening is known as the back hearth.

HEADWAY: Clear space or height under an arch, or over a stairway and the like.

HERRINGBONE BOND: Bricks laid in an angular or zigzag fashion resembling the bone structure of a herring.

HIGH-LIFT GROUTING: The technique of grouting masonry in lifts up to 12 ft.

HOLLOW MASONRY UNIT: One whose net cross-sectional area in any plane parallel to the bearing surface is less than 75 percent of the gross.

HOLLOW WALL: A wall built of masonry units arranged to provide an air space within the wall. The separated facing and backing are bonded together with masonry units.

HYDRATED LIME: Quicklime treated with sufficient water to satisfy its chemical needs and then processed for use. Hydrated lime is the usual material used to add lime to mortar.

INCISE: To cut inwardly or engrave, as in an inscription.

INITIAL SET: The first setting action of mortar, the beginning of the set.

INTERLOCKING: The binding of particles one with another.

INTRADOS: The concave curve which bounds the lower side of the arch. (See soffit.) The distinction between soffit and intrados is that the intrados is linear, while the soffit is a surface.

JACK ARCH: One having horizontal or nearly horizontal upper and lower surfaces. Also called flat or straight arch.

JAMB: The side of an opening, such as a window or door.

JAMB BLOCK: A concrete block especially formed with a slot for holding the jambs of window or door frames.

JOINT: The narrow space between adjacent stones, bricks or other building blocks usually filled with mortar.

JOINTER: A tool used for smoothing or indenting the surface of a mortar joint.

JOINTING: The process of facing or tooling the mortar joints.

JOISTS, COMBINATION TILE AND CONCRETE: A floor or roof system consisting of reinforced concrete and structural clay tile.

JUMBO BRICK: A generic term indicating a brick larger in size than the standard. Some producers use this term to describe oversize brick of specific dimensions manufactured by them.

KEY: The relative position of the headers of various courses with reference to a vertical line.

KEYSTONE: The center masonry unit of the arch.

KILN RUN: Brick or tile from one kiln which have not been sorted or graded for size or color variation.

KING CLOSER: A brick cut diagonally to have one 2-in. end and one full-width end.

LAITANCE: An accumulation of fine particles on the surface of fresh concrete due to an upward movement of water.

LAP: The distance one brick extends over another.

LATERAL SUPPORT: Means whereby walls are braced either vertically or horizontally by columns, pilasters, crosswalls, beams, floors, roofs, etc.

LATERAL THRUST: The pressure of a load which extends to the sides.

LAYING OVERHAND: Building the further face of a wall from a scaffold on the other side of the wall.

LAYING TO BOND: Laying the brick of the entire course without a cut brick.

LEAD: The section of a wall built up and racked back on successive courses. A line is attached to leads as a guide for constructing a wall between them.

LEAN MORTAR: Mortar which is deficient in cementitious components. It is usually harsh and difficult to spread.

LIGHT HARD: A term applied to red brick that are not the hardest in the kiln. Although suitable for carrying moderate loads, they are not able to withstand alternate freezing and thawing as well as the hard brick.

LIGHTWEIGHT AGGREGATE: Aggregate made up of granular, puffy materials such as cinders, pumice, or expanded shale.

LIME: The base of mortar and the result of limestone burned in a kiln until the carbon dioxide has been driven off.

LIME, HYDRATED: Quicklime to which sufficient water has been added to convert the oxides to hydroxides.

LIME PUTTY: Slaked lime in a soft puttylike condition before sand or cement is added.

LINE: The string stretched taut from lead to lead as a guide for laying the top edge of a brick course.

LINE PIN: A metal pin used to attach line used for alignment of masonry units.

LINEAR FOOT: A foot measurement along a straight line.

LINETL: A beam placed over an opening in a wall.

LINTEL BLOCK: U or W shaped concrete block used in construction of horizontal bond beams and lintels.

LOAD-BEARING TILE: Tile for use in masonry walls carrying superimposed loads.

LOCK: Any special device or method of construction used to secure a bond in masonry.

MANTEL: A shelf projecting beyond the chimney breast above the fireplace opening.

MASON: A worker skilled in laying brick, block or stone; as a brickmason, blockmason or stonemason.

MASONRY: Brick, tile, stone, etc., or combination thereof, bonded with mortar. Also: That branch of construction dealing with plaster, concrete construction and the laying up of stone, brick, tile and other such units with mortar.

MASONRY CEMENT: A mill-mixed mortar to which sand and water must be added.

MASONRY UNIT: Natural or manufactured building units of burned clay, stone, glass and gypsum.

MODULAR MASONRY UNIT: One whose nominal dimensions are based on the 4 in. module.

MORTAR: A plastic mixture of cementitious materials, fine aggregate and water.

MORTAR BOARD: A board about 3 ft. square laid on the scaffold to receive the mortar ready for the use of a bricklayer.

MORTAR BOX: The box in which the mortar is mixed and softened by water for use.

NATURAL BED: The surface of a stone parallel to the stratification.

NOMINAL DIMENSION: The dimension equal to the actual masonry dimension plus the thickness of one mortar joint.

NONCOMBUSTIBLE: Any material which will neither ignite nor actively support combustion in air at a temperature of 1200 deg. F when exposed to fire.

NON-LOAD-BEARING TILE: Tile designed for use in masonry walls carrying no superimposed loads.

NORMAN BRICK: A brick whose nominal dimensions are 2 2/3" by 4" by 12".

OFFSET: A course that sets in from the course directly under it. Also called setoff, setback, etc.; the opposite of corbel.

OPEN END BLOCK: A concrete block with an end web removed for placing the block around vertical steel reinforcement.

OUTRIGGER: A joist projecting out of a window to support an outside scaffold.

OVERHAND WORK: An entire wall built with a staging located on only one side of the wall.

OVERHANG: A face of the wall leaning from the vertical away from the wall.

PANEL WALL: A non-load-bearing wall in skeleton form construction, wholly supported at each story.

PARAPET: A wall or barrier on the edge of an elevated structure for the purpose of protection or ornament.

PARAPET WALL: That part of any wall entirely above the roof line.

PARGING: The process of applying a coat of cement mortar on masonry.

PARTITION: An interior wall, one story or less in height.

PARTITION TILE: Tile for use in interior partitions.

PARTY WALL: A wall used for joint service by adjoining buildings.

PAVING: Regularly placed stones or bricks forming a floor.

PAVING BRICK: Vitrified brick especially suitable for use in pavements where resistance to abrasion is important.

PEACH BASKET: A template against which the entire head of a tall chimney is built.

PERFORATED WALL: One which contains a considerable number of relatively small openings. Often called pierced wall or screen wall.

PERPEND BOND: Signifies that a header extends through the whole thickness of the wall.

PICK AND DIP: A method of laying brick whereby the bricklayer simultaneously picks up a brick with one hand and, with the other hand, enough mortar on a trowel to lay the brick. Sometimes called the Eastern or New England method.

PIER: An isolated column of masonry. Also: A block of brickwork usually between two openings which is built to support arches, or to carry beams or girders.

PILASTER: A wall portion projecting from either or both wall faces and serving as a vertical column and/or beam. Also: A pillar of brickwork, rectangular in form, used as a supplement to a pier, usually projecting one-third of the thickness of the wall.

PITCH: To square a stone.

PITCH STONE: Stone having the arris clearly define by a line beyond which the rock is cut away by the pitching chisel, so as to make approximately true edges.

PLASTIC: In the form of a sticky paste.

PLINTH: The square block at the base of a column or pedestal. In a wall, the term plinth is applied to the projecting base or water table, generally at the level of the first floor.

PLUMB BOB: The lead weight to make taut the plumb line.

PLUMB RULE: A tool (mason's level) used to aid in building surface in a vertical plane.

PLYFORM: A special plywood used to make forms for

concrete.

POINTING: Troweling mortar into a joint after the masonry units are laid.

POINTING TROWEL: A small tool used for filling joints on the exposed surface of the wall.

PORTLAND CEMENT: A cement made by a mixture of various clays, chalk, limestone, river mud, slate and the like, which are mixed together, burned, ground into powder and put through a sieve with fine meshes.

PORTLAND CEMENT: A type of cement manufactured by combining, burning and finely grinding a combination of lime, silica, alumina, and iron oxide. It is capable of hardening through a chemical reaction when mixed with water.

PRESSED BRICKS: Those that are pressed in the mold by mechanical power before they are burned or baked.

PUGGING: A course kind of mortar laid on the boarding between floor joists to prevent the passage of sound; also called deafening.

PUTLOG: The cross support of the scaffold which holds the scaffold planks or platform.

QUEEN CLOSER: A cut brick having a nominal 2 in. horizontal face dimension.

QUOIN: A projecting right angle masonry corner. Also: Projecting courses of brick at the corners of building as ornamental features.

RACKING: Laying the lead or end of the wall with a series of steps so that when work is resumed, the bond can be easily continued. More convenient and structurally better than toothing.

RAGGLE: A groove in a joint or special unit to receive roofing or flashing.

RAKE: The end of a wall that racks back.

RAKING BOND: Brick laid in an angular or zigzag fashion.

RANGE WORK: In this construction, a course of any thickness, once started, is continued across the entire face, but all courses need not be of the same thickness.

RBM: Reinforced brick masonry. Also: Reinforced clay masonry.

RECESS: A depth of some inches in the thickness of a wall such as a niche, etc.

REGLET: A recess to receive and secure metal flashing.

REINFORCED CONCRETE: Concrete which has iron and steel rods and pieces to enable it to withstand greater stress and strain.

REINFORCED MASONRY: Masonry units, reinforcing steel, grout and/or mortar combined to act together in resisting forces.

RELIEVING ARCH: One built over a lintel, flat arch, or smaller arch to divert loads, thus relieving the lower member from excessive loading. Also known as discharging or safety arch.

RETURN: Any surface turned back from the face of a principal surface.

REVEAL: That portion of a jamb or recess which is visible from the face of a wall back to the frame placed between jambs.

RIPRAP: Rough stones of various sizes placed irregularly and compactly to prevent scour by water.

RISE: The distance at the middle of the arch between the springing line and intrados or soffit.

RODDING CONCRETE: An up-and-down action with a tamping rod.

ROLLED: A brick laid with an overhanging face.

ROMAN BRICK: Brick whose nominal dimensions are 2" by 4" by 12".

ROWLOCK: A brick laid on its face edge so that the normal bedding area is visible in the wall face. Frequently spelled rolok.

ROWLOCK COURSE: Bricks set on edge.

RUBBLE: Field stone or rough stone as it comes from the quarry.

RUNNING BOND: Same as stretcher bond.

SAG: A depression in a horizontal line, meaning that there is a slight fall below the level. Usually refers to the bricklayer's line which in a long distance, will fall below the level because of its own weight, no matter how tightly it is stretched.

SALMON BRICK: Relatively soft, under-burned brick, or named because of color. Sometimes called chuff or place brick.

SALT GLAZE: A gloss finish obtained by thermochemical reaction between silicates of clay and vapors of salt or chemicals.

SAND: Small grain of mineral, largely quartz, which is the result of disintegration of rock.

SCAFFOLD HEIGHT: The height of the unfinished wall which requires another raising of the scaffold to continue the building.

SCALE BOX: A derrick box made with an open top and one open end.

SCANT: A slight slope inwards from the plumb line.

SCR ACOUSTILE: A side-construction two-celled facing tile, having a perforated face backed with glass wool for acoustical purposes.

SCR BRICK: Brick whose nominal dimensions are 2 2/3" by 6" by 12".

SCR BUILDING PANEL: Prefabricated, structural ceramic panels, approximately 2 1/2 in. thick.

SCR MASONRY PROCESS: A construction aid providing greater efficiency, better workmanship and increased production in masonry construction. It utilizes story poles, marked lines and adjustable scaffolding.

SCREEDING (STRIKE-OFF): The operation of leveling a concrete surface. Performed by moving a straightedge across the top of the side forms.

SCUTCH: A tool resembling a pick on a small scale with flat cutting edges, for trimming bricks for particular uses.

SEGMENTAL ARCH: An arch whose intrados and extrados make the line of a half circle.

SEGREATION: The tendency of coarse aggregate (stone) to separate from the mortar (cement paste and sand) as concrete is placed.

SELECTS: The bricks accepted as the best after culling.

SET: A change from a plastic to a hard state, also a name given for the chisel used for cutting bricks, also called a "bolster."

SET-IN: The amount that the lower edge of a brick on the face tier is back from the line of the top edge of the brick directly below it.

SEWER BRICK: Low absorption, abrasive-resistant brick intended for use in drainage structures.

SHANK: That part of the trowel between the blade and the handle or hold.

SHOT SAWED: Description of a finish obtained by using steel shot in the gang sawing process to produce random markings for a rough surface texture.

SHOVED JOINTS: Vertical joints filled by shoving a brick against the next brick when it is being laid in a bed of mortar.

SIDE-CONSTRUCTION TILE: Tile intended for placement with axes of cells horizontal.

SILICA: A mineral contained in the clay used for brick-making.

SILL BLOCK: A solid concrete masonry unit used for sills of openings.

SILL HIGH: The height for the window will upon which the window frame rests.

SKEWBACK: The inclined surface on which the arch joins the supporting wall. For jack arches the skewback is indicated by a horizontal dimension.

SMOKE CHAMBER: The space in a fireplace immediately above the throat where the smoke gathers before passing into the flue and narrowed by corbeling to the size of the flue lining above.

SLUMP MOLD OR CONE: A standard metal mold in the form of a truncated cone with a base diameter of 8 in., a top diameter of 4 in., and a depth of 12 in., used to fabricate a concrete specimen for the slump test.

SLUMP OF CONCRETE: A measure of consistency or fluidity of concrete equal to the number of inches of subsidence of a truncated cone of concrete released immediately after molding in a standard slump cone.

STRIKE-OFF: Action of removing concrete in excess of that which is required to fill the form evenly. Also, the name applied to the timber straightedge used to level the concrete with form edges.

SLUSHED JOINTS: Vertical joints filled, after units are laid, by "throwing" mortar in with the edge of a trowel. (Generally, not recommended.)

SOAP: A brick or tile of normal face dimensions, having a nominal 2 in. thickness.

SOFFIT: The underside of a beam, lintel or arch.

SOFT-BURNED: Clay products which have been fired at low temperature ranges, producing relatively high absorption and low compressive strengths.

SOFT-MUD BRICK: Brick produced by molding relatively wet clay (20 to 30 percent moisture). Often a hand process. When insides of molds are sanded to prevent sticking of clay, the product is sand-struck brick. When molds are wetted to prevent sticking, the product is water-struck brick.

SOLAR SCREEN: A perforated wall used as a sunshade.

SOLAR SCREEN TILE: Tile manufactured for masonry screen construction.

SOLDIER: A stretcher set on end with face showing on the wall surface.

SOLID MASONRY UNIT: One whose net cross-sectional area in every plane parallel to the bearing surface is 75 percent or more of the gross.

SOLID MASONRY WALL: A wall built of solid masonry units, laid contiguously, with joints between units completely filled with mortar.

SPALL: A small fragment removed from the face of a masonry unit by a flow or by action of the elements.

SPAN: The distance to be covered by an arch, lintel, beam, girder or the like, between two abutments or supports; the width of an opening.

SPANDREL: The triangular portion of the wall contained between the arches where a horizontal line is drawn from crown to crown.

SPANDREL WALL: That part of a curtain wall above the top of a window in one story and below the sill of the window in the story above.

SPALY: A slope or bevel, particularly at the sides of a window or door.

SPLIT BLOCK: Concrete masonry units with one or more faces having a rough surface from being split during manufacture.

SPRINGER: The stone from which an arch springs. In some cases this is a capital or import; in other cases the moldings continue down the pier. The lowest stone of a gable is sometimes called a springer.

SPRINGING COURSE: The course from which an arch springs.

SPRINGING LINE: The upper and inner edge of the line of skewbacks on an abutment.

STACK: Any structure or part thereof which contains a flue or flues for the discharge of gases.

STIFF-MUD BRICK: Brick produced by extruding a stiff but plastic clay (12 to 15 percent moisture) through a die.

STORY HIGH: The height for the floor joists.

STORY POLE: A marked pole for measuring masonry coursing during construction.

STRAIGHTEDGE: A board having an edge trued and straight, used for leveling and plumbing.

STRETCHER: A masonry unit laid with its greatest dimension horizontal and its face parallel to the wall face.

STRINGING MORTAR: The name of a method where a bricklayer picks up mortar for a large number of brick and spreads it before laying the brick.

STRUCK JOINT: Any mortar joint which has been finished with a trowel.

STRUCTURAL STEEL: Steel beams, girders, and columns used for building purposes, particularly for high buildings known as skyscrapers.

SUPERSTRUCTURE: That part of a building which is above ground.

TEMPER: To mix so as to get the mortar in the proper condition for use.

TEMPLATE: Any form or pattern, such as centering, over which brickwork may be formed.

TENDER: A laborer who tends masons. A general name covering hod and pack carriers and wheelbarrow handlers.

TERMITE SHIELD: A barrier, usually of sheet metal, placed in masonry work to prevent the passage of termites.

THREE-QUARTER: A brick with one end cut off, usually measures about 6 in. in length.

THROAT: An opening at the top of a fireplace through which the smoke passes to the smoke chamber and chimney.

THROUGH BONDS: Bonds which extend clear across from face to back.

TIE: Any unit of material which connects masonry to masonry or other materials.

TIER: One of the 4 in., or one-brick layers in the thickness of a wall.

TILE, STRUCTURAL CLAY: Hollow masonry building units composed of burned clay, shale, fire clay or mixtures thereof.

TOOLING: Compressing and shaping the face of a mortar joint with a special tool other than a trowel.

TOOTHING: Constructing the temporary end of a wall with the end stretcher of every alternate course projecting. Projecting units are toothers.

TRIG: The bricks laid in the middle of a wall between the two main leads to overcome the sag in the line, and also to keep the center plumb in case there is a wind bearing upon the line.

TRIM: Stone used as sills, copings, enframenets, etc. with the facing of another material.

TRIMMER ARCH: An arch, usually a low rise arch of brick, used for supporting a fireplace hearth.

TROWEL: A flat, broad-bladed steel hand tool used in the final stages of finishing operations to impart a relatively smooth surface to concrete slabs or other unformed concrete surfaces.

TUCK POINTING: The filling in with fresh mortar of cut out or defective mortar joints in masonry.

TWO-INCH PIECE: A closer about one-quarter of a brick in length used to start the bond from the corner.

VENEER: A single wythe of masonry for facing purposes, not structurally bonded.

VENEERED WALL: A wall having a masonry facing which is attached to the backing but not bonded to exert common action under load.

VERTICAL JOINTS: Same as head joints.

VITRIFICATION: The condition resulting when kiln temperatures are sufficient to fuse grains and close pores of a clay product, making the mass impervious.

VOUSSOIR: One of the wedge-shaped masonry units which forms an arch ring.

WALL: A vertical platelike member, enclosing or dividing spaces and often used structurally.

WALL PLATE: A horizontal member anchored to a masonry wall to which other structural elements may be attached. Also called head plate.

WALL TIE: A bonder or metal piece which connects wythes of masonry to each other or to other materials.

WALL TIE, CAVITY: A rigid, corrosion-resistant metal tie which bonds two wythes of a cavity wall. It is usually steel, 2/16 in. in diameter and formed in a "z" shape or a rectangle.

WALLS, BEARING: A wall supporting a vertical load in addition to its own weight.

WALLS, CAVITY: A wall in which the inner and outer wythes are spearated by an air space, but tied together with metal ties.

WASHING DOWN: Cleaning the surface of the brick wall with a mild solution of muriatic acid after it is completed and pointed.

WATER RETENTIVITY: That property of a mortar which prevents the rapid loss of water to masonry units of high suction. It prevents bleeding or water gain when mortar is in contact with relatively impervious units.

WATER TABLE: A projection of lower masonry on the outside of the wall slightly above the ground. Often a damp course is placed at the level of the water table to prevent upward penetration of ground water.

WEATHERING: The process of decay brought about by the effect of weather conditions.

WEB: The cross wall connecting the face shells of a hollow concrete masonry unit.

WEEP HOLES: Openings in mortar joints of facing material at the level of flashing, to let moisture escape.

WIRE CUT BRICK: A brick having its surfaces formed by wires cutting the clay before it is baked.

WYTHE: Each continuous vertical section of masonry one unit in thickness. Also: The thickness of masonry separating flues in a chimney. Also called withe or tier.

WORKABILITY: That property of fresh concrete or mortar which determines the ease with which it can be mixed, placed and finished.

WORKING DRAWINGS: Drawings that show sufficient detailed information including sizes and shapes from which sufficient interpretation can be obtained to properly build the object described by the drawings.

INDEX

Index

Index